Environmental Science and Technology

Concepts and Applications

Frank R. Spellman

and

Nancy E. Whiting

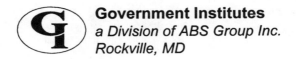

Government Institutes
a Division of ABS Group Inc.
Rockville, MD

Government Institutes, a Division of ABS Group Inc.
4 Research Place, Rockville, Maryland 20850, USA
Phone: (301) 921-2300
Fax: (301) 921-0373
Email: giinfo@govinst.com
Internet: http://www.govinst.com

Library of Congress Cataloging-in-Publication Data

Spellman, Frank R.
    Environmental Science and Technology: Concepts and Applications / by Frank R. Spellman, Nancy E. Whiting.
       p.   cm.
    Includes index.
    ISBN 0-86587-644-4
    1. Environmental sciences. 2. Green technology. 3. Ecological engineering.
    I. Whiting, Nancy E.   II. Title.
GE105.S73   1999
628.—dc21

                        98-51821
                        CIP

to

Rachel Morgan Spellman

and

Wilson Bates Whiting

and

Katherine Grace Whiting

Hopefully, when it comes to our environment, you will be good stewards.

Summary of Contents

Contents

PART II
AIR QUALITY 157

CHAPTER 9
THE ATMOSPHERE: BASIC AIR QUALITY 159

CHAPTER 10
METEOROLOGY 173

CHAPTER 21
SOIL POLLUTION CONTROL TECHNOLOGY 341

List of Figures and Tables

PREFACE

You have heard about, read about, or witnessed the horrific impact of technology upon the environment. You may have observed or heard about rivers so oil-laden that they actually burned. Or perhaps you've observed, heard, or read about skies above metropolitan areas that were red with soot. Some of you may have breathed air you could actually see. You may have seen lakes choked with algae, or lakes so poisoned that they could no longer support life (even though they might still be a source of drinking water). Or you may have seen the flip side of this coin—you may have seen a place so pristine, so unlittered, so filled with abundant life that you recognize it as special, as different, as a priceless treasure.

You may have seen earth-moving machines enter what appeared to be pristine wilderness areas to gouge out huge, gaping masses from the face of the earth—to remove the soil to a segregated holding area because the soil was contaminated, polluted by hazardous materials dumped there by environmental criminals—criminals who, by wont of their greed, their unconcern, and their thoughtlessness have committed an ultimate crime—they have poisoned us all by poisoning what we all must have to live. They fouled our earth.

You've read about, heard about, or actually seen such travesties, such total disregard for the source of our being, for the basis of our very survival. And, in the past and in the present (we hope not in the future), we have directed the blame to a consistent culprit—people using (without complete knowledge of the consequences), misusing, or abusing technology and technological advances.

In *Environmental Science: Concepts and Applications* we examine the many environmental problems that beset us, the dimensions of those problems, and their varied and interrelated causes. We examine in detail the three environmental media we are wholly dependent upon: air, water, and soil. We look at data on the ecological relationships of endangered species (are we one of them?), the impact of toxic materials on air, water, and soil (and ultimately on human health), the dispersal of pollutants in the atmosphere, bodies of water, and soil, the accumulation of persistent chemicals in aquatic food webs, and the control of agricultural pests with pesticides. We look at acid rain and management of hazardous wastes.

We also look at wastewater treatment plants that have significantly lessened the impact of some water pollutants. We look at the groundwater problem and the ozone depletion problem. We look at advancements made in reusing and recycling contaminated soils. All of these problems can, will, and do affect each of us, and all are related to the environment.

Are you wondering "So what else is new? We've heard all this before."

You have seen, heard, read about, and studied these topics. But consider this: a fundamental knowledge of environmental science is a prerequisite to meeting the environmental and natural resource challenges that will face us all in the 21st century. Granted, this is an old concept. But we must go beyond concepts to the connections those concepts can lead us to—a realistic view, clearly presented, which shows the environmental effects of human activity (primarily technological advances) that, when constructively directed, can have a beneficial impact on the earth.

This text analyzes environmental problems related to air, water, and soil as objectively as possible. We leave you to form your own opinion on many of the controversial issues. That means you're going to have to think about them—a worthy goal for any learning activity.

In short, the primary message in this text is that while we have environmental problems and challenges, the news is not all bad. Remember, every problem has a solution. This solution is critically important. How technology can remediate environmental problems will come up again and again. For environmental challenges, the solution is the use of the proper technology to minimize environmental disruption.

Organization and Content

Environmental Science and Technology: Concepts and Applications is divided into five parts and twenty-four chapters, and is organized to provide an even and logical flow of concepts. It provides the reader with a clear and thoughtful picture of this complex field.

Part I provides the foundation for the underlying theme of this book—the connections between environmental science and technology. We augment the philosophical with more practical aspects, which include a presentation of fundamentals: Energy, Materials Balance, and Units of Measurement. We present our materials simply, in a down-to-earth, user-friendly format. Concepts of environmental chemistry, biology/ecology, toxicology, geology and groundwater hydrology, environmental processes, and technology and engineering are also presented as part of the introduction, all germane to the foundation-building process, and are presented in an understandable format.

Part II develops the air quality principles basic to an understanding of air quality. Meteorology, air pollution, atmospheric air dispersion, atmospheric change (greenhouse effect and global change), and air pollution control are discussed in detail.

Part III focuses on water quality and the characteristics of water and water bodies, water sciences, water pollution, and water treatment.

Part IV deals with soil science and emphasizes soil as a natural resource, highlighting the many interactions between soil and other components of the ecosystem.

Part V is devoted to showing how decisions regarding handling solid and hazardous waste have or can have profound impact on the environment and the three media discussed in this text: air, water, and soil.

Finally, the epilogue looks at the state of the environment, past, present, and future. The emphasis here is on mitigating present and future environmental concerns by incorporating technology into the remediation process—not by blaming technology for the problem.

Frank R. Spellman

Nancy E. Whiting

PART I

INTRODUCTION

Fundamentals: Energy, Materials Balance, and Units of Measurement

CHAPTER 1

ENVIRONMENTAL SCIENCE AND TECHNOLOGY: THE CONNECTION

One of the marvels of early Wisconsin was the Round River, a river that flowed into itself, and thus sped around and around in a never-ending circuit. Paul Bunyan discovered it, and the Bunyan saga tells how he floated many a log down its restless waters.

No one has suspected Paul of speaking in parables, yet in this instance he did. Wisconsin not only had a round river, Wisconsin is one. The current is the stream of energy which flows out of the soil and into the plants, thence into the animals, thence back into the soil in a never ending circuit of life. —Leopold

CHAPTER OUTLINE
- Introduction
- What is Environmental Science?
- Key Terms
- Science and Environmental Science
- Environmental Science and Technology: The Connection
- When Properly Connected, Science and Technology Offer Solutions

INTRODUCTION

You don't need to be a practitioner of rocket science (or any other science, for that matter) to ascertain the message implicit in the opening to this chapter. Leopold develops the idea of the entire earth as the round river, and we, as representatives of but a few of the organisms who depend on the river for life and for survival, are tied to it, component and portion of the environment that we influence and are influenced by.

Most environmental science texts examine four areas: air, water, soil, and biota. This text focuses on only three, because without air, water, or soil, no biota can exist. Without them, the planet would be a sterile hunk of orbiting rock. Without air, water, and soil, there is nothing we could relate to.

Along with the standard focus on air, water, soil, and biota, many environmental science texts leave out a crucial element that is related to and impacts all of them: Technology. This should come as no surprise to anyone concerned about the environment and environmental issues, especially since technology and our use of technology is usually blamed for the degradation of the earth's environment.

We damage the environment through use, misuse, and abuse of technology. Frequently we use technological advances before we fully understand their long-term effects. We weigh the advantages a technological advance can give us against the environment and discount the importance of the environment, through greed, hubris, lack of knowledge, or stupidity. We often only examine short-term plans without fully developing how problems will be handled years later. We assume that when the situation becomes critical, technology will be there to fix it. We believe the scientists will be able to figure it out, and we therefore ignore the immediate consequences of our technological abuse.

Consider this: while technological advances have provided us with nuclear power, the light bulb and its energy source, plastics, the internal combustion engine, air conditioning and refrigeration (and scores of other advances that make our modern lives pleasant and comfortable), these advances have affected the earth's environment in ways we did not expect, in ways we deplore, and in ways we may not be able to live with. In this text, the argument is made that this same technology can be used to mitigate the consequences of technology's misuse.

In this chapter we set out on a journey that begins with the basics, the building blocks that enable us to pursue and understand difficult concepts. We will develop clearer perceptions about ideas some view as controversial concerns that impact us all (although almost everyone will agree that our environment is not as pristine and unspoiled as we want and expect it to be). Most importantly, we will offer insights that will enable you to make up your mind on what needs to be done, on what needs to be undone, and on what needs to be mankind's focus on maintaining life as we know, want, and deserve it. Have a smooth and enlightening journey.

WHAT IS ENVIRONMENTAL SCIENCE?

To define *environmental science,* we must first break down the phrase and look at both terms separately. The *environment* includes all living and nonliving (such as air, soil, and water) things that influence organisms. *Science* is the observation, identification, description, experimental investigation, and theoretical explanation of natural phenomena. When we combine the two, we are left with a complex interdisciplinary study that must be defined both narrowly and broadly—and then combined—to allow us an accurate definition.

The narrow definition of *environmental science* is the study of the human impact on the physical and biological environment of an organism. In this sense, environmental scientists are interested in determining the effects of pesticides on croplands, in learning how acid rain affects vegetation, or determining the impact of introducing an exotic species of game fish into a pond or lake.

Beginning in the early 1960s, environmental science evolved out of the studies of natural science, biology, ecology, conservation, and geography. Increasing awareness of the interdependence that exists among all the disparate elements that make up our environment led to the field of study that contains aspects of all of those elements. The biological and physical ideas were combined, along with ideas from the social sciences—sociology, economics, and political science—into the new, interdisciplinary field of environmental science (see Figure 1.1).

Many environmental scientists have widely varying, diverse backgrounds. A well-trained environmentalist is a generalist, trained as a biologist, ecologist (formerly known as natural scientist), geologist, or environmental engineer—or in many other related areas. While environmental scientists are generalists, and while they may have concentrated their study on a particular specialty, solidly trained environmental scientists have one thing in common: they are well grounded in several different branches of science.

In its broadest sense, environmental science also encompasses the social and cultural aspects of the environment. As a mixture of several traditional sciences, political awareness, and societal

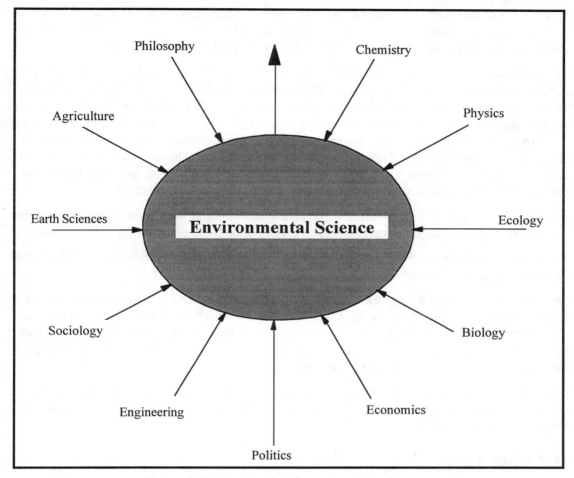

Figure 1.1 Decisions related to environmental science

values, environmental science demands examination of more than the concrete physical aspects of the world around us—and many of those political, societal, and cultural aspects are far more slippery than what we can prove as scientific fact.

In short, we can accurately say that environmental science is a pure science, because it includes study of all the mechanisms of environmental processes: the study of the air, water, and soil. But ·it is also an applied science, because it examines problems with the goal of contributing to their solution: the study of the effects of technology thereon. Obviously, to solve environmental problems and understand the issues, environmental scientists need a broad base of information upon which to draw.

The environment in which we live has been affected irreversibly by advancements in technology; it has been affected for as long as humans have wielded tools to alter their circumstances. We will continue to alter our environment to suit ourselves as long as we remain a viable species. But to do so wisely, we need to closely examine what we do and how we do it.

We build a bridge between science and technology. On one side, science, and on the other side, technology.

Why is science necessary? We can all see the signs of decay around us. The air we breath is filthy, the water we drink has a foul odor and taste, the landfill (not in my backyard!—NIMBY) is overflowing with God only knows what—why do we need science to tell us that?

We need science, first, for quantitative analysis. We use science to obtain basic information of existing conditions of air, water, and soil. We need to know what is causing the problem and to define its severity. We also need science to show us the hidden problems—the ones we can't see: the lake that appears normal on the surface, but in fact is sterile from acid rain (and yet may remain a source of drinking water). We rely on scientific measurements and computer models to help us define, understand, and effect change on these problems. The environmental challenges we face today are often less visible to the unaware, and more global in scope—and have longer response times. We can solve these problems only through scientific methods.

We use technology to address environmental problems of air and water quality, soil contamination, and of solid and hazardous waste—the technology that will clean up the environment.

But didn't technology cause our present environmental problems in the first place? Why do we want to make a bad situation worse by throwing more technology at it?

For better or worse, technology has changed our environment. And while technology has contributed to our environmental problems, remember that the human element must bear the brunt of the blame. People must use technology to repair the damage as well. For example, water and wastewater treatment technologies have made enormous strides in the task of purifying the water we drink and treating the water we waste.

Whether we agree or disagree with the advantages or disadvantages of technology, to sustain life on our planet we must learn that the marriage between science and technology is not only compatible, but is critically important.

KEY TERMS

To understand the basic concepts of environmental science, you'll need to learn the core vocabulary. Here are some of the key terms used in this chapter.

Scientists gather information and draw conclusions about the workings of the environment by applying the *scientific method*, which is a way of gathering and evaluating information. It involves observation, speculation (hypothesis formation), and reasoning.

Environmental science may be divided among the study of air (atmosphere), water (hydrosphere), soil (geosphere), and life (biosphere). Again, the emphasis in this text is on the first three—air, water, and soil—because without any of these, life as we know it is impossible.

The *atmosphere* is the envelope of thin air around the earth. The role of the atmosphere is multifaceted: (1) it serves as a reservoir of gases, (2) it moderates the earth's temperature, (3) it absorbs energy and damaging ultraviolet (UV) radiation from the sun, (4) it transports energy away from equatorial regions, and (5) it serves as a pathway for vapor-phase movement or water in the hydrologic cycle. *Air*, the mixture of gases that constitutes the earth's atmosphere, is, by volume at sea level, 78.0% nitrogen, 21.0% oxygen, 0.93% argon, and 0.03% carbon dioxide, together with very small (trace) amounts of numerous other constituents.

The *hydrosphere* is the water component of the earth, encompassing the oceans, seas, rivers, streams, swamps, lakes, groundwater, and atmospheric water vapor. *Water* (H_2O) is a liquid without color, taste, or odor, which covers 70% of the earth's surface and occurs as standing (oceans, lakes) and running (rivers, streams) water, rain, and vapor. It supports all forms of the earth's life.

The *geosphere* consists of the solid earth, including *soil*—the *lithosphere*, the topmost layer of decomposed rock and organic matter, which usually contains air, moisture, and nutrients, and can therefore support life.

The *biosphere* is the region of the earth and its atmosphere in which life exists, an envelope extending from up to 6000 meters above to 10,000 meters below sea level. Living organisms and the aspects of the environment pertaining directly to them are called *biotic* (biota), and other portions, the nonliving part of the physical environment, are *abiotic*.

A series of biological, chemical, and geological processes by which materials cycle through ecosystems is called a *biogeochemical cycle*. We are concerned with two types, the *gaseous* and the *sedimentary*. Gaseous cycles include the carbon and nitrogen cycles. The main sink (the main receiving area for material: e.g., plants are sinks for carbon dioxide) of nutrients in the gaseous cycle is the atmosphere and the ocean. The sedimentary cycles include the sulfur and phosphorous cycles. The main sink for sedimentary cycles is soil and rocks of the earth's crust.

Formerly known as natural science, *ecology (*as it is now more commonly known) is critical to the study of environmental science, and is the study of the structure, function, and behavior of the natural systems that comprise the biosphere. The terms "ecology" and "interrelationship" are interchangeable; they mean the same thing. In fact, ecology is the scientific study of the interrelationships among organisms and between organisms, and all aspects, living and non-living, of their environment.

Ecology is normally approached from two viewpoints: (1) the environment and the demands it places on the organisms in it, and (2) organisms and how they adapt to their environmental conditions. An *ecosystem* (a cyclic mechanism) describes the interdependence of species in the living world (the biome or community) with one another and with their non-living (abiotic) environment. An ecosystem has physical, chemical, and biological components along with energy sources and pathways.

An ecosystem can be analyzed from a functional viewpoint in terms of several factors. The factors important in this discussion include *biogeochemical cycles*, *energy*, and *food chains* (all discussed in detail in Chapter 2). Each ecosystem is bound together by biogeochemical cycles through which living organisms use energy from the sun to obtain or "fix" non-living inorganic elements such as carbon, oxygen, and hydrogen from the environment, and transform them into vital food, which is then used and recycled. The environment in which a particular organism lives is called a *habitat*. The role of an organism in a habitat is called its *niche*.

Figure 1.2 depicts an ecosystem where biotic and abiotic materials are constantly exchanged. *Producers* construct organic substances through photosynthesis and chemosynthesis. *Consumers* and *decomposers* use organic matter as their food and convert it into abiotic components—that is, they dissipate energy fixed by producers through food chains. The abiotic part of the pond in Figure 1.2 is formed of inorganic and organic compounds such as carbon, oxygen, nitrogen, sulfur, calcium, hydrogen, and humic acids. The biotic part is represented by producers—rooted plants and phytoplankton. Fish, crustaceans, and insect larvae make up the consumers. Mayfly nymphs, for example, are detrivores, which feed on organic detritus. Decomposers (aquatic bacteria and fungi) make up the final biotic element.

In short, while many branches of science help us understand the physical, chemical, and biological processes of our environment, ecology concentrates on the way these processes interact as systems. A well-grounded knowledge of ecology is fundamental to gaining knowledge of environmental science. Ecology and the holistic approach are also interchangeable because ecologists study nature as a functioning system instead of as a collection of distinct, unrelated parts.

SCIENCE AND ENVIRONMENTAL SCIENCE

When you view Figure 1.1, you see immediately that the environmental practitioner, whether scientist or technologist, is really an interdisciplinarian who must have a wide range of scientific knowledge. The field also demands knowledge beyond the various scientific fields or specialties. The environmentalist must understand and attempt to solve the problems caused by the interaction of natural and cultural systems. A working environmental scientist or technologist can't focus strictly on "pure science." The interdisciplinary nature of the field itself prevents it. But what are the differences between "pure science" and "environmental science"? The best way to differentiate them is to compare and contrast them.

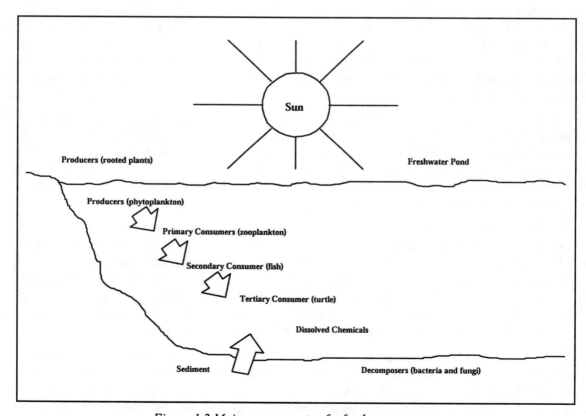

Figure 1.2 Major components of a freshwater ecosystem

In science, the scientist uses the scientific method, grounded in experimentation. Scientists conduct controlled experimentation, which tends to be reductive (they isolate the problem to a single variable, sometimes missing the big picture). Experimentation takes time. Rushing it can, and often will, invalidate the results. Effective scientists must remain objective, value-free, and bias-free. On the other hand, environmental scientists use problem-solving techniques. They start with a human-caused problem and take into account the human values pertinent to it. The environmentalist considers human values, which are neither objective nor bias-free.

Scientists define natural system structure, function, and behavior, which may or may not have direct application to a particular environmental problem. Environmental scientists define a process for solving environmental problems.

Scientists propose hypotheses based on past observations, and use the scientific method to continue questioning and testing to establish the validity of hypotheses. Environmental scientists use problem solving to propose future-directed solutions, and continually evaluate and monitor situations to improve the solutions.

Scientists are interested in knowledge for its own sake, or in applied sciences, in applications of knowledge found through a precise and thorough process that may or may not solve environmental problems. Environmental scientists are interested in finding the best solution (sometimes before all the facts are in) to actual environmental problems within a particular social setting.

Scientists who conduct studies to determine and understand how the *biosphere* creates and supports all life and environmentalists who solve environmental problems that arise from the degrading effects of human activity strive to accomplish two very different undertakings. Consider Case Study 1.1.

In short, scientists study to find the answer to a problem through scientific analysis and study. Their interest is in pure science. The environmentalists are capable of arriving at the same causal conclusions as the scientists, but they are also able to factor in socio-economic, political, and cultural influences as well.

Case Study 1.1
Salmon and the Rachel River

The Rachel River, a hypothetical river system in the northwestern United States, flows through an area that includes an Indian Reservation. The river system empties into the Pacific Ocean, and the headwaters begin within the Cascade Mountain Range of Washington State. For untold centuries this river system provided a natural spawning area for salmon. The salmon fry thrived in the river, eventually growing the characteristic dark blotches on their bodies, and transformed from fry to parr. Then the salmon, now called smolt, their bodies now larger and covered with silver pigment, migrated to the ocean, where they thrived until it was time to return to the river and spawn (about four years later). In spawning season, the salmon instinctively homed their way toward the odor generated by the Rachel River, and up the river to their home waters.

Before the non-Native Americans arrived in this pristine wilderness region, nature, humans, and salmon lived in harmony and provided for each other. Nature gave the salmon the perfect habitat; the salmon provided Native Americans with sustenance. But after the non-Native Americans came to the Rachel River Valley, changes began. The salmon still ran the river, humans still fed on the salmon, but circumstances quickly altered. The non-Native Americans wanted more land, and Native Americans were forced to give way. As the area became more crowded, the Rachel River started to show the affects of civilization's influence.

In its natural course, sometimes the river flooded, creating problems for the settler populations. And everyone wanted power to maintain modern lifestyles—and hydropower poured down the Rachel River to the ocean constantly. So they built flood control systems and a dam to convert hydropower to hydroelectric power. The salmon still ran, but in reduced numbers and size. Soon local inhabitants couldn't catch the quantity and quality of salmon they had in the past. When the inconvenience finally struck home, they began to ask, "Where are the salmon?"

No one seemed to know. The time had come to call in the scientists. The scientists came and studied the situation, conducted testing, and decided that the salmon population needed to increase. They determined increased population could be achieved by building a fish hatchery, which would take the eggs from spawning salmon, raise the eggs to fingerling-sized fish, release them into specially built basins, and later, release them to restock the river.

A lot of science goes into the operation of a fish hatchery. It can't operate successfully on its own, but must be run by trained scientists and technicians following a proven protocol based on biological studies of salmon life-cycles. When the time was right, the salmon were released into the river. Other scientists and engineers realized that some mechanism had to be installed in the dam to allow the salmon to swim downstream to the ocean. In salmon lives (they are an anadromous species and therefore spend their adult lives at sea but return to freshwater to spawn), what goes down must go up (upstream). Those salmon would eventually need some way of getting back up past the dam and into home water, their spawning grounds. So the scientists and engineers devised, designed, built, and installed fish ladders in the dam so that the salmon could climb the ladders, scale the dam, and return to their native waters to spawn and die.

In a few seasons, the salmon again ran strong in the Rachel River. The scientists had temporarily solved the problem. But nothing in Nature is static or permanent. Things shift from static to dynamic in natural cycles that defy human intervention relatively quickly and without notice. In a few years, local Rachel River residents noticed an alarming trend. Studies over a five-year period showed that no matter how many salmon were released into the river, fewer and fewer returned to spawn each season. They called in the scientists again.

The scientists came in, analyzed the problem, and came up with five conclusions:

1) The Rachel River was extremely polluted both from point and nonpoint sources.
2) The Rachel River Dam had radically reduced the number of returning salmon to the spawning grounds.
3) Foreign fishing fleets off the Pacific Coast were depleting the salmon.
4) Native Americans are removing salmon downstream, before they reached the fish ladder.
5) The large volume of water withdrawn each year from the river for cooling machinery in local factories was having an effect. Large rivers with rapid flow rates usually can dissipate heat rapidly and suffer little ecological damage unless their flow rates are sharply reduced by seasonal fluctuations. This is not the case with the Rachel River. The large input of heated water from Rachel River area factories back into the Rachel River is having an adverse effect called *thermal pollution*. Thermal pollution and salmon do not mix. Increased water temperatures lower the dissolved oxygen (DO) content by decreasing the solubility of

oxygen in the river water, and warmer river water also causes aquatic organisms to increase their respiration rates and consume oxygen faster, increasing their susceptibility to disease, parasites, and toxic chemicals. Although salmon to some extent can survive in heated water, many other fish (the salmon's food supply) cannot. The heated discharge water from the factories also disrupts the spawning process, killing the young fry.

They had the causal factors defined—but what was the solution? Within days, the city officials hired an environmental engineering firm to study the salmon depletion problem. The environmentalists came up with the same causal conclusions as the scientists, which they related to the city officials, but they also related the political, economic, and philosophical implications of the situation to the city powers. The environmentalists explained that most of the pollution constantly pouring into the Rachel River would soon be eliminated when the city's new wastewater treatment plant came on line, and that point source pollution was eliminated. They explained that the state agricultural department and their environmental staff were working with farmers along the lower river course to modify farming practices and pesticide treatment regimes to help control nonpoint source pollution. The environmentalists explained that the Rachel River dam's present fish ladder was incorrectly configured, but could be modified with minor retrofitting. They explained that the over-fishing by the foreign fishing fleets off the Pacific Coast was a problem that the Federal government was working to resolve with the governments involved. The environmentalists explained that the State of Washington and the Federal Government were addressing the problem with the Native Americans fishing the down-river locations before the salmon ever reached the dam. Both governmental entities were negotiating with the local tribes on this problem, and the local tribes had pending litigation against the state and the federal government over who actually owned fishing rights to the Rachel River and the salmon. For the final problem, thermal pollution from the factories making the Rachel River unfavorable for spawning and/or killing off the young salmon fry, the environmentalists explained that to correct this problem the outfalls from the factories would have to be relocated. The environmentalists also recommended construction of a channel basin whereby the ready-to-release salmon fry could be released in a favorable environment, at ambient stream temperatures, and would have a controlled one-way route to a safe downstream location where they could thrive until time to migrate to the sea.

ENVIRONMENTAL SCIENCE AND TECHNOLOGY: THE CONNECTION

As long as capitalism drives most modern economies, people will still want and desire material things—precipitating a high level of consumption. For better or for worse, the human desire to lead the "good life" (which Americans may interpret as a life enriched by material possessions) is a fact of life. Arguing against someone who wants to purchase a new, modern home with all the amenities, who wants to purchase the latest, greatest automobile is difficult. Arguing against the person wanting to make a better life for his or her children by making sure they have all they need and want to succeed in their chosen pursuit is even harder. How do you argue with someone who earns his or her own way and spends his or her hard-earned money at will? Look at the tradeoff, though. The tradeoff often affects the environment. That new house purchased with hard-earned money may sit in a field of radon-rich soil. That new automobile may get only eight miles to the gallon. The boat they use on weekends gets even worse mileage, and exudes wastes into the local lake, river, or stream. Their weekend retreat on the five wooded acres is part of the watershed of the local community.

The environmental tradeoff never enters the average person's mind. Most people don't commonly think about it. In fact, we usually don't think about the environment until we damage it, until it becomes unsightly, until it is so fouled that it offends us. We can put up with a lot of abuse, especially with our surroundings—until the surroundings no longer please us. We treat our resources the same way. How often do we think about the air we breathe, the water we drink, the soil we plant our vegetables in? Not often enough.

Our typical attitude toward natural resources is often deliberate ignorance. Only when this person must wait in line for hours to fill the car gas tank does gasoline become a concern. Only when he can see—and smell—the air he breathes, and when he coughs when he inhales, does air become a visible resource. Water, the universal solvent, causes no concern (and very little thought) until shortages occur, or until it is so foul that nothing can live in it or drink it. Only when we lack water or the quality is poor do we think of water as a resource to "worry" about. Is soil a resource or is it "dirt"? Unless you farm or plant a garden, soil is only "dirt." It depends on what you use soil for—and on how hungry you are.

Resource utilization and environmental degradation are tied together. While we are dependent on resources and must use them, our use impacts the environment. A *resource* is usually defined as anything of use to mankind that is obtained from the physical environment. Some resources, such as edible growing plants, water (in many places), and fresh air, are directly available to us. But most resources, like coal, iron, oil, groundwater, game animals, and fish are not. They become resources only when we use science and technology to find them, extract them, process them, and convert them, at a reasonable cost, into usable and acceptable forms. Natural gas, found deep below the earth's surface, was not a resource until the technology for drilling a well and installing pipes to bring it to the surface became available. For centuries people stumbled across stinky, messy pools of petroleum and had no idea of its potential uses or benefits. When its potential was realized, we exploited petroleum by learning how to extract it and convert (refine) it into heating oil, gasoline, sulfur extract, road tar, and other products.

Excluding *perpetual resources* (solar energy, tides, wind, and flowing water), two different classes (types) of resources are available to us: renewable and nonrenewable (see Figure 1.3). *Renewable resources* (fresh air, fresh water, fertile soil, plants and animals (via genetic diversity) can be depleted in the short run if used or contaminated too rapidly, but normally will be replaced through natural processes in the long run. Because renewable resources are relatively plentiful, we often ignore, overlook, destroy, contaminate, and/or mismanage them.

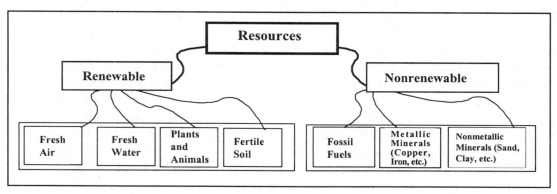

Figure 1.3 Perpetual and renewable resources

Classifying anything as "renewable" is a double-edged sword. Renewable resources are renewable only to a point. Timber or grass used for grazing must be managed for *maximum sustainable yield* (the highest rate at which a renewable resource can be used without impairing or damaging its ability to be fully renewed). If timber or grass yield exceed this rate, the system gives ever-diminishing returns. Recovery is complicated by the time-factor, which is life-cycle dependent. Grass can renew itself in a season or two. Timber takes decades. Any length of time is problematic when we get impatient.

One of the contributing factors of the plight of the Rachel River salmon was over-fishing. When a fishery is pushed past its limit, if the catch is maintained by collecting greater and greater numbers of younger salmon, no increase is possible. If the same practices are used on a wild species, extinction can result. We have no more passenger pigeons, dodos, solitaires, or great auks.

Exceeding maximum sustainable yield is only the tip of the iceberg—other environmental, social, and economic problems may develop. Let's look at *overgrazing* (depleting) grass on livestock lands. The initial problem occurs when the grass is depleted. But secondary problems kick in fast. Without grass, the soil erodes quickly. With time, so much soil is gone that the land is no longer capable of growing grass—or anything else. Productive land converted to non-productive deserts (*desertification*) is a process of *environmental degradation*—and it impacts social and economic factors. Those who depend on the grasslands must move on, and moving on costs time, energy, and money.

Environmental degradation is not limited to salmon and grass. Let's look at a few other examples. Along with over-fishing and overgrazing, land can also be over-cultivated. Intense over-cultivation reduces soil nutrients and increases erosion to the point where agricultural productivity is reduced. If irrigation of agricultural lands proceeds without proper drainage, the excessive accumulation of water or salts in the soil decreases productivity. Environmental degradation takes place when trees are removed from large areas without adequate re-planting. The results are destruction of wildlife habitat, increased soil erosion, and flooding. Land is often environmentally degraded when a metropolitan area expands. Often in growth areas, productive land is covered with concrete, asphalt, buildings, water, or silt to such an extent that agricultural productivity declines and wildlife habitat is lost. See Case Study 1.2.

Case Study 1.2
The Amish and Lancaster County, PA

Lancaster County, Pennsylvania is a case in point. Fortunate enough to possess some of the best farmland in the country, Lancaster also possesses a higher than average dewfall, and in years when counties around it are in drought, Lancaster can still bring in a reasonable crop. For many years, farming was the primary industry in the region, primarily accomplished by Amish farmers, working with minimum technology, teams of mules, and their neighbors and families. Lancaster County still has a high, steadily expanding Amish farmer population. Those who farm typically raise high-profit, labor-intensive crops, and garden produce for their own use. But farming is no longer their primary local industry. More than 50% of the Amish work in something other than farming because property values around Lancaster County have risen steeply in the last 20 years. Because developers, seeing the tourist trade increase, wish to increase the number of attractions that tourists can visit, the increased per-acre cost of good farmland means that a small farm that goes on the market now sells for an astronomical price.

This problem is compounded by local growth. Industrial growth and a heavy influx of population over the last 20 years have made the county expand quickly, especially in and around Lancaster City itself. Many of these people don't want to live in the city in row houses, though hundreds of homes in the city are for sale. They want new houses with yards, and good schools. They don't want to shop downtown, where businesses are dying. They want to shop at the outlet malls or in the strip malls in their own neighborhoods. This causes problems:

- The Amish can't afford to buy land for their sons to start farming on. They now often band together and purchase what land they can cooperatively, but even though their farming methods allow them to farm incredibly profitably, they can't compete with the developers for the per-acre prices.
- Every year more acres of prime farmland are bulldozed for health campuses, outlet stores, entertainment complexes, and sub-developments, while the tax base of Lancaster City moves to the suburbs and the city dies a slow death.
- Lancaster's urban sprawl is encompassing what only a few years ago were separate towns, 5 or 6 miles away. Now developments full of houses, strip malls, and car dealerships fill in those spaces.
- All this growth is springing up on what used to be farmland.
- Those people who move out into the developments built on former farms realize quickly that they don't like the smells associated with the farms still in their neighborhoods. They sometimes go so far as to take these cases to court.
- Low-lying areas suffer from flooding. When up-hill land that used to be open to the rain is covered with asphalt, run-off quickly overwhelms the creeks below, creating new flood plains in areas that weren't at risk before.
- Construction has fouled local creeks, releasing quantities of silt and dirt into the streams, altering habitat.
- The system of roads designed for the expected lower rates of growth predicted when the roads were built 30 or so years ago are clogged and congested with traffic. Route 30 from Gap to Lancaster (15 miles) can take 2 hours to travel in tourist season. Exhaust fumes build up until the air is grey, and people keep their car windows closed and air conditioners going so they don't have to breath it.

People everywhere are blind to their own conditions in some respects. Lancaster County is clearly changed—vastly changed from what it was twenty years ago. Lancastrians allowed free growth, free expansion for years. Now, Lancaster does recognize and work to combat these problems. They have conservancy programs designed to keep farmland in farming, but funds are limited—and the developers get the bank loans. The local residents recognize that a valuable natural resource is slipping away, but the problem is complicated by local economics, politics, sociology, religion, government, and business, as well as money interests outside the community. Any successful action taken to preserve farmland in one place is immediately countered by a loss somewhere else. Is there a solution? Yes. These problems could be solved. But to solve them, people who don't want to agree with one another will have to work together—people in local politics, business, and religion.

Nonrenewable resources (copper, coal, tin, and oil, among many others) have built up or evolved in a geological time-span. They can't be replaced at will—only over a similar time-scale. In this age of advanced technology, we often hear, for example, that when high-grade tin ore runs out (when 80% of its total estimated supply has been removed and used) low-grade tin ore (the other 20%) will become economically workable. This erroneous view neglects the facts of energy resource depletion and increasing pollution with lower grade burdens. In short, to find, extract, and process the remaining 20% generally costs more than the result is worth. Even with unlimited supplies of energy (impossible according to the *Laws of Thermodynamics*, which we will discuss later), what if we could extract that last 20%? When it is gone, nothing is going to bring it back except time measured in centuries and millennia.

Advances in technology have allowed us to make great strides in creating "the good life." These same technological advances have increased environmental degradation. But not all the news is bad. Technological advances have also let us (via recycling and reuse) conserve finite resources—aluminum, copper, iron, and glass, for example. *Recycling* involves collecting household waste items (aluminum beverage cans, for example) and reprocessing usable portions. *Reuse* involves using a resource over and over in the same form (refillable beverage bottles, water).

We discussed the so-called "good life" earlier—modern homes, luxury cars, boats, and that second home in the woods. With the continuing depletion of natural resources, prices must be forced upward until economically, attaining the "good life," or even gaining a foothold toward it, becomes difficult or impossible—and maintaining it becomes precarious. Ruthless exploitation of natural resources and the environment—overfishing a diminishing species (look at shark populations, for example), intense exploitation of energy and mineral resources, cultivation of marginal land without proper conservation practices, and the problems posed by further technological advances—will result in environmental degradation that will turn the "good life" into something we don't even want to think about.

To solve our problems, some would have us all "return to nature." These people suggest returning to Thoreau's Walden Pond on a large scale, giving up the "good life" to which we have become accustomed. They think that giving up the cars, boats, fancy homes, bulldozers that make construction and farming easier, pesticides that protect our crops, and medicines that improve our health and save our lives—the myriad material improvements that make our lives comfortable and productive—will solve the problem. Is this approach the answer, or even realistic? To a small minority, it is.

But not to the rest of us. To us it is a pipe dream, founded in romance, not logic. It cannot, should not, and will not happen. We can't abandon ship—we must prevent the need for abandoning our society from ever happening. Technological development is a boon to civilization, and will continue to be. Technological development isn't the problem—improper use of technology is. But we must continue to make advances in technology, we must find further uses for technology, and we must learn to use technology for the benefit of mankind and the environment. Technology and the environment must work hand in hand, not stand opposed.

Just how bad are the problems of misuse of technology on the environment? Let's take a closer look.

Major advances in technology have provided us with enormous transformation and pollution of the environment. While transformation is generally glaringly obvious (damming a river system, for example), "polluting" or "pollution" is not always as clear. What do we mean by pollution?

To *pollute* means to impair the purity of some substance or environment. *Air pollution* and *water pollution* refer to alteration of the normal compositions of air and water (their environmental quality) by the addition of foreign matter (gasoline, sewage).

Ways technology has contributed to environmental transformation and pollution include:

- The extraction, production, and processing of raw natural resources, such as minerals, with accompanying environmental disruption
- The manufacturing of enormous quantities of industrial products that consume huge amounts of natural resources and produce large quantities of hazardous waste and water/air pollutants
- Agricultural practices that have resulted in intensive cultivation of land, irrigation of arid lands, drainage of wetlands, and application of chemicals
- Energy production and utilization accompanied by disruption of soil by strip mining, emission of air pollutants, and pollution of water by release of contaminants from petroleum production
- Transportation practices (particularly reliance on the airplane) that cause scarring of land surfaces from airport construction, emission of air pollutants, and greatly increased demands for fuel (energy) resources

Throughout this text, we discuss the important aspects of the impact of technology on the environment.

WHEN PROPERLY CONNECTED, SCIENCE AND TECHNOLOGY OFFER SOLUTIONS

When technology is based on a sound foundation of environmental science and common sense, it can be used to solve environmental problems. In short, the goal is to produce manufacturing processes that have minimum environmental impact. This procedure has already been aptly demonstrated in the redesign of standard manufacturing processes. In these new environmentally friendly designs, the focus has been to minimize raw material and energy consumption and waste production. In this redesign process, one of the things that can be done is to construct a manufacturing process that can use raw materials and energy sources in ways that minimize environmental impact. When processing chemicals, reactions can be modified in such a way that the process is much more environmentally friendly. Another key change is in raw material and water usage. An environmentally friendly manufacturing process should be designed so that raw materials and water may be recycled. State-of-the-art technologies should be employed to minimize air, water, and solid waste emissions. A few of the ways in which technology can be applied to minimize environmental impact are listed in the following:

- Use of waste heat recovery systems to achieve maximum energy use, increase efficiency, and maximum utilization of fuel
- Use of precision machining and processing systems (e.g. lasers) to minimize waste production
- Optimization of process operations to increase efficiency
- Use of materials that minimize pollution
- Use of computerized control systems to achieve maximum energy efficiency, maximum utilization of raw materials, and minimum production of pollutants
- Application of processes that enable maximum materials recycling and minimum waste production.

- Application of advanced technologies to treat waste products efficiently

Advancements in technology are evolutionary. One advancement breeds other advancements. Each advancement builds on its predecessor to produce (evolve) a technology that is an improvement on its predecessors. The applications of technology to environmental improvements (the connection) that are available to us today are addressed throughout this text.

SUMMARY

When you throw a stone into a pool of quiet water, the ensuing ripples move out in concentric circles from the point of impact. Eventually, those ripples, much dissipated, reach the edge of the pond, where they break, disturbing the shore environment. When we alter our environment, similar repercussions affect the world around us—and some of these actions can be felt across the world. We use technology to alter our environment to suit our needs. That same technology can be put into effect so that our environment is protected from unrecoverable losses. Environmental scientists must maintain an acute sense of awareness of the global repercussions of the problems we create for the environment—to extend the boundaries of the problem beyond our own backyard.

Cited References

Leopold, A., *A Sand County Almanac*. New York: Ballentine Books, 1970.

Suggested Readings

Allaby, A. and Allaby, M., *The Concise Dictionary of Earth Sciences*. Oxford: Oxford University Press, 1991.

Arms, K., *Environmental Science*, 2nd ed., HBJ College and School Division, Saddle Brook, NJ, 1994.

Baden, J., and Stroup, R. C., (eds.). *Bureaucracy vs. Environment*. Ann Arbor: University of Michigan, 1981.

Botkin, D. B., *Environmental Science: Earth as a Living Planet*, Wiley, New York, NY, 1995.

Cobb, R. W., and Elder, C.D. *Participation in American Politics,* 2nd ed. Baltimore: John Hopkins, 1983.

Downing, P. B. *Environmental Economics and Policy.* Boston: Little, Brown, 1984.

Easterbrook, G., *A Moment on the Earth: The Coming Age of Environmental Optimism*, Viking Penguin, Bergenfield, NJ, 1995.

Field, B. C., *Environmental Economics: An Introduction,* 2nd ed., McGraw-Hill, New York, NY, 1996.

Franck, I. and Brownstone, D., *The Green Encyclopedia*. New York: NY, Prentice Hall, 1992.

Henry, J. G. and Heinke, G. W., *Environmental Science and Engineering*, 2nd Ed., Prentice Hall, New York, NY, 1995.

Jackson, A. R. and Jackson, J. M., *Environmental Science: The Natural Environment and Human Impact,* Longmand, New York, NY, 1996.

Lave, L. B., *The Strategy of Social Regulations: Decision Frameworks for Policy.* Brookings, Washington, D.C., 1981.

McHibben, B., *Hope, Human and Wild: True Stories of Living Lightly on the Earth,* Little Brown and Company, Boston, MA, 1995.

Miller, G. T., *Environmental Science: Working with the Earth,* 5th ed., Wadsworth Publishing Co., Belmont, CA, 1997.

Ophuls, W., *Ecology and the Politics of Scarcity.* New York: W. H. Freeman, 1977.

Pepper, I. L., Gerba, C. P., and Brusseau, M. L., *Pollution Science,* Academic Press Textbooks, San Diego, CA, 1996.

Spellman, F. R., *Stream Ecology and Self Purification: An Introduction for Wastewater and Water Specialists.* Lancaster, PA: Technomic Publishing, 1996.

Tower, E., *Environmental and Natural and Natural Resource Economics,* Eno River Press, New York, NY, 1995.

Walker, M., *The Nature of Scientific Thought.* Englewood Cliffs, N.J., Prentice-Hall, Spectrum Book, 1963.

CHAPTER 2

ENVIRONMENTAL SCIENCE: THE FUNDAMENTALS

*When you can measure what you are speaking about, and express it in numbers,
you know something about it; but when you cannot measure it, when you cannot
express it in numbers, your knowledge is of a meager and unsatisfactory kind; it
may be the beginning of knowledge, but you have scarcely...advanced to the
state of science. —Lord Kelvin*

CHAPTER OUTLINE

- Introduction
- Biogeochemical Cycles
- Energy Flow through an Ecosystem and the Biosphere
- Units of Measurement

INTRODUCTION

In this chapter we discuss fundamental concepts foundational to more complex material that
follows in subsequent chapters. The first area discussed covers the basics involved with the
circulation of matter through the ecosystem—the *biogeochemical cycles*. Because biogeochemical
cycles (and most other processes on the earth) are driven by energy from the sun, we present
energy and energy transfer next. The final section deals with units and measurement.

BIOGEOCHEMICAL CYCLES

To find out about our physical world, you must understand the natural biogeochemical cycles
that take place in our environment. Biogeochemical cycles are categorized into two types, the
gaseous and the *sedimentary*. Gaseous cycles include the carbon and nitrogen cycles. The atmosphere and the ocean are the main sinks of nutrients in the gaseous cycle. The sedimentary
cycles include the sulfur and phosphorous cycles. Soil and rocks of the earth's crust are the main
sinks for sedimentary cycles. These cycles are ultimately powered by the sun and are fine-tuned
and directed by energy expended by organisms. Another important cycle, the *hydrological cycle*
(to be discussed later), is also solar-powered and acts like a continuous conveyor system that
moves through the ecosystem materials essential for life.

Between twenty and forty elements of the earth's ninety-two naturally occurring elements are
ingredients that make up living organisms. The chemical elements carbon, hydrogen, oxygen,
nitrogen, and phosphorous are critical in maintaining life as we know. Of the elements needed
by living organisms to survive, oxygen, hydrogen, carbon, and nitrogen are needed in larger
quantities than some of the other elements. The point is that no matter what elements are needed
to sustain life, these elements exhibit definite biogeochemical cycles. For now, let's cover the
life-sustaining elements in greater detail.

The elements needed to sustain life are products of the global environment. The global environment consists of three main subdivisions:

1)   Hydrosphere—includes all the components formed of water bodies on the earth's surface.

2)   Lithosphere—comprises the solid components such as rocks.

3)   Atmosphere—is the gaseous mantle that envelops the hydrosphere and lithosphere.

To survive, organisms require inorganic metabolites from all three parts of the biosphere. For example, the hydrosphere supplies water as the exclusive source of needed hydrogen. Essential elements (calcium, sulfur, and phosphorus) are provided by the lithosphere. Finally, oxygen, nitrogen, and carbon dioxide are provided by the atmosphere.

Within the biogeochemical cycles, all the essential elements circulate from the environment to organisms and back to the environment. Because these elements are critically important for sustaining life, you can easily understand why the biogeochemical cycles are readily and realistically labeled "*nutrient cycles*."

Through these biogeochemical (or nutrient) cycles, nature processes and reprocesses the critical life-sustaining elements in definite inorganic-organic phases. Some cycles (the carbon cycle, for example) are more perfect than others—that is, the cycle loses no material in the process for long periods of time. Others are less perfect, but one essential point to keep in mind is that energy flows through an ecosystem (we'll explain how later), but nutrients are cycled and recycled.

Because humans need almost all the elements in our complex culture, we have speeded up the movement of many materials so that the cycles tend to become imperfect, or what Odum (1971) calls *acyclic*. One example of a somewhat imperfect (acyclic) cycle is demonstrated by our use of phosphate, which, of course, affects the phosphorus cycle. Phosphate rock is mined and processed with careless abandon, which leads to severe local pollution near mines and phosphate mills. We also increase the input of phosphate fertilizers in agricultural systems without controlling in any way the inevitable increase in run-off output that severely stresses our waterways and reduces water quality through *eutrophication*, the natural aging of a land-locked body of water.

In agricultural ecosystems, we often supply necessary nutrients in the form of fertilizer to increase plant growth and yield. In natural ecosystems, however, these nutrients are recycled naturally through each trophic level (feeding level). Elemental forms are taken up by plants. The consumers ingest these elements in the form of organic plant material. They cycle through the food chain from producer to consumer, and eventually the nutrients are degraded back to the inorganic form again.

The following sections present and discuss the nutrient cycles for carbon, nitrogen, phosphorus, and sulfur.

Carbon Cycle

Carbon, an essential ingredient for all living things and the basic building block of the large organic molecules necessary for life, is cycled into food webs from the atmosphere (see Figure 2.1).

Green plants obtain carbon dioxide (CO_2) from the air (Figure 2.1) and, through photosynthesis—probably the most important chemical process on earth—produce the food and oxygen that all organisms live on. Part of the carbon produced remains in living matter; the other part is released as CO_2 in cellular respiration and returned to the atmosphere.

Some carbon is contained in buried dead plant and animal materials. Much of these buried plant and animal materials were transformed into fossil fuels (coal, oil, and natural gas), which contain large amounts of carbon. When fossil fuels are burned, stored carbon combines with oxygen in the air to form carbon dioxide, which enters the atmosphere.

In the atmosphere, carbon dioxide acts as a beneficial heat screen—it does not allow the radiation of earth's heat into space. This balance is important. As more carbon dioxide is released into the atmosphere, that balance can (and is) being altered. Massive increases of carbon dioxide into the atmosphere tend to increase the possibility of global warming. The consequences of global warning might be catastrophic, and the resulting climate change may be irreversible. We'll discuss carbon dioxide and global warning more fully later in this text.

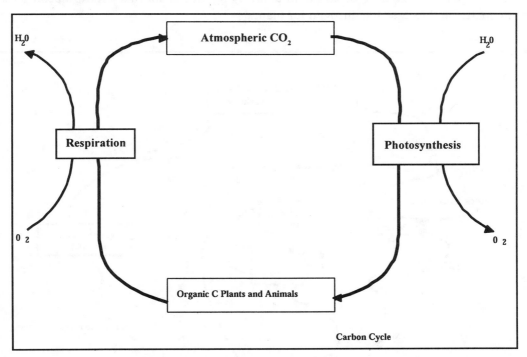

Figure 2.1 The carbon cycle

Nitrogen Cycle

The atmosphere contains 78% by volume of nitrogen. Nitrogen, an essential element for all living matter, constitutes 1-3% dry weight of cells, yet nitrogen is not a common element on earth. Although it is an essential ingredient for plant growth, nitrogen is chemically very inactive, and before it can be incorporated by the vast majority of the biomass it must be fixed.

Though nitrogen gas does make up about 78% of the volume of the earth's atmosphere, in that form it is useless to most plants and animals. Fortunately, nitrogen gas is converted into compounds containing nitrate ions, which are taken up by plant roots as part of the nitrogen cycle (shown in simplified form in Figure 2.2).

Aerial nitrogen is converted into nitrates mainly by microorganisms, bacteria, and blue-green algae. Lightning also converts some aerial nitrogen gas into forms that return to the earth as nitrate ions in rainfall and other types of precipitation. Ammonia plays a major role in the nitrogen cycle (see Figure 2.2). Excretion by animals and aerobic decomposition of dead organic matter by bacteria produce ammonia. Ammonia, in turn, is converted by nitrification bacteria into nitrites and then into nitrates. This process is known as *nitrification*. Nitrification bacteria are *aerobic*. Bacteria that convert ammonia into nitrites are known as *nitrite bacteria* (*Nitrosococcus* and *Nitrosomonas*); they convert nitrites into nitrates and nitrate bacteria (*Nitrobacter*).

Because nitrogen is often a *limiting factor* in naturally occurring soil, it can inhibit plant growth. To increase yields, farmers often provide extra sources of nitrogen by applying inorganic fertilizers or by spreading manure on the field and relying on the soil bacteria to decompose the organic matter and release the nitrogen for plant use.

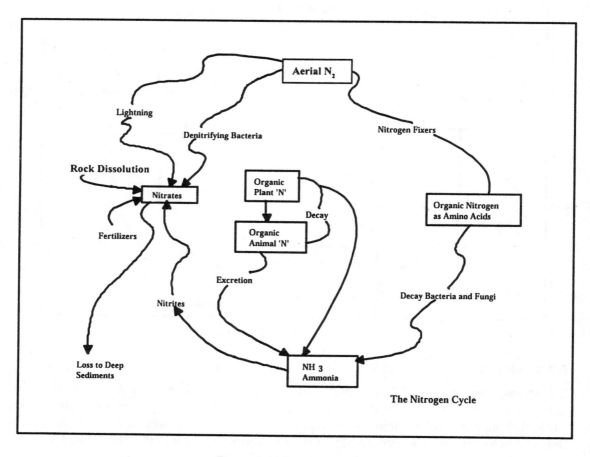

Figure 2.2 The nitrogen cycle

Phosphorus Cycle

Phosphorus is another element common in the structure of living organisms. The ultimate source of phosphorus is rock (see Figure 2.3). Phosphorus occurs as phosphate or other minerals formed in past geological ages. These massive deposits are gradually eroding to provide phosphorus to ecosystems. A large amount of eroded phosphorus ends up in deep sediments in the oceans and in lesser amounts in shallow sediments. Some phosphorus reaches land when marine animals are brought out. Birds also play a role in phosphorus recovery. The great guano deposit (bird excreta) of the Peruvian coast is an example. Humans have hastened the rate of phosphorus loss through mining and the production of fertilizers, which are washed away and lost.

Phosphorus has become very important in water quality studies, since it is often a limiting factor. Phosphates, upon entering a stream, act as fertilizer, which promotes the growth of undesirable algae blooms. As the organic matter decays, dissolved oxygen levels decrease, and fish and other aquatic species die.

Sulfur Cycle

Sulfur, like nitrogen, is characteristic of organic compounds. The sulfur cycle (see Figure 2.4) is both sedimentary and gaseous. Bacteria play a major role in the conversion of sulfur from one form to another. In an *anaerobic* environment, bacteria break down organic matter, thereby producing hydrogen sulfide with its characteristic rotten-egg odor. A bacterium called *Beggiatoa* converts hydrogen sulfide into elemental sulfur. An aerobic sulfur bacterium, *Thiobacillus thiooxidans,* converts sulfur into sulfates. Other sulfates are contributed by the dissolving of rocks and some sulfur dioxide. Sulfur is incorporated by plants into proteins. Some of these plants are then consumed by organisms. Sulfur from proteins is liberated as hydrogen sulfide by many heterotrophic anaerobic bacteria.

ENERGY FLOW THROUGH AN ECOSYSTEM AND THE BIOSPHERE

We often take energy for granted through a deceptive familiarity, because we think of it in so many different ways: atomic energy, food energy, cheap energy, abundant energy, and so on. This presents a huge double irony, because on one hand most people know that without energy our energy-dependent industrialized society would grind to a halt. On the other hand, energy is more than just the force that powers our machines and civilization; it powers hurricanes, the movement of the planets—the entire universe. Despite its pervasiveness and its familiarity, energy is a complex and puzzling concept. It cannot be seen, tasted, smelled, or touched. What is it? To answer this question we must first gain an understanding of *materials balance*.

Materials Balance

Probably the simplest way to express one of the most important but fundamental scientific principles is to point out that everything has to go somewhere. According to the *law of conservation of mass*, when chemical reactions take place, matter is neither created nor destroyed (exception: in a nuclear reaction, mass can be converted to energy). The importance of this concept in environmental science is that it allows us to track pollutants from one location to another using *mass balance equations*.

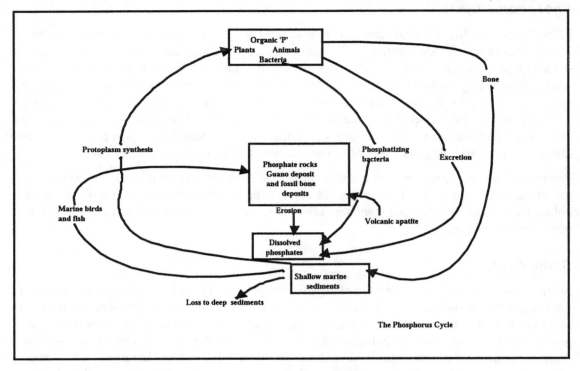

Figure 2.3 The phosphorus cycle

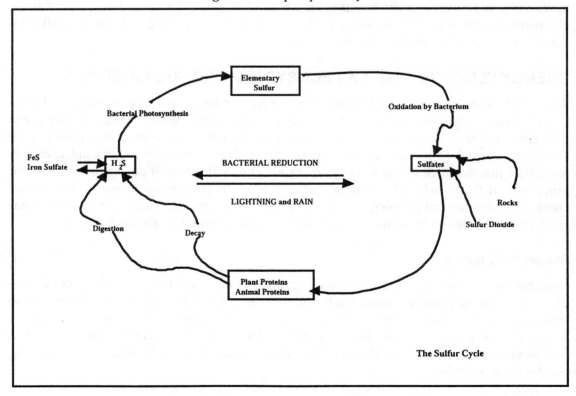

Figure 2.4 The sulfur cycle

To perform mass balance analysis, you must first define the particular region to be analyzed. The region you select could include anything—from a lake, a stretch of river or stream, an air basin above a city or factory, a chemical mixing vat, a coal-fired power plant, or the earth itself. Whatever region you select for analysis, you must confine the region with an imaginary boundary (see Figure 2.5). From such a region we can begin to identify the flow of materials across the boundary as well as the accumulation of materials within the region.

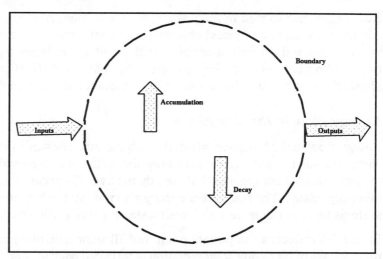

Figure 2.5 Materials balance diagram

When a material enters the region it has three possible fates: some of it may enter and slip through the region unchanged, some of it may accumulate within the boundary, and some of it may be converted—for example, CO to CO_2—to some other material. If we use Figure 2.5 as a guide, a materials balance equation can be written as follows:

Input rate = Output rate + Decay rate + Accumulation rate   (2.1)

Note that the decay rate in (2.1) does not imply a violation of the law of conservation of mass. There is no constraint on the change of one substance to another (chemical reactions), but atoms are conserved.

> NOTE: In practice, (2.1) often is simplified by assuming a steady state of equilibrium conditions (i.e., that nothing changed with time), but discussion of this practice is beyond the scope of this text and is generally presented in environmental engineering studies.

Let's get back to our discussion of energy. First of all, what is energy? Energy is often defined as the capacity for doing work, and work is often described as the product of force and the displacement of some object caused by that force.

Along with understanding and analyzing the flow of materials through a particular region, we can also determine and analyze the flow of energy. Using the *first law of thermodynamics*, we can write *energy balance equations*. The first law of thermodynamics states that energy cannot be created or destroyed. In short, energy may change forms in a given process, but we should be able to account for every bit of energy as it takes part in the process. In simplified form this relationship is shown in (2.2).

Energy in = Energy out   (2.2)

Equation (2.2) may give you the false impression that the transfer of energy in a process is 100% efficient. This is not the case. In a coal-fired electrical power generating plant, for example, only a portion of the energy from the burned coal is converted directly into electricity. A large portion of the coal-fired energy ends up as waste heat that is given off to the environment. This is the case, of course, because of the *second law of thermodynamics*, which states there will always be some waste heat. Devising a process or machine that can convert heat to work with 100% efficiency is impossible.

Heat can be transferred in three ways: by conduction, convection, and radiation. When direct contact between two physical objects at different temperatures occurs, heat will be transferred via *conduction* from the hotter object to the colder one. When a gas or liquid is placed between two solid objects, heat can be transferred by *convection*. Heat can also be transferred when no physical medium exists by *radiation* (e.g., radiant energy from the sun).

Energy Flow in the Biosphere

Energy flow in the biosphere all starts with the sun. The sun's radiant energy sustains all life on earth. The sun not only lights and warms the earth but also provides energy used by green plants to synthesize the compounds that keep them alive. These compounds serve as food for almost all other organisms. The sun's solar energy also powers the biochemical cycles and drives climate systems that distribute heat and fresh water over the earth's surface.

Figure 2.6 reflects an important point: not all solar radiant energy reaches the earth. Approximately 34% of incoming solar radiation is reflected back to space by clouds, dust, and chemicals in the atmosphere, and by the earth's surface. Most of the remaining 66% warms the atmosphere and land, evaporates water and cycles it through the biosphere, and generates winds. Surprisingly, only a small percentage (about 0.022%) is captured by green plants and used to make the glucose essential to life.

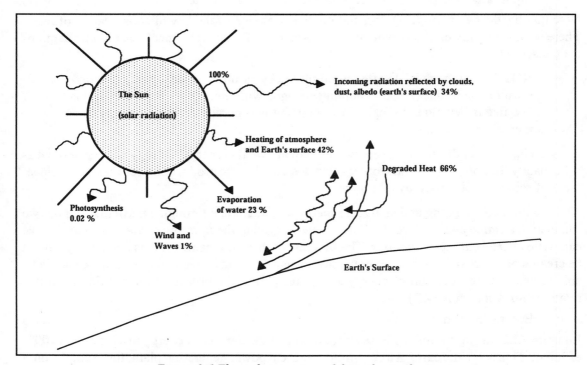

Figure 2.6 Flow of energy to and from the earth

Most of the incoming solar radiation not reflected away is degraded (or wasted) into longer-wavelength heat (in accordance with the second law of thermodynamics) and flows into space. The actual amount of energy that returns to space is affected by the presence of molecules of water, methane, carbon dioxide, and ozone, and by various forms of particulate matter in the atmosphere. Many of these barriers are created by man-made activities and might affect global climate patterns by disrupting the rate at which incoming solar energy flows through the biosphere and returns to space. We'll discuss the possible effects of human activities on climate in a later chapter.

Energy Flow in the Ecosystem

For an ecosystem to exist and to maintain itself, it must have energy. All activities of living organisms involve work—the expending of energy. This means the degradation of a higher state of energy to a lower state. The flow of energy through an ecosystem is governed by the two laws mentioned earlier: the first and second laws of thermodynamics.

Remembering that the first law, sometimes called the *conservation law*, states that energy may not be created or destroyed, and that the second law states that no energy transformation is 100% efficient sets the stage for a discussion of energy flow in the ecosystem. Hand in hand with the second law (some energy is always lost, dissipated as heat) is another critically important concept: *entropy*. Used as a measure of the non-availability of energy to a system, entropy increases with an increase in heat dissipation. Because of entropy, input of energy into any system is higher than the output or work done; the resultant efficiency is less than 100%.

Environmental scientists and technicians are primarily concerned with the interaction of energy and materials in the ecosystem. Earlier we discussed biogeochemical nutrient cycles and pointed out that the flow of energy drives these cycles. Energy does not cycle as nutrients do in biogeochemical cycles. For example, when food passes from one organism to another, energy contained in the food is reduced step by step until all the energy in the system is dissipated as heat.

This process has been referred to as a *unidirectional flow* of energy through the system, with no possibility for recycling of energy. When water or nutrients are recycled, energy is required. The energy expended in the recycling is not recyclable. As Odum (1975) points out, this is a "fact not understood by those who think that artificial recycling of man's resources is somehow an instant and free solution to shortages" (p. 61).

The principal source of energy for any ecosystem is sunlight. *Producers* (green plants: flowers, trees, ferns, mosses, and algae), through the process of *photosynthesis*, transform the sun's energy into carbohydrates, which are consumed by

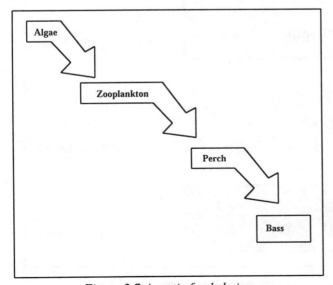

Figure 2.7 Aquatic food chain

animals. This transfer of energy, as stated earlier, is unidirectional—from producers to consumers. Often the transfer of energy to different organisms is called a *food chain*. Figure 2.7 shows a simple aquatic food chain.

All organisms, alive and dead, are potential sources of food for other organisms. All organisms that share the same general type of food in a food chain are said to be at the same *trophic level* (feeding level). Since green plants use sunlight to produce food for animals, they are called producers of the first trophic level. The herbivores eat plants directly, and are called the second trophic level or the *primary consumers*. The carnivores are flesh-eating consumers; they include several trophic levels from the third on up (see Figure 2.8). At each transfer, a large amount of energy (about 80 to 90%) is lost as heat and wastes. Nature normally limits food chains to four or five links. Note, however, that in aquatic food chains, the links are commonly longer than they are on land, because several predatory fish may be feeding on the plant consumers. Even so, the built-in inefficiency of the energy transfer process prevents development of extremely long food chains.

Only a few simple food chains are found in nature and most are interlocked. This interlocking of food chains forms a *food web*. A food web can be characterized as a map that shows what eats what. An organism in a food web may occupy one or more trophic levels. Food chains and webs help to explain how energy moves through the ecosystem.

An important trophic level of the food web that has not been discussed thus far is comprised of *decomposers*. The decomposers feed on dead plants or animals and play an important role in

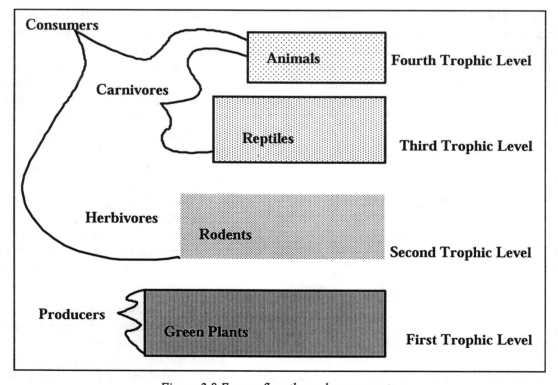

Figure 2.8 Energy flow through an ecosystem

recycling nutrients in the ecosystem. In an ecosystem, there is no waste. All organisms, alive or dead, are potential sources of food for other organisms.

UNITS OF MEASUREMENT

A basic knowledge of units of measurement and how to use them is essential for students of environmental science. Environmental science students and practitioners should be familiar both with the *U.S. Customary System (USCS) or English System* and the *International System of Units (SI)*. Some of the important units are summarized here to enable better understanding of material covered later in the text. Table 2.1 gives conversion factors between USCS and SI systems for some of the most basic units that will be encountered.

Table 2.1: Commonly Used Units and Conversion Factors

Quantity	SI units	SI symbol x	Conversion factor =	USCS units
Length	meter	m	3.2808	ft
Mass	kilogram	kg	2.2046	lb.
Temperature	Celsius	C	1.8 (C) + 32	F
Area	square meter	m^2	10.7639	ft^2
Volume	cubic meter	m^3	35.3147	ft^3
Energy	kilojoule	Kj	0.9478	Btu
Power	watt	W	3.4121	Btu/hr
Velocity	meter/second	m/s	2.2369	mi/hr

In the study of environmental science, you will commonly encounter both extremely large quantities and extremely small ones. The concentration of some toxic substance may be measured in parts per million or billion (ppm or ppb), for example. *PPM* may be roughly described as an amount contained in a shot glass in the bottom of a swimming pool. To describe quantities that may take on such large or small values, it is useful to have a system of prefixes that accompany the units. Some of the more important prefixes are presented in Table 2.2.

Table 2.2: Common Prefixes

Quantity	Prefix	Symbol	Quantity	Prefix	Symbol
10^{-12}	pico	p	10^{-1}	deci	d
10^{-9}	nano	n	10	deca	da
10^{-6}	micro	μ	10^2	hecto	h
10^{-3}	milli	m	10^3	kilo	
10^{-2}	centi	c	10^6	mega	M

Units of Mass

Simply defined, *mass* is a quantity of matter and measurement of the amount of inertia that a body possesses. Defining mass in a different way may help in understanding it. Mass expresses the degree to which an object resists a change in its state of rest or motion and is proportional to the amount of matter in the object. An even simpler way to understand mass is to think of it as the quantity of matter an object contains.

Beginning science students often confuse mass with *weight*, but they are different. Weight, for example, is the gravitational force action upon an object and is proportional to mass. In the SI system (a modernized metric system) the fundamental unit of mass is the *gram* (g). How does this stack up against weight?

To show the relationship between mass and weight, consider that there are 452.6 grams per pound. In laboratory-scale operations, the gram is a convenient unit of measurement. However, in real world applications the gram is usually prefixed with one of the prefixes shown in Table 2.2. For example, human body mass is expressed in kilograms (1 kg = 2.2 pounds). In everyday terms, a kilogram is the mass of one liter of water. When dealing with units of measurement pertaining to environmental conditions such as air pollutants and toxic water pollutants, they may be measured in teragrams (1 x 10^{12} grams) and micrograms (1 x 10^{-6} grams) respectively. When dealing with large-scale industrial commodities, the mass units may be measured in units of megagrams (Mg), which is also known as a metric ton.

Often mass and *density* are mistaken as signifying the same thing—they are not. Where mass is the quantity of matter and measurement of the amount of inertia that a body contains, density refers to how compacted a substance is with matter, or the mass per unit volume of an object, and its formula can be written:

$$\text{density} = \frac{\text{mass}}{\text{volume}} \tag{2.3}$$

Something with a mass of 25 kg that occupies a volume of 5 m³ would have a density of 25 kg/ 5 m³ = 5 kg/ m³. In this example the mass was measured in kilograms and the volume in cubic meters.

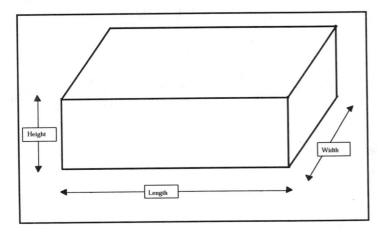

Figure 2.9 Length

Units of Length

In measuring locations and sizes we use the fundamental property of *length*, defined as the measurement of space in any direction. Space has three dimensions, each of which can be measured by length. This can be easily seen by considering the rectangular object shown in Figure 2.9. It has length, width, and height, but each of these dimensions is a length.

In the metric system, length is expressed in units based on the *meter* (m), which is 39.37 inches long. A kilometer (km) is equal to 1000 m and is used to measure relatively great distances. In practical laboratory applications the centimeter (cm = 0.01 m) is often used. There are 2.540 cm per inch, and the cm is employed to express lengths that would be given in inches in the English system. The micrometer (μm) is also commonly used to express measurements of bacterial cells and wavelengths of infrared radiation by which the earth re-radiates solar energy back to outer space. The nanometer (nm) (10^{-9}) is often used for measuring visible light (400 to 800 nm).

Units of Volume

The easiest way to approach measurements involving *volume* is to remember that volume is surface area multiplied by a third dimension. The *liter* is the basic metric unit of volume and is the volume of a decimeter cubed (1 L = 1 dm^3). A milliliter (ml) is the same volume as a cubic centimeter, cm^3.

Units of Temperature

Temperature is a measure of how 'hot' something is, or how much thermal energy it contains. Temperature is a fundamental measurement in environmental science, especially in most pollution work. The temperature of a stack gas plume, for example, determines its buoyancy and how far the plume of effluent will rise before attaining the temperature of its surroundings. This in turn determines how much it will be diluted before traces of the pollutant reach ground level.

Temperature is measured on several scales. For example, the *Celsius (centigrade)* and *Fahrenheit* scales are both measured from a reference point—the freezing point of water—which is taken as 0°C (or 32°F). The boiling point of water is taken as 100°C (or 212°F). Thermodynamic devices usually work in terms of absolute or "thermodynamic temperature," where the reference point is absolute zero, the lowest possible temperature attainable. For absolute temperature measurement, the thermodynamic unit or *Kelvin* (K) scale—which uses centigrade divisions for which zero is the lowest attainable measurement—is used. A unit of temperature on this scale is equal to a Celsius degree, and is not called a degree, but is called a Kelvin and designated as K, not °K. The value of absolute zero on the Kelvin scale is -273.15°C, so that the Kelvin temperature is always a number 273 (rounded) higher than the Celsius temperature. Thus water boils at 373 K and freezes at 273 K.

To convert from the Celsius scale to the Kelvin scale, simply add 273 to the Celsius temperature and you have the Kelvin temperature. Mathematically,

$$K = C + 273 \tag{2.4}$$

where   K = temperature on the Kelvin scale
          C = temperature on the Celsius scale

Converting from Fahrenheit to Celsius or vice versa is not so easy. The equations used are

$$C = 5/9(F - 32) \tag{2.5}$$

and

$$F = 9/5C + 32 \tag{2.6}$$

where   C = temperature on the Celsius scale
          F = temperature on the Fahrenheit scale

As examples, 15°C = 59°F and 68°F = 20°C. F or C, or both, of course, can be negative numbers.

Units of Pressure

Pressure is force per unit area and can be expressed in a number of different units, including the *atmosphere* (atm), the average pressure exerted by air at sea level, or the *pascal* (Pa), usually expressed in kilopascal (1 Kpa = 1000 Pa, and 101.3 Kpa = 1 atm). Pressure can also be given as millimeters of mercury (mm Hg), which is based on the amount of pressure required to hold up a column of mercury in a mercury barometer. 1 mm of mercury is a unit called the *torr* and 760 torr equal 1 atm.

Units Often Used in Environmental Studies

In environmental studies, often the concentration of some substance (foreign or otherwise) in air or water is of interest. In either medium, concentrations may be based on volume or weight, or a combination of the two (which may lead to some confusion). To understand how weight and volume are used to determine concentrations when studying liquids or gases/vapors, study the following explanations.

Liquids

Concentrations of substances dissolved in water are usually expressed in terms of weight of substance per unit volume of mixture. In environmental science, a good practical example of this weight per unit volume is best observed whenever a contaminant is dispersed in the atmosphere in solid or liquid form as a mist, dust, or fume. When this occurs, its concentration is usually expressed on a weight-per-volume basis. Outdoor air contaminants and stack effluents are frequently expressed as grams, milligrams, or micrograms per cubic meter, ounces per thousand cubic feet, pounds per thousand pounds of air, and grains per cubic foot. Most measurements are expressed in metric units. However, the use of standard U.S. units is justified for purposes of comparison with existing data, especially those relative to the specifications for air-moving equipment.

Alternatively, concentrations in liquids are expressed as weight of substance per weight of mixture, with the most common units being parts per million (ppm), or parts per billion (ppb).

Since most concentrations of pollutants are very small, one liter of mixture weighs essentially 1000 g, so that for all practical purposes we can write

$$1 \text{ mg/l} = 1 \text{ g/m}^3 = 1 \text{ ppm (by weight)} \tag{2.7}$$
$$1 \text{ μg/l} = 1 \text{ mg/m}^3 = 1 \text{ ppb (by weight)} \tag{2.8}$$

The environmental science practitioner may also be involved with concentrations of liquid wastes that may be so high that the *specific gravity* (the ratio of an object's or substance's weight to that of an equal volume of water) of the mixture is affected, in which case a correction to (2.7) and (2.8) may be required:

$$\text{mg/l} = \text{ppm (by weight)} \times \text{Specific gravity} \tag{2.9}$$

Gases/Vapors

For most air pollution work, by custom, we express pollutant concentrations in volumetric terms. For example, the concentration of a gaseous pollutant in parts per million (ppm) is the volume of pollutant per million volumes of the air mixture. That is:

$$ppm = \frac{\text{parts of contaminant}}{\text{million parts of air}} \qquad (2.10)$$

Calculations for gas and vapor concentrations are based on the gas laws. Briefly, these are as follows:

- The volume of gas under constant temperature is inversely proportional to the pressure.
- The volume of a gas under constant pressure is directly proportional to the Kelvin temperature. The Kelvin temperature scale is based on absolute zero (O°C = 372 K).
- The pressure of a gas of a constant volume is directly proportional to the Kelvin temperature.

When measuring contaminant concentrations, you must know the atmospheric temperature and pressure under which the samples were taken. At standard temperatures and pressure (STP), 1 gm-mol of an ideal gas occupies 22.4 liters (l). The STP is 0°C and 760 mmHg. If the temperature is increased to 25°C (room temperature) and the pressure remains the same, 1 g-mol of gas occupies 24.45 liters.

Sometimes you'll need to convert milligrams per cubic meter (mg/m³) (weight-per-volume ratio) into a volume-per-unit-volume ratio. If one gram-mole of an ideal gas at 25°C occupies 24.45 l, the following relationships can be calculated.

$$ppm = \frac{24.45 \ mg/m^3}{\text{molecular wt}} \qquad (2.11)$$

$$mg/m^3 = \frac{\text{molecular wt}}{24.45} \quad ppm \qquad (2.12)$$

SUMMARY

Environmental science, like any other true science, has a foundation in observation and numerical analysis. The biogeochemical cycles that allow and sustain life on our planet operate on levels beyond ordinary observation—a common enough problem in scientific analysis. Without the fundamental knowledge of the foundational biogeochemical cycles, energy's position and importance in all life-cycles, and the basics of how to quantify information gathered from the environments studied, you have not "advanced to the state of science" (Lord Kelvin).

Cited References

Odum, E. P., *Fundamentals of Ecology*. Philadelphia: Saunders College Publishing, 1971.

Odum, E. P., *Ecology: The link between the natural and the social sciences*. New York: Holt, Rinehart and Winston, Inc., 1975.

Suggested Readings

Bolin, B., and Cook, R. B., *The Major biogeochemical cycles and their interactions*. New York: Wiley, 1983.

Colinvaux, P., *Ecology*. New York: John Wiley and Sons, 1986.

Ehrlich, P. R., Ehrlich, A. H., and Holdren, J. P., *Ecoscience: Population, Resources, and Environment*. San Francisco: W. H. Freeman, 1977.

Kormondy, E. J., *Concepts of Ecology*, 3rd ed. Englewood Cliffs, NJ: Prentice-Hall, 1984.

Manahan, S. E., *Environmental Science and Technology*, Boca Raton, Fl: Lewis Publishers, 1997.

Odum, E. P., *Basic Ecology*. Philadelphia: Saunders College Publishing, 1983.

Porteous, A., *Dictionary of Environmental Science and Technology*. New York: John Wiley, 1992.

Ramade, F., *Ecology of Natural Resources*. New York: John Wiley, 1984.

Smith, R. E., *Ecology and Field Biology*. New York: Harper & Row, 1980.

Spellman, F. R., *Stream Ecology and Self-Purification for Wastewater and Water Specialists*. Lancaster, PA: Technomic Publishing Company, 1996.

Wanielista, M. P., Yousef, Y. A., Taylor, J. S., and Cooper, C.D., *Engineering and the Environment*. Monterey, CA: Brooks/Cole Engineering Division, 1984.

CHAPTER 3

ENVIRONMENTAL CHEMISTRY

This grand show is eternal. It is always sunrise somewhere; the dew is never all dried at once; a shower is forever falling; vapor is ever rising. Eternal sunrise, eternal sunset, eternal dawn and gloaming, on sea and continents and islands, each in its turn, as the round earth rolls. —John Muir

CHAPTER OUTLINE

- Introduction
- What is Chemistry?
- Elements and Compounds
- Classification of Elements
- Physical and Chemical Changes
- The Structure of the Atom
- Periodic Classification of the Elements
- Molecules and Ions
- Chemical Bonding
- Chemical Formulas and Equations
- Molecular Weights, Formulas, and the Mole
- Physical and Chemical Properties of Matter
- States of Matter
- The Gas Laws
- Liquids and Solutions
- Thermal Properties
- Acid + Bases ⇒ Salts
- Organic Chemistry
- Environmental Chemistry

INTRODUCTION

Why do we need to study chemistry for environmental science? In fact, a study of chemistry is important for environmental science or for any other science. In a general sense, consider that on foundational levels chemistry affects everything we do. Not a single moment of time goes by during which we are not affected in some way by a chemical substance, chemical process, or chemical reaction. Chemistry affects every aspect of our daily lives.

In a specific sense, consider that almost every environmental and pollution problem we face has a chemical basis. In short, in environmental studies, to examine such problems as the greenhouse effect, ozone depletion, groundwater contamination, toxic wastes, air pollution, stream pollution, and acid rain without some fundamental understanding of basic chemical concepts would be difficult, if not impossible. And of course, an environmental practitioner who must likewise solve environmental problems and understand environmental remediation clean-up processes, such as emission control systems or waste treatment facilities, must be well grounded in chemical principles and the techniques of chemistry in general, because many of these techniques are being used to solve environmental problems.

The environmental science student or interested reader who uses this text may or may not have some fundamental knowledge of chemistry. In this chapter, the topics have been selected with the goal of reviewing or providing only the essential chemical principles required to understand the nature of the environmental problems we face, and the chemistry involved with scientific and technological approaches to their solutions.

WHAT IS CHEMISTRY?

Chemistry is the science concerned with the composition of matter (gas, liquid, or solid) and of the changes that take place in it under certain conditions.

Every substance, material, and/or object in the environment is either a chemical substance or mixture of chemical substances. Your body is made up of chemicals, literally thousands of them. The food we eat, the clothes we wear, the fuel we burn, and the vitamins we take in from natural or synthetic sources are all products of chemistry wrought either by the forces of nature or the hand of man. Chemistry is about matter—its actual makeup, constituents, and consistency. It is about measuring and quantifying matter.

All matter can exist in three states: gas, liquid, or solid. It is composed of minute particles termed *molecules*, which are constantly moving, and may be further divided into *atoms*.

Molecules that contain atoms of one kind only are known as *elements*; those that contain atoms of different kinds are called *compounds*.

Chemical compounds are produced by a chemical action that alters the arrangements of the atoms in the reacting molecules. Heat, light, vibration, catalytic action, radiation, or pressure, as well as moisture (for ionization) may be necessary to produce a chemical change. Examination and possible breakdown of compounds to determine their components is *analysis*, and the building up of compounds from their components is *synthesis*. When substances are brought together without changing their molecular structures, they are said to be *mixtures*.

Organic substances consist of virtually all compounds that contain carbon. All other substances are *inorganic substances*.

ELEMENTS AND COMPOUNDS

A *pure substance* is a material that has been separated from all other materials. Examples of such substances (which are indistinguishable from each other no matter what procedures are used to purify them or what their origin is) are copper metal, aluminum metal, distilled water, table sugar, and oxygen. All samples of table sugar are alike and indistinguishable from all other sugar samples.

Usually expressed in terms of percentage by mass, a substance is characterized as a material having a fixed composition. Distilled water, for example, is a pure substance consisting of approximately 11% hydrogen and 89% oxygen by mass. By contrast, a lump of coal is not a pure substance because its carbon content may vary from 35% to 90% by mass. Materials (like coal) that are not pure substances are mixtures. When substances can be broken down into two or more simpler substances, they are called compounds.

When substances cannot be broken down or decomposed into simpler forms of matter, they are called elements. The elements are the basic substances of which all matter is composed. At the present time there are only 100+ known elements, but there are well over a million known compounds. Of the 100+ elements, only 88 are present in detectable amounts on Earth, and many of these 88 are rare. Ten elements make up approximately 99% by mass of the Earth's crust, including the surface layer, the atmosphere, and the bodies of water (see Table 3.1). From this table it is apparent that the most abundant element on Earth is oxygen, which is found in the free state in the atmosphere, as well as in combined form with other elements in numerous minerals and ores.

Table 3.1 also lists the symbols and atomic number of the ten chemicals listed. The symbols consist of either one or two letters, with the first letter capitalized. The *atomic number* of an element is the number of protons in the nucleus.

Table 3.1: ***Elements Making Up 99% of the Earth's Crust, Oceans, and Atmosphere***

Element	Symbol	% of Composition	Atomic number
Oxygen	O	49.5%	8
Silicon	Si	25.7%	14
Aluminum	Al	7.5%	13
Iron	Fe	4.7%	26
Calcium	Ca	3.4%	20
Sodium	Na	2.6%	11
Potassium	K	2.4%	19
Magnesium	Mg	1.9%	12
Hydrogen	H	1.9%	1
Titanium	Ti	0.58%	22

CLASSIFICATION OF ELEMENTS

Each element may be classified as a metal, nonmetal, or metalloid. *Metals*, elements that are typically lustrous solids, are good conductors of heat and electricity, melt and boil at high temperatures, possess relatively high densities, and are normally malleable (can be hammered into sheets) and ductile (can be drawn into a wire). Examples of metals are copper, iron, silver, and platinum. Almost all metals are solids (none are gaseous) at room temperature (mercury being the only exception).

Elements that do not possess the general physical properties just mentioned (i.e., they are poor conductors of heat and electricity, boil at relatively low temperatures, do not possess luster, and are less dense than metals) are called *nonmetals*. Most nonmetals (the exception is bromine, a liquid) are either solids or gases at room temperature. Nitrogen, oxygen, and fluorine are examples of gaseous nonmetals, while sulfur, carbon, and phosphorus are examples of solid nonmetals.

Several elements have properties resembling both metals and nonmetals. They are called *metalloids* (semimetal). The metalloids are boron, silicon, germanium, arsenic, tellurium, antimony, and polonium.

PHYSICAL AND CHEMICAL CHANGES

Internal linkages among a substance's units (between one atom and another) maintain their constant composition. These linkages are called *chemical bonds*. When a particular process occurs that involves the making and breaking of these bonds, we say that a *chemical change* or *chemical reaction* has occurred. In environmental science, combustion and corrosion are common examples of chemical changes that impact our environment.

Let's briefly consider a couple of examples of chemical change. When a flame is brought into contact with a mixture of hydrogen and oxygen gases, a violent reaction takes place. The covalent bonds in the hydrogen (H_2) molecules and of the oxygen (O_2) molecules are broken and new bonds are formed to produce molecules of water, H_2O.

The key point to remember is that whenever chemical bonds are broken, formed, or both, a chemical change takes place. The hydrogen and oxygen undergo a chemical change to produce water, a substance with new properties.

When mercuric oxide, a red powder, is heated, small globules of mercury are formed and oxygen gas is released. This mercuric oxide is changed chemically to form molecules of mercury and molecules of water.

By contrast, a *physical change* (nonmolecular change) is one in which the molecular structure of a substance is not altered. When a substance freezes, melts, or changes to vapor, the composition of each molecule does not change. For example, ice, steam, and liquid water all are made up of molecules containing two atoms of hydrogen and one atom of oxygen. A substance can be ripped or sawed into small pieces, ground into powder, or molded into a different shape without changing the molecules in any way.

The types of behavior that a substance exhibits when undergoing chemical changes are called its *chemical properties*. The characteristics that do not involve changes in the chemical identity of a substance are called its *physical properties*. All substances may be distinguished from one another by these properties, in much the same way that certain features (DNA, for example) distinguish one human being from another.

THE STRUCTURE OF THE ATOM

If a small piece of an element, say copper, is hypothetically divided and subdivided and subdivided into the smallest piece possible, the result would be one particle of copper. This smallest unit of the element, which is still representative of the element, is called an *atom*.

Although infinitesimally small, the atom is composed of particles, principally electrons, protons, and neutrons. The simplest atom possible consists of a *nucleus* having a single *proton* (positively charged particles) with a single *electron* (negatively charged particles) traveling around it. This is an atom of hydrogen, which, we say, has an atomic weight of one because of the single proton. The *atomic weight* of an element is equal to the total number of protons and *neutrons* (neutral particles) in the nucleus of an atom of an element. Electrons and protons bear the same magnitude of charge, but are of opposite polarity.

The hydrogen atom also has an atomic number of one because of its one proton. The atomic number of an element is equal to the number of protons in its nucleus. A neutral atom has the same number of protons and electrons. Therefore, in a neutral atom the atomic number is also equal to the number of electrons in the atom. The number of neutrons in an atom is always equal to or greater than the number of protons, except in the atom of hydrogen.

The protons and neutrons of an atom reside in the nucleus. Electrons reside primarily in designated regions of space surrounding the nucleus, called *atomic orbitals* or *electron shells*. Only a prescribed number of electrons may reside in a given type of electron shell. With the exception of hydrogen (which only has one electron), two electrons are always close to the nucleus in an atom's innermost electron shell. In most atoms, other electrons are located in electron shells some distance from the nucleus.

While neutral atoms of the same element have an identical number of electrons and protons, they may differ by the number of neutrons in their nuclei. Atoms of the same element having different numbers of neutrons are called *isotopes* of that element.

PERIODIC CLASSIFICATION OF THE ELEMENTS

Through experience, scientists discovered that the chemical properties of the elements repeat themselves. Chemists summarize all such observations in the *periodic law*: The properties of the elements vary periodically with their atomic numbers.

In 1869, Dimitri Mendeleev, using relative atomic masses, developed the original form of what today is known as the *periodic table*. The periodic table is a chart of elements arranged in order of increasing proton number to show the similarities of chemical elements with related electronic configurations. The elements fall into vertical columns, known as *groups*. Going down a group, the atoms of the elements all have the same outer shell structure, but have an increasing number of inner shells. Traditionally, the alkali metals are shown on the left of the table and the groups are numbered IA to VIIA, IB to VIIB, and 0 (for noble gases). Now we more commonly classify all the elements in the middle of the table as transition elements and regard the nontransition elements as *main-group* elements, numbered from I to VII, with the noble gases in group 0. Horizontal rows in the table are *periods*. The first three are called *short periods*; the next four (which include transition elements) are *long periods*. Within a period, the atoms of all the elements have the same number of shells, but with a steadily increasing number of electrons in the outer shell.

The periodic table is an important tool for learning chemistry because it tabulates a variety of information in one spot. For example, we can immediately determine the atomic number of the elements because they are tabulated on the periodic table (see Figure 3.1). We can also readily identify which elements are metals, nonmetals, and metalloids. Usually a bold zigzag line separates metals from nonmetals, while those elements lying to each immediate side of the line are metalloids. Metals fall to the left of the line, and nonmetals fall to the right of it.

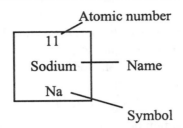

Figure 3.1: The element sodium as it is commonly shown in one of the horizontal boxes in the periodic table

MOLECULES AND IONS

When elements other than noble gases (which exist as single atoms) exist in either the gaseous or liquid state of matter at room conditions, they consist of units containing pairs of like atoms. These units are called *molecules*. For example, we generally encounter oxygen, hydrogen, chlorine, and nitrogen as gases. Each exists as a molecule having two atoms. These molecules are symbolized by the notations O_2, H_2, Cl_2, and N_2, respectively.

The smallest particle of many compounds is also the molecule. Molecules of compounds contain atoms of two or more elements. The water molecule, for example, consists of two atoms of hydrogen and one atom of oxygen (H_2O). The methane molecule consists of one carbon atom and four hydrogen atoms (CH_4).

Not all compounds occur naturally as molecules. Many of them occur as aggregates of oppositely charged atoms or groups of atoms called *ions*. Atoms become charged by gaining or losing some of their electrons. Atoms of metals, for example, that lose their electrons become positively charged, and atoms of nonmetals that gain electrons become negatively charged.

CHEMICAL BONDING

When compounds form, the atoms of one element become attached to, or associated with, atoms of other elements by forces called *chemical bonds*. Chemical bonding is a strong force of attraction holding atoms together in a molecule. There are various types of chemical bonds. *Ionic* bonds can be formed by transfer of electrons. For instance, the calcium atom has an electron configuration of two electrons in its outer shell. The chlorine atom has seven outer electrons. If the calcium atom transfers two electrons, one to each chlorine atom, it becomes a calcium ion with the stable configuration of an inert gas. At the same time, each chlorine, having gained one electron, becomes a chlorine ion, also with an inert-gas configuration. The bonding in calcium chloride is the electrostatic attraction between the ions.

Covalent bonds are formed by sharing of *valence* (the number of electrons an atom can give up or acquire to achieve a filled outer shell) electrons. Hydrogen atoms, for instance, have one outer electron. In the hydrogen molecule, H_2, each atom contributes 1 electron to the bond. Consequently, each hydrogen atom has control of two electrons—one of its own and the second from the other atom—giving it the electron configuration of an inert gas. In the water molecule, H_2O, the oxygen atom, with six outer electrons, gains control of two extra electrons supplied by the two hydrogen atoms. Similarly, each hydrogen atom gains control of an extra electron from the oxygen.

Chemical compounds are often classified into either of two groups based on the nature of the bonding between their atoms. As you might expect, chemical compounds consisting of atoms bonded together by means of ionic bonds are called *ionic compounds*. Compounds whose atoms are bonded together by covalent bonds are called *covalent compounds*.

There are some interesting contrasts between most ionic and covalent compounds. For example, ionic compounds have higher melting points, boiling points, and solubility in water than covalent compounds. Ionic compounds are nonflammable, while covalent compounds are flammable. Ionic compounds that are molten in water solutions conduct electricity. Molten covalent compounds do not conduct electricity. Ionic compounds generally exist as solids at room temperature, while covalent compounds exist as gases, liquids, and solids at room temperature.

CHEMICAL FORMULAS AND EQUATIONS

Chemists have developed a shorthand method of writing *chemical formulas*. Elements are represented by groups of symbols called formulas. A common compound is sulfuric acid; its formula is H_2SO_4. The formula indicates that the acid is composed of two atoms of hydrogen, one atom of sulfur, and four atoms of oxygen. However, this is not a recipe for making the acid. The formula does not tell you how to prepare the acid, only what is in the acid.

A *chemical equation* tells what elements and compounds are present before and after a chemical reaction. Sulfuric acid poured over zinc will cause the release of hydrogen and the formation of zinc sulfate. This is shown by the following equation:

$$Zn \quad + \quad H_2SO_4 \quad \Rightarrow \quad ZnSO_4 \quad + \quad H_2$$
(Zinc) (sulfuric acid) (zinc sulfate) (hydrogen) $\qquad\qquad$ (3.1)

One atom (also one molecule) of zinc unites with one molecule of sulfuric acid giving one molecule of zinc sulfate and one molecule (two atoms) of hydrogen. Notice that the same number of atoms of each element still exists on each side of the arrow. However, the atoms are combined differently.

MOLECULAR WEIGHTS, FORMULAS, AND THE MOLE

The relative weight of a compound that occurs as molecules is called the *molecular weight*, the sum of the atomic weights of each atom that comprises the molecule. Consider the water molecule. Its molecular weight is determined as follows:

$$
\begin{aligned}
&\text{2 hydrogen atoms} = 2 \times 1.008 \quad = 2.016 \\
&\text{1 oxygen atom} = 1 \times 15.999 \quad = 15.999 \\
&\overline{\qquad\qquad\qquad\qquad\qquad\qquad\qquad} \\
&\text{Molecular weight of } H_2O \quad\;\; = 18.015
\end{aligned}
$$
(3.2)

Thus, the molecular weight of a molecule is simply the sum of the atomic weights of all of the constituent atoms. If we divide the mass of a substance by its molecular weight, the result is the mass expressed in *moles* (mol). Usually the mass is expressed in grams, in which case the moles are written as *g-moles*; in like fashion, if mass is expressed in pounds, the result would be *lb.-*

moles. One g-mole contains 6.022 x 10²³ molecules (*Avogadro's number*, in honor of the scientist who first suggested its existence), and one lb.-mole about 2.7 x 10²⁶ molecules.

$$\text{Moles} = \frac{\text{Mass}}{\text{Molecular Weight}} \tag{3.3}$$

The relative weight of a compound that occurs as formula units is called the *formula weight*. It is the sum of the atomic weights of all atoms that comprise one formula unit. Consider sodium fluoride. Its formula weight is determined as follows:

$$
\begin{aligned}
1 \text{ sodium ion} &= 22.990 \\
1 \text{ fluoride ion} &= \underline{18.998} \\
\text{Formula weight of NaF} &= 41.988
\end{aligned}
\tag{3.4}
$$

PHYSICAL AND CHEMICAL PROPERTIES OF MATTER

Two basic types of properties (characteristics) of matter exist: physical and chemical. *Physical properties* of matter are those that do not involve a change in the chemical composition of the substance. Among these properties are hardness, color, boiling point, electrical conductivity, thermal conductivity, specific heat, density, solubility, and melting point. These properties may change with a change in temperature or pressure. Those changes that do not alter chemical composition of the substance are called *physical changes*. When heat is applied to solid ice to convert it to liquid, no new substance is produced, but the appearance has changed—melting is a physical change. Other examples of physical changes are dissolving sugar in water, heating a piece of metal, and evaporating water.

The physical properties that are most commonly used in describing and identifying particular kinds of matter are density, color, and solubility. *Density* (d) is mass per unit volume and is expressed by the equation

$$d = \frac{\text{mass}}{\text{volume}} \tag{3.5}$$

All matter has weight and takes up space; it also has density, which depends on weight and space. We commonly say that a certain material will not float in water because it is *heavier* than water. What we really mean is that a particular material is *denser* than water. The density of an element differs from the density of any other element. The densities of liquids and solids are normally given in units of grams per cubic centimeter (g/cm³), which is the same as grams per milliliter (g/ml). The advantage of using the physical property of *color* is that no chemical or physical tests are required. *Solubility* refers to the degree to which a substance dissolves in a liquid, such as water. In environmental science, the density, color, and solubility of a substance are important physical properties that aid in the determination of various pollutants, stages of pollution or treatment, the remedial actions required to clean up toxic/hazardous waste spills, and other environmental problems.

The properties involved in the transformation of one substance into another are known as *chemical properties*. For example, when a piece of wood burns, oxygen in the air unites with the different substances in the wood to form new substances. When iron corrodes, during the corrosion process, oxygen combines with the iron and water to form a new substance commonly known as rust. Changes that result in the formation of new substances are known as *chemical changes*.

STATES OF MATTER

The three common states (or phases) of matter (solid state, liquid state, gaseous state) each have unique characteristics. In the *solid state*, the molecules or atoms are in a relatively fixed position. The molecules are vibrating rapidly, but about a fixed point. Because of this definite position of the molecules, a solid holds its shape. *A solid occupies a definite amount of space and has a fixed shape.*

When the temperature of a gas is lowered, the molecules of the gas slow down. If the gas is cooled sufficiently, the molecules slow down so much that they lose the energy needed to move rapidly throughout their container. The gas may turn into *liquid*. Common liquids are water, oil, and gasoline. *A liquid is a material that occupies a definite amount of space, but which takes the shape of the container.*

In some materials, the atoms or molecules have no special arrangement at all. Such materials are called *gases*. Oxygen, carbon dioxide, and nitrogen are common gases. *A gas is a material that takes the exact volume and shape of its container.*

Although the three states of matter discussed above are familiar to most students and others, the change from one state to another is of primary interest to environmentalists. Changes in matter that include water vapor changing from the gaseous state to liquid precipitation or a spilled liquid chemical changed to a semi-solid substance (by addition of chemicals, which aids in the cleanup effort) are just two ways changing from one state to another has impact on environmental concerns.

THE GAS LAWS

The atmosphere is composed of a mixture of gases, the most abundant of which are nitrogen, oxygen, argon, carbon dioxide, and water vapor (gases and the atmosphere are addressed in greater detail later). The *pressure* of a gas is the force that the moving gas molecules exert upon a unit area. A common unit of pressure is newton per square meter, N/m_2, called a pascal (Pa). An important relationship exists among the pressure, volume, and temperature of a gas. This relation is known as the *ideal gas law* and can be stated as

$$\frac{P_1 V_1}{T_1} = \frac{P_2 V_2}{T_2} \tag{3.6}$$

where P_1, V_1, T_1 are pressure, volume, and absolute temperature at time 1, and P_2, V_2, T_2, are pressure, volume, and absolute temperature at time 2. A gas is called perfect (or ideal) when it obeys this law.

A temperature of 0°C (273 K) and a pressure of 1 atmosphere (atm) have been chosen as *standard temperature and pressure (STP)*. At STP the volume of 1 mole of ideal gas is 22.4 l.

LIQUIDS AND SOLUTIONS

The most common solutions are liquids. However, solutions, which are homogenous mixtures, can be solid, gaseous, or liquid. The substance in excess in a solution is called the *solvent*. The substance dissolved is the *solute*. Solutions in which water is the solvent are called *aqueous solutions*. A solution in which the solute is present in only a small amount is called a *dilute solution*. If the solute is present in large amounts, the solution is *concentrated solution*. When the maximum amount of solute possible is dissolved in the solvent, the solution is called a *saturated solution*.

The concentration, or amount of solute dissolved, is frequently expressed in terms of the molar concentration. The *molar concentration*, or *molarity*, is the number of moles of solute per liter of solution. Thus a one molar solution, written 1.0M, has one gram formula weight of solute dissolved in one liter of solution. In general

$$\text{Molarity} = \frac{\text{moles of solute}}{\text{number of liters of solution}} \tag{3.7}$$

Note that the *number of liters of solution*, not the number of liters of solvent, is used.

Example: Exactly 40 g of sodium chloride (NaCl), or table salt, were dissolved in water and the solution was made up to a volume 0.80 liter of solution. What was the molar concentration, M, of sodium chloride in the resulting solution?

Answer: First find the number of moles of salt.

$$\text{Number of moles} = \frac{40 \text{ g}}{58.5 \text{ g/mole}} = 0.68 \text{ mole} \tag{3.8}$$

$$\text{Molarity} = \frac{0.68 \text{ mole}}{0.80 \text{ liter}} = 0.85 \text{M} \tag{3.9}$$

THERMAL PROPERTIES

Thermal properties of chemicals and other substances are important to the environmental practitioner. Such knowledge is used in hazardous materials spill mitigation and in solving many other complex environmental problems. *Heat* is a form of energy. Whenever work is performed, usually a substantial amount of heat is caused by friction. The conservation of energy law tells us the work done plus the heat energy produced must equal the original amount of energy available. That is,

$$\text{Total energy} = \text{work done} + \text{heat produced} \tag{3.10}$$

As environmental scientists, technicians, and/or practitioners, we are concerned with several properties related to heat for particular substances. Those thermal properties that we need to be familiar with are discussed in the following sections.

A traditional unit for measuring heat energy is the calorie. A *calorie* (cal) is defined as the amount of heat necessary to raise one gram of pure liquid water by one degree Celsius at normal atmospheric pressure.

In SI units

$$1 \text{ cal} = 4.186 \text{ J (Joule)} \qquad\qquad (3.11)$$

The calorie we have defined should not be confused with the one used when discussing diets and nutrition. A *kilocalorie* is 1000 calories as we have defined it. That is, a kilocalorie is the amount of heat necessary to raise the temperature of one kilogram of water by 1°C.

In the British system of units, the unit of heat is the British thermal unit, or *Btu*. One Btu is the amount of heat required to raise one pound of water one degree Fahrenheit at normal atmospheric pressure (1 atm).

Specific Heat

Earlier we pointed out that one kilocalorie of heat is necessary to raise the temperature of one kilogram of water one degree Celsius. Other substances require different amounts of heat to raise the temperature of one kilogram of the substance one degree. The *specific heat* of a substance is the amount of heat in kilocalories necessary to raise the temperature of one kilogram of the substance one degree Celsius.

The units of specific heat are Kcal/kg°C, or, in SI units, J/kg°C. The specific heat of pure water, for example, is 1.000 kcal/kg°C. This is 4186 J/kg°C.

The greater the specific heat of a material, the more heat is required. Also, the greater the mass of the material or the greater the temperature change desired, the more the heat that is required.

The amount of heat necessary to change one kilogram of a solid into a liquid at the same temperature is called the *latent heat of fusion* of the substance. The temperature of the substance at which this change from solid to liquid takes place is known as the *melting point*. The amount of heat necessary to change one kilogram of a liquid into a gas is called the *latent heat of vaporization*. When this point has been reached, the substance is all in the gas state. The temperature of the substance at which this change from liquid to gas occurs is known as the *boiling point*.

ACIDS + BASES ⇒ SALTS

When acids and bases are combined in the proper proportions, they neutralize each other, each losing its characteristic properties and forming a salt and water.

$$NaOH + Hcl \Rightarrow NaCl + H_2O \hspace{4cm} (3.12)$$

which is

sodium hydroxide + hydrochloric acid ⇒ sodium chloride + water

The acid-base-salt concept originated with the beginning of chemistry and is very important in the environment and the life processes, and as industrial chemicals.

The word acid is derived from the Latin *acidus*, which means sour. The sour taste is one of the properties of acids (however, you should never actually taste an acid in the laboratory or anywhere else). An *acid* is a substance that, in water, produces hydrogen ions, H^+, and has the following properties:

1)   conducts electricity

2)   tastes sour

3)   changes the color of blue litmus paper to red

4)   reacts with a base to neutralize its properties

5)   reacts with metals to liberate hydrogen gas

A *base* is a substance that produces hydroxide ions, OH^-, and/or accepts H^+, and when dissolved in water has the following properties:

1)   conducts electricity

2)   changes the color of red litmus paper to blue

3)   tastes bitter and feels slippery

4)   reacts with an acid to neutralize the acid's properties.

pH Scale

A common way to determine whether a solution is an acid or a base is to measure the concentration of hydrogen ions (H^+) in the solution. The concentration can be expressed in powers of 10, but is more conveniently expressed as *pH*. For example, pure water has 1×10^{-7} grams of hydrogen ions per liter. The negative exponent of the hydrogen ion concentration is called the pH of the solution. The pH of water is 7, a neutral solution. A concentration of 1×10^{-12} has a pH of 12. A pH less than 7 indicates an acid solution and a pH greater than 7 indicates a basic solution (see Table 3.2).

Table 3.2 Standard pH Scale. The "p" is for the German word "poentz," and the "H" stands for hydrogen.

pH	Concentration of H- Ions	Acidic/Basic
1	1.0×10^{-1} mole/liter	Very Acidic
2	1.0×10^{-2} mole/liter	
3	1.0×10^{-3} mole/liter	
4	1.0×10^{-4} mole/liter	
5	1.0×10^{-5} mole/liter	
6	1.0×10^{-6} mole/liter	Acidic
7	1.0×10^{-7} mole/liter	Neutral
8	1.0×10^{-8} mole/liter	Basic
9	1.0×10^{-9} mole/liter	
10	1.0×10^{-10} mole/liter	
11	1.0×10^{-11} mole/liter	
12	1.0×10^{-12} mole/liter	
13	1.0×10^{-13} mole/liter	
14	1.0×10^{-14} mole/liter	Very Basic

Table 3.3: Approximate pH of Common Substances

SUBSTANCE	pH
Battery acid	0.0
Gastric juice	1.2
Lemons	2.3
Vinegar	2.8
Soft drinks	3.0
Apples	3.1
Grapefruit	3.1
Wines	3.2
Oranges	3.5
Tomatoes	4.2
Beer	4.5
Bananas	4.6
Carrots	5.0
Potatoes	5.8
Coffee	6.0
Milk (cow)	6.5
Pure Water —NEUTRAL—	7.0
Blood (human)	7.4
Eggs	7.8
Sea water	8.5
Milk of magnesia	10.5
Oven cleaner	13.0

The pH of substances found in our environment varies in value. Acid-base reactions are among the most important in environmental science. In diagnosing various environmental problems (acid rain problems, hazardous materials spills into lakes and ponds), pH value is critical. Remediation or prevention are also important. To protect local ecosystems, wastes often require neutralization before being released into the environment. Another important aspect of pH control is seen in waste treatment (wastewater treatment), where removing nitrogen is essential. If not removed, nitrogen stimulates growth of algae in the receiving body of water. (Table 3.3 gives an approximate pH of some common substances.)

ORGANIC CHEMISTRY

Organic chemistry is the branch of chemistry concerned with compounds of carbon. The science of organic chemistry is incredibly complex and varied. Millions of different organic compounds are known today, and 100,000+ of these are products of synthesis, unknown in nature. In this text, about all we can do is provide a very basic introduction to some of the most common organic substances important to environmental science (because of their toxicities as pollutants and other hazards) so they will be more familiar when we encounter them later in the text.

Before 1828, scientists thought that organic compounds could only be made by plants and animals. In that year, Friedrich Wohler made urea from ammonium cyanate. Wohler's discovery disproved the theory that stated that urea (and thus all organic compounds) could only be made by living things. Because of his discovery, the science of organic chemistry was born.

Organic compounds are components of all the familiar commodities that our technological world requires—motor and heating fuels, adhesives, cleaning solvents, paints, varnishes, plastics, refrigerants, aerosols, textiles, fibers and resins, among many others.

From an environmental science perspective, the principal concern about organic compounds is that they are pollutants of water, air, and soil environments. As such, they are safety and health hazards. They are also combustible or flammable substances, with few exceptions. From a health standpoint, they have the ability to cause a wide range of detrimental health effects. In humans, some of these compounds damage the kidneys, liver, and heart, others depress the central nervous system, and several are suspected to cause cancer. If human beings are subject to such health hazards from these compounds, for the environmental scientist, the logical question that follows is: What about their impact on delicate ecosystems?

Organic Compounds

The molecules of organic compounds have one common feature: one or more carbon atoms that covalently bond to other atoms; that is, pairs of electrons are shared between atoms. A carbon atom may share electrons with other nonmetallic atoms and also with other carbon atoms. As Figure 3.2 shows, methane, carbon tetrachloride, and carbon monoxide are compounds having moles in which the carbon atom is bonded to other nonmetallic atoms.

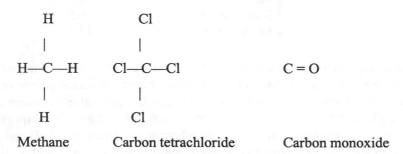

Figure 3.2 Carbon atoms sharing their electrons with the electrons of other nonmetallic atoms, like hydrogen, chlorine, and oxygen. The compounds that result from such electron sharing are methane, carbon tetrachloride, and carbon monoxide, respectively.

When carbon atoms share electrons with other carbon atoms, we find that two carbon atoms may share electrons in such a manner that they form either of the following: carbon-carbon single bonds (C - C), carbon-carbon double bonds (C =), or carbon-carbon triple bonds (C ≡). Each bond written here as a dash (—) is a shared pair of electrons. Figure 3.3 illustrates the bonding of molecules of ethane, ethylene, and acetylene—compounds having molecules with only two carbon atoms. Molecules of ethane possess carbon-carbon single bonds, molecules of ethylene possess carbon-carbon double bonds, and molecules of acetylene possess carbon-carbon triple bonds.

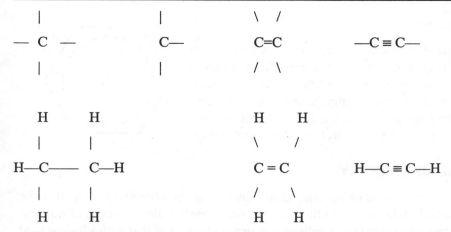

Figure 3.3 Two carbon atoms may share their own electrons in any of the three ways noted. When these carbon atoms further bond to hydrogen atoms, the resulting compounds are ethane, ethene, and acetylene, respectively.

Covalent bonds between carbon atoms in molecules of more complex organic compounds may be linked into chains, including branched chains, or into rings.

Hydrocarbons

The simplest organic compounds are the *hydrocarbons*—compounds whose molecules are composed only of carbon and hydrogen atoms. All hydrocarbons are broadly divided into two groups: aliphatic and aromatic hydrocarbons.

Aliphatic hydrocarbons

Aliphatic hydrocarbons are those that can be characterized by the chain arrangements of their constituent carbon atoms. They are divided into the alkanes, alkenes, and alkynes.

The *alkanes*, also called *paraffins* or *aliphatic hydrocarbons*, are saturated hydrocarbons (its hydrogen content is at maximum) with the general formula C_nH_{2n+2}. In systematic chemical nomenclature, alkane names end in the suffix *-ane*. They form the alkane series methane (CH_4), ethane (C_2H_6), propane (C_3H_8), butane (C_4H_{10}), etc. The lower members of the series are gases; the high-molecular weight alkanes are waxy solids. Alkanes are present in natural gas and petroleum.

Alkenes (olefins) are unsaturated hydrocarbons (can take on hydrogen atoms to form saturated hydrocarbons) that contain one or more double carbon-carbon bonds in their molecules. In systematic chemical nomenclature, alkene names end in the suffix *-ene*. Alkenes that have only one double bond form the alkene series starting with ethene—the gas that is liberated when food rots—(ethylene), $CH_2:CH_2$; propene, $CH_3CH: CH_2$, etc.

Alkynes (acetylenes) are unsaturated hydrocarbons that contain one or more triple carbon-carbon bonds in their molecules. In systematic chemical nomenclature alkyne names end in the suffix *-yne*—acetylene, $H—C \equiv C—H$.

Aromatic Hydrocarbons

Aromatic hydrocarbons are those unsaturated organic compounds that contain a benzene ring in their molecules or that have chemical properties similar to benzene, a clear, colorless, water-insoluble liquid that rapidly vaporizes at room temperature, whose molecular formula is C_6H_6. The molecular structure of benzene is commonly represented by a hexagon with a circle inside as shown here:

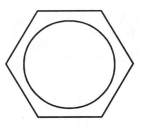

ENVIRONMENTAL CHEMISTRY

Environmental chemistry is a blend of aquatic, atmospheric, and soil chemistry, as well as the "chemistry" generated by human activities thereon. We stated earlier that the focus of this text deals with the three major environmental media: air, water, and soil, and that with a loss or total damage to any of these three there could be no life, as we know it, on earth. This is the case, of course, and the environmental effects brought about by humans will not be ignored here.

Actually, understanding that the three environmental media are what they are because of many interwoven scientific principles (including chemistry) is most important. We are interested in the chemistry that makes up these mediums as well as the chemical reactions that take place to preserve or destroy them. As we proceed, we concern ourselves with the environmental impact of human activities—mining, acid rain, erosion from poor cultivation practices, disposal of hazardous wastes, photochemical reactions (smog), air pollutants from particulate matter to the greenhouse effect, and ozone and water degradation problems related to organic, inorganic, and biological pollutants. All of these activities and problems have something to do with chemistry. The remediation and/or mitigation processes to repair them are also tied to chemistry.

To say that environmental science, environmental studies, and environmental engineering are built upon a strong foundation of chemistry is to more than mildly understate chemistry's real importance and relevance in the field.

SUMMARY

Chemical actions and reactions accompany everything we do, every day. For environmental scientists to perform their work properly, they must possess more than just the chemical basis of how mankind alters and affects his environment. Chemistry is part of the material at the foundation of environmental science—without understanding and mastery of the science of chemistry, you will not master the science of our environment.

Suggested Readings

A Concise Dictionary of Chemistry, Oxford: Oxford University Press, 1990.

Andrews, J. E., *Environmental Chemistry*, Cambridge, MA: Blackwell Science Publications, 1996.

Bohn, H. L., McNeal, B. L., and O'Connor, G. A., *Soil Chemistry*, 2nd ed. New York: John Wiley and Sons, 1985.

Henry, J. G., and Heinke, G. W., *Environmental Science and Engineering*, Englewood Cliffs, NJ: Prentice-Hall, Inc., 1989.

Manahan, S. E., *Fundamentals of Environmental Chemistry*, Boca Raton, FL: CRC Press/Lewis Publishers, 1993.

Masters, G. M., *Introduction to Environmental Engineering and Science*, Englewood Cliffs, NJ: Prentice-Hall, 1991.

Meyer, E., *Chemistry of Hazardous Materials*, 2nd ed., Englewood Cliffs, NJ: Prentice-Hall, 1989.

Peavy, H. S., Rowe, D. R., and Tchobanoglous, G., *Environmental Engineering*, New York: McGraw-Hill, 1985.

Shipman, J. T., Adams, J. L., and Wilson, J.D., *An Introduction to Physical Science*, 5th ed. Lexington, Massachusetts: D.C. Heath and Company, 1987.

Snoeyink, V. L., and Jenkins, D., *Water Chemistry*, New York: Wiley, 1980.

Spiro, T. G. and Stigliani, W. M., *Chemistry of the Environment*, Upper Saddle River, NJ: Prentice-Hall, 1996.

Stumm, W. and Morgan, J. J., *Aquatic Chemistry: Chemical Equilibria and Rates in Natural Waters*, New York: Wiley, 1996.

CHAPTER 4

ENVIRONMENTAL BIOLOGY

TINY ENEMY, BIG WORRIES

The rugged residents of [Saxis, Virginia]—population: "about 300, give or take a few," says Mayor Charles Tull—are worried they might be wiped out.

Erosion from waves and storms off the Chesapeake Bay may get them in the long run. But the source of their anxiety these days is a deadly microbe, believed to be hanging around in the neighboring Pocomoke River—and, more important, the consumer fears now spreading across the nation about eating seafood pulled from waters here.

Saxis...relies almost exclusively on the bounty of Pocomoke Sound and the Pocomoke River to make ends meet. "It's a scary thought if they close down Pocomoke Sound," Aubrey Justice, president of the Pocomoke Sound Watermen's Association, said as he leaned against the sun-drenched dock here... "You take away the crabs and you can pretty much take Saxis off the map. Just take the scissors and cut us off the map."—Scott Harper, The Virginian-Pilot, *Norfolk, 1997*

CHAPTER OUTLINE
- Introduction
- Microbiology
- The Cell
- Bacteria
- Viruses
- Fungi
- Algae
- Enzymes
- Metabolic Transformations
- Pathogenicity

INTRODUCTION

The preceding newspaper article points to an environmental problem caused by an infestation of the microbe *pfiesteria*, which has killed a billion fish in the coastal waterways of North Carolina and has been discovered as far north as Virginia and Maryland. On the surface, the microbe *pfiesteria*—which is a common organism, but can take on a toxic form under certain environmental conditions—is the problem. But there is more to this problem. *Pfiesteria* seem to thrive where manure washes into the waterways. Scientists believe that pollution triggers the toxic transformation. What we are left with is a compound environmental problem, a chain of events leading to a chain of conditions—manure rich waters, which provide a habitat for the fish-killing microbe *pfiesteria*.

Just as the science of chemistry is essential to environmental science, so is the science of biology. In addition to a fundamental knowledge of chemistry, this text assumes readers have a rudimentary knowledge of biological sciences—in particular, microbiology. We emphasize microbiology because of the positive and negative influences tiny microbes have on our environment. We also include some of the basic tenets of biochemistry that deal with enzymes and metabolic processes. These compliment our discussion of environmental microbiology.

Biology is the science of life. Strictly speaking, biology includes all the life sciences—anatomy and physiology, cytology, zoology, botany, ecology, genetics, biochemistry and biophysics, animal behavior, embryology, and microbiology. In this text we are concerned with the micro-life—the study of microbiology.

We focus on microbiology for several reasons. One is that we are concerned with those "things" (the micro-life) that impact our environment—our air, our water, our soil—our lives. Another important reason has to do with the new research topics that have emerged and the emphasis now being increasingly placed on the biological treatment of hazardous wastes and the detection and control of new pathogens. For example, the field of water/wastewater microbiology has blossomed during the last two decades as new modern tools have been developed to study the role of microorganisms in the treatment of water and wastewater.

Another important reason for emphasizing microbiology is that we have witnessed dramatic advances in the methodology for detection of pathogenic microorganisms and parasites in various environmental samples. Environmental practitioners and microbiologists are increasingly interested in toxicity (to be discussed in the next chapter) and the biodegradation of *xenobiotics* (defined as any chemical that is present in a natural environment but that does not normally occur in nature, e.g., pesticides and/or industrial pollutants) by aerobic and anaerobic biological processes in wastewater treatment plants. Thus, the essence of this chapter is an exploration of the interface between environmental studies and microbiology, which will hopefully lead to fruitful interactions between microbiologists and environmental practitioners.

MICROBIOLOGY

Microbiology is the study of organisms of microscopic (cannot be seen without the aid of a microscope) dimensions. *Microbiologists* are scientists concerned with studying the form, structure, reproduction, physiology, metabolism, and identification of microorganisms. The microorganisms they study generally include bacteria, fungi, protozoa, algae, and viruses. These tiny organisms make up a large and diverse group of free-living forms that exist either as single cells or cell bunches or clusters.

Microscopic organisms can be found in abundance almost anywhere on earth. The vast majority of microorganisms are not harmful. Many microorganisms, or microbes, occur as single cells (unicellular); others are multicellular; still others (viruses) do not have a true cellular appearance.

Because microorganisms exist as single cells or cell bunches, they are unique and distinct from the cells of animals and plants, which are not able to live alone in nature but can exist only as part of multicellular organisms. A single microbial cell, for the most part, exhibits the characteristic features common to other biological systems, such as metabolism, reproduction, and growth.

Classification

For centuries, scientists classified the forms of life visible to the naked eye as either animal or plant. Much of the current knowledge about living things was organized by the Swedish naturalist Caroulus Linnaeus in 1735.

The importance of classifying organisms cannot be overstated. Without a classification scheme, how could we establish criteria for identifying organisms and arranging similar organisms into groups? The most important reason for classification is that a standardized system allows us to handle information efficiently—it makes the vastly diverse and abundant natural world less confusing.

Linnaeus's classification system was extraordinarily innovative. His *binomial system of nomenclature* is still with us today. Under the binomial system, all organisms are generally described by a two word scientific name, the g*enus* and *species*. Genus and species are groups that are part of a hierarchy of groups of increasing size, based on their nomenclature (taxonomy). This hierarchy is shown as follows:

Kingdom

    Phylum

       Class

          Order

            Family

              Genus

                Species

Using this hierarchy and Linnaeus's binomial system of nomenclature, the scientific name of any organism (as stated previously) includes both the genus and the species name. The genus name is always capitalized, while the species names begins with a lowercase letter. On occasion, when there is little chance for confusion, the genus name will be abbreviated with a single capital letter. The names are always in Latin, so they are usually printed in italics or underlined. Some organisms also have English common names. Some microbe names of interest, for example, are listed as follows:

- *Salmonella typhi* - the typhoid bacillus
- *Escherichia coli* - a coliform bacteria
- *Giardia lamblia* - a protozoan

Escherichia coli is commonly known as simply *E. coli*, while *Giardia lamblia* is usually referred to by only its genus name, *Giardia*.

Let's take a look, for example, at a simplified system of microorganism classification used in water and wastewater treatment. Classification is broken down into the kingdoms of animal, plant, and protista. As a general rule the animal and plant kingdoms contain all the multi-celled organisms, and the protists contain all single-celled organisms. Along with microorganism classification based on the animal, plant, and protista kingdoms, microorganisms can be further classified as being eucaryotic or procaryotic (See Table 4.1). A *eucaryotic* organism is characterized by a cellular organization that includes a well-defined nuclear membrane. A *procaryotic* organism is characterized by having a nucleus that lacks a limiting membrane.

Table 4.1: Simplified Classification of Microorganisms

Kingdom	Members	Cell Classification
Animal		Rotifers
	Crustaceans	
	Worms and larvae	Eucaryotic
Plant	Ferns	
	Mosses	
Protista	Protozoa	
	Algae	
	Fungi	
	Bacteria	Procaryotic
	Lower Algae Forms	

THE CELL

Since the 19th century, we have known that all living things, whether animal or plant, are made up of cells. The fundamental unit of all living things, no matter how complex, is the cell. A typical cell is an entity isolated from other cells by a membrane or cell wall. The cell membrane contains protoplasm (the living material found within them) and the nucleus.

In a typical mature plant cell, the cell wall is rigid and is composed of non-living material, while in the typical animal cell, the wall is an elastic living membrane. Cells exist in a very great variety of sizes and shapes, as well as functions. Their average size ranges from bacteria too small to be seen with the light microscope to the largest single cell known, the ostrich egg. Microbial cells also have an extensive size range, some being larger than human cells.

Cell Structure

The cell is the unit of paramount importance to all living organisms and is the fundamental unit of life. Cells consist of a small body on the order of micrometers in size. Just as there are two classifications of microorganisms, *procaryotic* and *eucaryotic*, there are two kinds of cells classified as *procaryotic* and *eucaryotic*. Single cell bacteria are composed of prokaryotic cells, primitive and relatively simpler than the eucaryotic cells, which comprise all organisms other than bacteria. A simplified representation of a cell is shown in Figure 4.1.

Eucaryotic cells are described in the following. Prokaryotic (bacteria) cells are described in detail later in this chapter.

Eucaryotic cells can contain a cell membrane, cell nucleus, cytoplasm, mitochondria, ribosomes, cell walls, vacuoles, and chloroplasts.

- *Cell membrane* (cytoplasmic membrane)—the lipid- and protein-containing selectively permeable membrane that surrounds the cytoplasm in procaryotic and eucaryotic cells. In most types of microbial cells, the cell membrane is bordered externally by the cell wall. In microbial cells, the precise composition of the cell membrane depends on the species, on growth conditions, and on the age of the cell

- *cell nucleus*—a distinct region not delimited by a membrane in which at least some species of RNA are synthesized and assembled into ribonucleoprotein subunits of ribosomes
- *cytoplasm*—the ground substance of cells located between the nucleus and the cell membrane
- *mitochondria*—commonly called the "power house" of a cell, an *organelle* (a specialized structure within cells) in which aerobic respiration produces the energy molecule, ATP

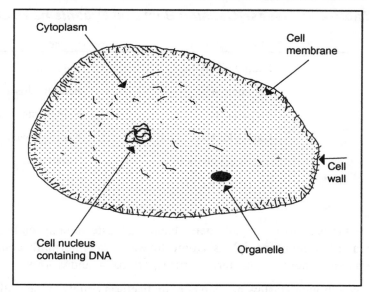

Figure 4.1 Basic features of a cell

- *ribosomes*—minute particles found attached to endoplasmic reticulum or loose in the cytoplasm that are the site of protein synthesis
- *endoplasmic reticulum*—a complex system of tubules, vesicles, and sacs in cells, sometimes having attached ribosomes
- *cell walls*—in plants, strong structures composed mostly of cellulose that provide stiffness and strength
- *vacuoles*—various membrane-delimited compartments within a cell
- *chloroplasts*—in plants, the sites for photosynthesis

BACTERIA

Of all microorganisms, bacteria are the most widely distributed, the smallest in size, the simplest in morphology (structure), the most difficult to classify, and the hardest to identify. Because of their considerable diversity, even providing a descriptive definition of what a bacterial organism is can be difficult. About the only generalizations that can be made for the entire group is that they are single-celled plants, prokaryotic, are seldom photosynthetic, and reproduce by binary fission.

Bacteria are found everywhere in our environment. They are present in soil, water, and the air. Bacteria are also present in and on the bodies of all living creatures, including man. Most bacteria do not cause disease. They are not pathogenic. Many bacteria carry on useful and necessary functions related to the life of larger organisms.

However, when we think about bacteria in general terms, we usually think of the damage they cause. In water, for example, the form of water pollution that poses the most direct menace to human health is *bacteriological contamination*. This is partly the reason that bacteria are of great significance to water and wastewater specialists. For water treatment personnel tasked with providing the public with safe potable water, disease-causing bacteria pose a constant challenge (See Table 4.2).

Table 4.2: Disease-Causing Bacterial Organisms Found in Polluted Water

Microorganism	Disease
Salmonella typhi	Typhoid fever
Salmonella sp.	Salmonellosis
Shigella sp.	Shigellosis
Campylobacter jejuni	Campylobacter enteritis
Yersinia entercolitice	Yersiniosis
Escherichia coli	

As far as controlling pathogenic bacteria, wastewater specialists share the same challenge as do water treatment specialists. Domestic wastewater normally contains huge quantities of microorganisms, including bacteria, viruses, protozoa, and worms.

Even though wastewater can contain bacteria counts in the millions per ml, in wastewater treatment under controlled conditions, bacteria can help to destroy pollutants in wastewater. In such a process, bacteria function to stabilize organic matter (e.g., activated biosolids [sludge] processes) and thereby assist the treatment process in performing as designed—to produce effluent that does not impose an excessive oxygen demand on the receiving body.

How Well Do We Know Bacteria?

The conquest of disease has placed bacteria high on the list of microorganisms of great interest to the scientific community. There is more to this interest and accompanying large research effort than just an incessant search for understanding and the eventual conquest of disease-causing bacteria. Not all bacteria are harmful to man. Some, for example, produce substances (antibiotics) which have helped in the fight against disease. Others are used to control insects that attack crops. Bacteria also have an impact on the natural cycle of matter. Bacteria work to increase soil fertility, which increases the potential for more food production. With the burgeoning world population, increasing future food productivity is no small matter.

We still have a lot to learn about bacteria, because we are still principally engaged in making observations and collecting facts, trying wherever possible to relate one set of facts to another but still lacking much of a basis for grand unifying theories. Like most learning processes, gaining knowledge about bacteria is a slow and deliberate process. With more knowledge about bacteria, we can minimize their harmful potential and exploit their useful activities.

Shapes, Forms, Sizes, and Arrangements of Bacterial Cells

Bacteria come in three shapes: elongated rods called *bacilli*, rounded or spherical cells called *cocci*, and spirals (helical and curved) called *spirilla* (for the less rigid form) and *spirochaete* (for those that are flexible). Elongated rod-shaped bacteria may vary considerably in length, have square, round, or pointed ends, and may be motile (possess the ability to move) or nonmotile. The spherical-shaped bacteria may occur singly, in pairs, in tetrads, in chains, and in irregular masses. The helical and curved spiral-shaped bacteria exist as slender spriochaetes, spirillum, and bent rods (See Figure 4.2).

Bacterial cells are usually measured in microns (μ) or micrometers (μm; 1 μm = 0.001 or 1/1000 of a millimeter, mm). A typical coliform bacterial cell that is rod-shaped is about 2 μm long and about 0.7 microns wide. The size of each cell changes with time during growth and death.

The arrangement of bacterial cells, viewed under the microscope, may be seen as separate (individual) cells or as cells in groupings. Within their species, cells may appear in pairs (diplo), chains, groups of four (tetrads), cubes (Sarcinae), and clumps. Long chains of cocci result when cells adhere after repeated divisions in one plane; this pattern is seen in the genera *Enterococcus* and *Lactococcus*. In the genus *Sarcina*, cocci divide in three planes, producing cubical packets of eight cells (tetrads). The shape of rod-shaped cells varies, especially the rod's end, which may be flat, cigar-shaped, rounded, or bifurcated. While many rods do occur singly, they may remain together after division to form pairs or chains (See Figure 4.2). These characteristic arrangements are frequently useful in bacterial identification.

Bacterial Cell Surface Structures and Cell Inclusions

Cell structure can best be studied in the rod form (see Figure 4.3). Keep in mind that cells may differ greatly both in their structure and chemical composition; for this reason there is no typical bacterium.

> Note: Figure 4.3 shows a generalized bacterium; not all bacteria have all the features shown in the figure, and some bacteria have structures not shown.

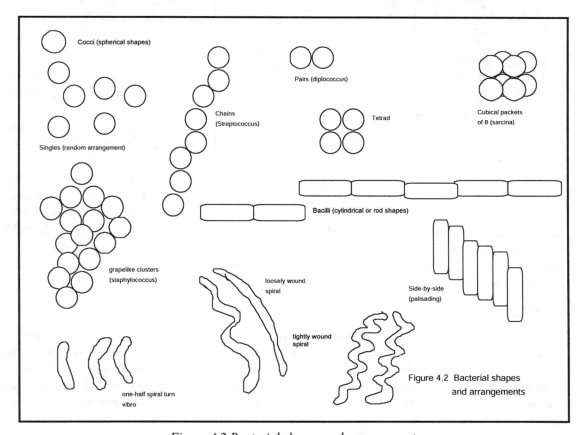

Figure 4.2 Bacterial shapes and arrangements

Capsule

Bacterial capsules (See Figure 4.3) are organized accumulations of gelatinous material on cell walls, in contrast to *slime layers* (a water secretion that adheres loosely to the cell wall and commonly diffuses into the cell), which are unorganized accumulations of similar material. The capsule is usually thick enough to be seen under the ordinary light microscope (*macrocapsule*), while thinner capsules (microcapsules) can only be detected by electron microscopy.

The production of capsules is determined largely by genetics, as well as by environmental conditions, and depends on the presence or absence of capsule-degrading enzymes and other growth factors. The capsules vary in composition and are mainly comprised of water; the organic contents are made of complex polysaccharides, nitrogen-containing substances, and polypeptides.

Capsules confer several advantages to bacteria growing in their normal habitat. For example, they (1) help to prevent desiccation, (2) help bacteria resist phagocytosis by host phagocytic cells, (3) prevent infection by bacteriophages, and (4) aid bacterial attachment to tissue surfaces in plant and animal hosts or to surfaces of solid objects in aquatic environments. Capsule formation often correlates with pathogenicity. On the positive side, capsule-secreted polysaccharides have been used for industrial purposes. In the food industry, for example, the polysaccharides have been used as gelling agents.

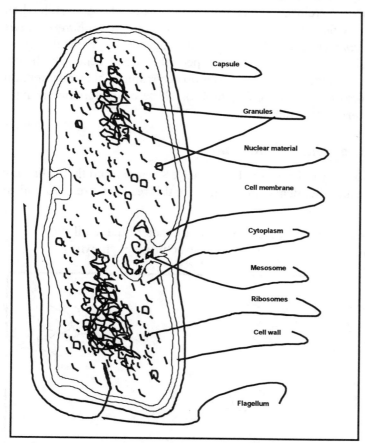

Figure 4.3 Bacterial cell

Flagella

Many bacteria are motile, and this ability to move independently is usually due to a special structure, the flagella (singular: flagellum). Depending on species, a cell may have a single flagellum (*Monotrichous* bacteria; trichous means hair), one flagellum at each end (*Amphitrichous* bacteria; amphi means on both sides), a tuft of flagella at one or both ends (*lophotrichous* bacteria; lopho means tuft), or flagella that arise all over the cell surface (*peritrichous* bacteria; peri means around).

Flagella are threadlike appendages extending outward from the plasma membrane and cell wall. They are slender, rigid locomotor structures, about 20 nm across and up to 15 or 20 µm long.

Bacterial cells benefit from flagella in several ways. They can increase the concentration of nutrients or decrease the concentration of toxic materials near the bacterial surfaces by causing a change in the flow rate of fluids. They can also disperse flagellated organisms to areas where colony formation can take place. The main benefit of flagellated organisms is their ability to flee areas that might be harmful.

Cell Wall

The rigid cell wall is the main structural component of most procaryotes. Some of the functions of the cell wall are (1) to provide protection for the delicate protoplast from osmotic lysis; (2) to determine a cell's shape; (3) to act as a permeability layer that excludes large molecules and various antibiotics, and also plays an active role in regulating the cell's intake of ions; and (4) to provide a solid support for flagella.

The cell walls of different species may differ greatly in structure, thickness, and composition. The cell wall accounts for about 20 to 40% of a bacterium's dry weight.

Plasma Membrane (Cytoplasmic membrane)

Bordered externally by the cell wall and composed of a lipoprotein complex, the plasma membrane is the critical barrier separating the inside from the outside of the cell. About 7-8 nm thick and comprising 10 to 20% of a bacterium's dry weight, the plasma membrane controls the passage of all material into and out of the cell. The inner and outer faces are embedded with water-loving (hydrophilic) lipids, while the interior is hydrophobic. Control of material into the cell is accomplished by screening, as well as by electric charge. The plasma membrane is the site of the surface charge of the bacteria.

In addition to serving as an osmotic barrier that passively regulates the passage of material into and out of the cell, the plasma membrane participates in the active transport of various substances into the bacterial cell. Inside the membrane many highly reactive chemical groups guide the incoming material to the proper points for further reaction. This active transport system provides bacteria with certain advantages, including the ability to maintain a fairly constant intercellular ionic state in the presence of varying external ionic concentrations. The cell membrane transport system also participates in waste excretion and protein secretions.

Cytoplasm

Within a cell and bounded by the cell membrane is a complicated mixture of substances and structures called the cytoplasm. The cytoplasm is a water-based fluid containing ribosomes, ions, enzymes, nutrients, storage granules (under certain circumstances), waste products, and various molecules involved in synthesis, energy metabolism, and cell maintenance.

Mesosome

A common intracellular structure found in the bacterial cytoplasm is the *mesosome*. Mesosomes are invaginations of the plasma membrane in the shape of tubules, vesicles, or lamellae. They are seen in both Gram-positive and Gram-negative bacteria, although they are generally more prominent in the former.

The exact function of mesosomes is still unknown. Many bacteriologists believe they are artifacts generated during the chemical fixation of bacteria for electron microscopy.

Nucleoid (nuclear body or region)

The nuclear region of the procaryotic cell is primitive, and is in striking contrast to that of the eucaryotic cell. Procaryotic cells lack a distinct nucleus, the function of the nucleus being carried out by a single long double-strand of deoxyribonucleic acid (DNA) that is efficiently packaged to fit within the nucleoid. The nucleoid is attached to the plasma membrane. A cell can have more than one nucleoid when cell division occurs after the genetic material has been duplicated.

Ribosomes

The bacterial cytoplasm is often packed with ribosomes. They are minute, rounded bodies, made of RNA (ribonucleic acid), and are loosely attached to the plasma membrane. Ribosomes are estimated to account for about 40% of a bacterium's dry weight; a single cell may have as many as 10,000 ribosomes. Ribosomes are the site of protein synthesis and are part of the translation apparatus.

Inclusions (storage granules)

Storage granules or other inclusions are often seen within bacterial cells. Some inclusion bodies are not bound by a membrane and lie free in the cytoplasm. Other inclusion bodies are enclosed by a single-layered membrane about 2.0 to 4.0 nm thick. Many bacteria produce polymers that are stored as granules in the cytoplasm.

Volutin or *polyphosphate* granules are inorganic inclusion bodies often seen in bacterial systems. They are believed to act as reservoirs of phosphate (an important component of nucleic acids) and appear to be involved with energy metabolism. These granules show the metachromatic effect; that is, they appear a different shade of color than the color they were stained with.

A variety of sulfur-metabolizing procaryotes are capable of oxidizing and accumulating free *elemental sulfur* within the cell. The elemental sulfur granules remain only under conditions when excess energy nutrients are present. As the sulfur is oxidized to sulfate, the granules slowly disappear.

Chemical Composition

The normal growth of a bacterial cell in excess nutrients results in a cell of definite chemical composition. This growth, however, involves a coordinated increase in the mass of its constituent parts and not solely an increase in total mass.

Bacteria, in general, are composed primarily of water (about 80%) and of dry matter (about 20%). The dry matter consists of both organic (90%) and inorganic (10%) components. All basic elements for protoplasm must be derived from the liquid environment, and if the environment is deficient in vital elements, the cell will show a characteristic lack of development.

Metabolism

Metabolism refers to the bacteria's ability to grow in any environment. The metabolic process refers to the chemical reactions that occur in living cells. In this process, *anabolism* works to build up cell components. Then *catabolism* breaks down or changes the cell components from one form to another.

Metabolic reactions require energy. So does locomotion and the uptake of nutrients. Many bacteria obtain their energy by processing chemicals from the environment through *chemosynthesis*. Other bacteria obtain their energy from sunlight through *photosynthesis*.

Chemosynthesis

The synthesis of organic substances (such as food nutrients) using the energy of chemical reactions is called chemosynthesis. A bacterium that obtains its carbon from carbon dioxide is called *autotrophic*. Bacteria that obtain carbon through organic compounds are called *heterotrophic* (See Figure 4.4).

Autotrophic Bacteria

Organisms that can synthesize organic molecules needed for growth from inorganic compounds using light or another source of energy are called *autotrophs*. For their carbon requirements, autotrophs are able to use ("fix") carbon dioxide to form complex organic compounds.

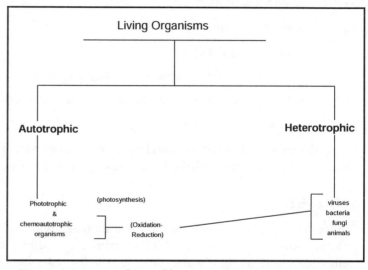

Figure 4.4 Autotrophic and heterotrophic organisms in relation to their means of obtaining energy

Heterotrophic Bacteria

Most bacteria are not autotrophic: they cannot use carbon dioxide as a major source of carbon, but must rely upon the presence of more reduced, complex molecules (mostly derived from other organisms) for their carbon supply. Bacteria that need complex carbon compounds are called *heterotrophs*. Heterotrophs use a vast range of carbon sources—including fatty acids, alcohols, sugars, and other organic substances. Heterotrophic bacteria are widespread in nature, and include all those species that cause disease in man, other animals, and plants.

Classification

Classifying microbes, including bacteria, is not always an easy undertaking. The classification process is complicated by the enormous variety of microorganisms, which differ widely in metabolic and structural properties. Some microorganisms are plant-like, others are animal-like, and still others are totally different from all other forms of life.

As an example of the classification process, consider bacteria in terms of activities: Bacteria can be classified as *aerobic, anaerobic,* or *facultative*. An *aerobe* must have oxygen to live. On the other extreme, the same oxygen would be toxic to an *anaerobe* (lives without oxygen). *Facultative* bacteria are capable of growth under aerobic or anaerobic conditions.

Like other microorganisms, there are so many different forms of bacteria that proper classification or identification of bacteria through a systematic application of procedures that are designed to grow, isolate, and identify the individual bacteria is required. These procedures are highly specialized and technical. Ultimately, bacteria are characterized based on observation and experience. Fortunately, certain classification criteria have been established based on observation and experience to help in the sorting out process:

1)   shape

2)   size and structure

3)   chemical activities

4)   types of nutrients they need

5)   form of energy they use

6)   physical conditions under which they can grow

7)   ability to cause disease (pathogenic or non-pathogenic)

8)   staining behavior

Using the above listed criteria, and based on observation and experience, bacteria can be identified from descriptions published in *Bergey's Manual of Determinative Bacteriology*.

VIRUSES

Viruses are parasitic particles that are the smallest living infectious agents known. They are not cellular—they have no nucleus, cell membrane, or cell wall. They multiply only within living cells (hosts) and are totally inert outside of living cells, but can survive in the environment. Just a single virus cell can infect a host. As far as measurable size goes, viruses range from 20-200 millimicrons in diameter, about 1-2 of magnitude smaller than bacteria. More than 100 virus types excreted from humans through the enteric tract could find their way into sources of drinking water. In sewage, these average between 100-500 enteric infectious units/100 ml. If the viruses are not killed in various treatment processes and become diluted by a receiving stream, for example, to 0.1-1 viral infectious units/100 ml, the low concentrations make it very difficult to determine virus levels in water supplies. Since tests are usually run on samples of less than 1 ml, at least 1,000 samples would have to be analyzed to detect a single virus unit in a liter of water. For this reason, viruses are usually concentrated by filtration or centrifugation prior to analysis.

Viruses differ from living cells in at least three ways: (1) they are unable to reproduce independently of cells and carry out cell division, (2) they possess only one type of nucleic acid, either DNA or RNA, and (3) they have a simple cellular organization. Viruses can be controlled by chlorination, but at much higher levels than those necessary to kill the bacteria. Some viruses that may be transmitted by water include hepatitis A, adeno virus, polio, coxsackie, echo, and Norwalk agent. A virus that infects a bacterium is called a *bacteriophage*.

Bacteriophage

Lewis Thomas (1974), in *The Lives of a Cell,* points out that when humans "catch diphtheria it is a virus infection, *but not of us.*" That is, when humans are infected by the virus causing diphtheria, it is the bacterium that is really infected—humans simply "blundered into someone else's accident" (p. 76). The toxin of diphtheria bacilli is produced when the organism has been infected by a bacteriophage.

A bacteriophage (phage) is any viral organism whose host is a bacterium. Most of the bacteriophage research that has been carried out has been on the bacterium *Escherichia coli,* one of the Gram-negative bacteria that environmental specialists such as water and wastewater operators are concerned about because it is a dangerous typical coliform.

A *virus* does not have a cell-type structure from which it is able to metabolize or reproduce. However, when the *genome* (a complete haploid set of chromosomes) of a virus is able to enter into a viable living cell (a bacterium), it may "take charge" and direct the operation of the cell's internal processes. When this occurs, the genome, through the host's synthesizing process, is able to reproduce copies of itself, move on, and then infect other hosts. Hosts of a phage may involve a single bacterial species or several bacteria genera.

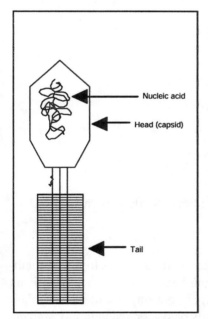

The most important properties used in classifying bacteriophages are nucleic acid properties and phage morphology. Bacterial viruses may contain either DNA or RNA; most phages have double stranded DNA.

Many different basic structures have been recognized among phages. Phages appear to show greater variation in form than any other viral group. (Basic morphological structure of the T-2 phage is shown in Figure 4.5.) The T-2 phage has two prominent structural characteristics: the head (a polyhedral capsid) and the tail.

The effect of phage infection depends on the phage and host, and to a lesser extent on conditions. Some phages multiply within and *lyse* (destroy) their hosts. When the host lyses (dies and breaks open), phage progeny are released.

FUNGI

The fungi (singular *fungus*) constitute an extremely important and interesting group of eucaryotic, aerobic microbes ranging from the unicellular yeasts to the extensively mycelial molds. Not considered plants, they are a distinctive life form

Figure 4.5 Simplified diagram of the major components of a T-2 phage

of great practical and ecological importance. Fungi are important because, like bacteria, they metabolize dissolved organic matter; they are the principal organisms responsible for the decomposition of carbon in the biosphere. Fungi, unlike bacteria, can grow in low moisture areas and in low pH solutions, which aids them in the breakdown of organic matter.

Fungi comprise a large group of organisms that include such diverse forms as the water molds, slime molds, other molds, mushrooms, puffballs, and yeasts. Because they lack chlorophyll (and thus are not considered plants), they must get nutrition from organic substances. They are either parasites, existing in or on animals or plants, or more commonly are *saporytes*, obtaining their food from dead organic matter. The fungi belong to the kingdom *Myceteae*. The study of fungi is called *mycology*.

McKinney (1962), in his well-known text *Microbiology for Sanitary Engineers*, complains that the study of mycology has been directed solely toward classification of fungi and not toward the actual biochemistry involved with fungi. McKinney goes on to point out that for those involved in the sanitary field it is important to recognize the "sanitary importance of fungi…and other steps will follow" (p. 40). For students of environmental science, it is important to understand the role of fungi as it relates to the water purification process. Moreover, environmental specialists need knowledge and understanding of the organism's ability to function and exist under extreme conditions, which make them important elements in biological wastestream treatment processes and in the degradation that takes place during waste-composting processes.

Fungi may be unicellular or filamentous. They are large (5-10 microns wide) and can be identified by a microscope. The distinguishing characteristics of the group, as a whole, include the following: (1) they are non-photosynthetic, (2) they lack tissue differentiation, (3) they have cell walls of polysaccharides (chitin), and (4) they propagate by spores (sexual or asexual).

Classification

Fungi are divided into five classes:

1)   Myxomycetes, or slime fungi

2)   Phycomycetes, or aquatic fungi (algae)

3)   Ascomycetes, or sac fungi

4)   Basidiomycetes, or rusts, smuts, and mushrooms

5)   Fungi imperfecti, or miscellaneous fungi

Although fungi are limited to only five classes, there are more than 80,000 known species.

Identification

Fungi differ from bacteria in several ways, including in their size, structural development, methods of reproduction, and cellular organization. They differ from bacteria in another significant way as well: their biochemical reactions (unlike the bacteria) are not important; instead, their structure is used to identify them. Fungi can be examined directly or suspended in liquid, stained, and dried to be observed under microscopic examination, where they can be identified by the appearance (color, texture, and diffusion of pigment) of their mycelia.

One of the tools available to environmental science students and specialists for use in the fungal identification process is the distinctive terminology used in mycology. Fungi go through several phases in their life cycle; their structural characteristics change with each new phase. Become familiar with the terms listed and defined in the following. As a further aid in learning how to identify fungi, relate the defined terms to their diagrammatic representations (Figure 4.6).

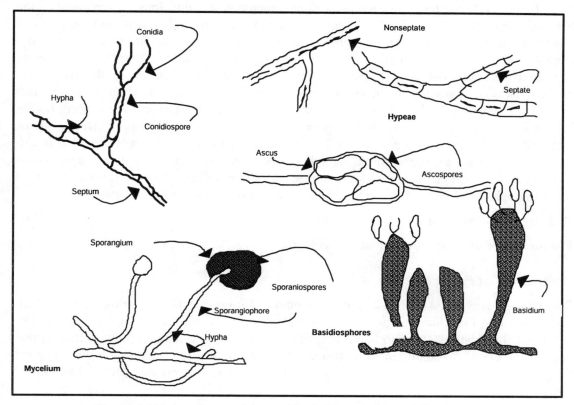

Figure 4.6 Nomenclature of fungi (Adapted from Mckinney, 1962, p. 36)

Definition of Key Terms

Hypha (pl. *hyphae*)—a tubular cell that grows from the tip and may form many branches. Probably the best known example of how extensive fungal hyphae can become is demonstrated in an individual honey fungus, *Armalloria ostoyae*, which was discovered in 1992 in Washington State. This particular fungus has been identified as the world's largest living thing; it is estimated to be 500 to 1,000 years old. Estimations have also been made about its individual network of hyphae: it covers almost 1,500 acres.

Mycelium—consists of many branched hypha and can become large enough to be seen with the naked eye.

Spore—reproductive stage of the fungi.

Septate hyphae—when a filament has crosswalls.

Nonseptate or *aseptate*—when crosswalls are not present.

Sporangiospores—spores that form within a sac called a *sporangium*. The sporangia are attached to stalks called *sporangiophores*.

Conidia—asexual spores that form on specialized hyphae called *conidiophores*. Large conidia are called *macroconidia,* and small conidia are called *microconidia.*

Sexual spores—In the fungi division Amastigomycota, four sub-divisions are separated on the basis of type of sexual reproductive spores present: (1) Subdivision *Zygomycotina*—consists of nonseptate hyphae and *Zygospores*. Zygospores are formed by the union of nuclear material from the hyphae of two different *strains*. (2) Subdivision *Ascomycotina*—fungi in this group are commonly referred to as the *ascomycetes*. They are also called *sac fungi*. They all have septate hyphae. *Ascospores* are the characteristic sexual reproductive spores and are produced in sacs called *asci* (ascus, singular). The mildews and *Penicillium* with asci in long fruiting bodies belong to this group. (3) Subdivision *Basidiomycotina*—consists of mushrooms, puffballs, smuts, rust, and shelf fungi (found on dead trees). The sexual spores of this class are known as *basidiospores*, which are produced on the club-shaped *basidia*. (4) Subdivision *Deutermycotina*—consists of only one group, the *Deuteromycetes*. Members of this class are referred to as the *fungi imperfecti* and include all the fungi that lack sexual means of reproduction.

Budding—process by which yeasts reproduce.

Blastospore or *bud*—spores formed by budding.

Cultivation of Fungi

Fungi can be grown and studied by cultural methods. However, when culturing fungi, use culture media that limit the growth of other microbial types—controlling bacterial growth is of particular importance. This can be accomplished by using special agars (culture media) that depress the pH of the culture medium (usually Sabouraud glucose or maltose agar) to prevent the growth of bacteria. Antibiotics can also be added to the agar to prevent bacterial growth.

Reproduction

As part of their reproductive cycle, fungi produce very small spores that are easily suspended in air and widely dispersed by the wind. Fungal spores are also spread by insects and other animals. The color, shape, and size of spores are useful in the identification of fungal species.

Reproduction in fungi can be either sexual or asexual. Sexual reproduction is accomplished by the union of compatible nuclei. Specialized asexual and/or sexual spore-bearing structures (fruiting bodies) are formed by most fungi. Some fungal species are self-fertilizing, and other species require outcrossing between different but compatible vegetative thalluses (mycelia).

Most fungi are asexual. Asexual spores are often brightly pigmented and give their colony a characteristic color (e.g., green, red, brown, black, blue—the blue spores of *Penicillium roquefort* are found in blue or Roquefort cheese).

Asexual reproduction is accomplished in several ways:

1) Vegetative cells may bud to produce new organisms. This is very common in the yeasts.

2) A parent cell can divide into two daughter cells.

3) The most common method of asexual reproduction is the production of spores (See Figure 4.7). There are several types of asexual spores:

   a) A hypha may separate to form cells (*arthrospores*) that behave as spores.

   b) If the cells are enclosed by a thick wall before separation, they are called *chlamydospores*.

c) If the spores are produced by budding, they are called *blastospores*.

d) If the spores develop within a sporangia (sac), they are called *sporangiospores*.

e) If the spores are produced at the sides or tips of the hypha, they are called *conidiospores*.

Nutrition and Metabolism

Fungi are found wherever organic material is available. They prefer moist habitats and grow best in the dark. Most fungi are *saprophytes*, acquiring their nutrients from dead organic matter, gained when the fungus secretes hydrolytic enzymes, which digest external substrates. They are able to use dead organic matter as a source of carbon and energy. Most fungi use glucose and maltose (carbohydrates) and nitrogenous compounds to synthesize their own proteins and other needed materials. Knowing what materials fungi use to synthesize their own protein and other needed materials in comparison to what bacteria use is important to those who work in the environmental disciplines for understanding the growth requirements of the different microorganisms.

ALGAE

You don't have to be an environmental specialist to understand that algae can be a nuisance. Many ponds, lakes, rivers and streams in the United States (and elsewhere) are currently undergoing eutrophication, the enrichment of an environment with inorganic substances (e.g., phosphorous and nitrogen). When eutrophication occurs, when filamentous algae like *Caldophora*

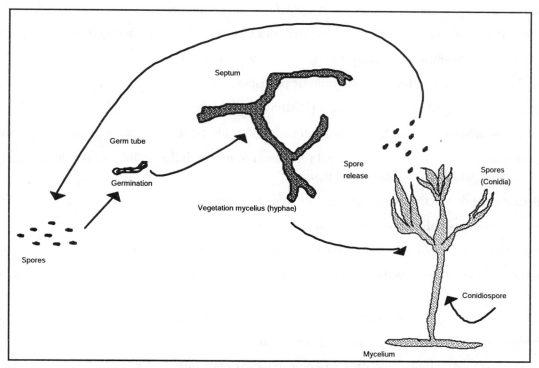

Figure 4.7 Asexual life of Penicilluum *sp. (adapted from Wistrecih and Lechtman, 1980, p. 163)*

break loose in a pond, lake, stream, or river and wash ashore, algae makes its stinking, noxious presence known.

For environmental specialists, algae are both a nuisance and a valuable ally. In water treatment, for example, although they are not pathogenic, algae are a nuisance. They grow easily on the walls of troughs and basins, and heavy growth can cause plugging of intakes and screens. Algae release chemicals that often give off undesirable tastes and odors. In wastewater treatment, on the other hand, controlled algae growth can be valuable in long-term oxidation ponds where they aid in the purification process by producing oxygen.

Before beginning a detailed discussion of algae, key terms are defined.

Definition of Key Terms

Algae—a large and diverse assemblage of eucaryotic organisms that lack roots, stems, and leaves but have chlorophyll and other pigments for carrying out oxygen-producing photosynthesis.

Algology or *Phycology*—the study of algae.

Antheridium—special male reproductive structures where sperm are produced.

Aplanospore—nonmotile spores produced by sporangia.

Benthic—algae attached and living on the bottom of a water body.

Binary fission—nuclear division followed by division of the cytoplasm.

Chloroplasts—packets that contain *chlorophyll a* and other pigments.

Chrysolaminarin—the carbohydrate reserve in organisms of division *Chrysophyta*.

Diatoms—photosynthetic, circular, or oblong chrysophyte cells.

Dinoflagellates—unicellular, photosynthetic protistan algae.

Epitheca—the larger part of the frustule (Diatoms).

Euglenids—contain chlorophylls *a* and *b* in their chloroplasts; representative genus is *Euglena*.

Fragmentation—a type of asexual algal reproduction in which the thallus breaks up and each fragmented part grows to form a new thallus.

Frustule—the distinctive two-piece wall of silica in diatoms.

Hypotheca—the small part of the frustule (Diatoms).

Neustonic—algae that live at the water-atmosphere interface.

Oogonia—vegetative cells that function as female sexual structures in algal reproductive systems.

Pellicle—a *Euglena* structure that allows for turning and flexing of the cell.

Phytoplankton—made up of algae and small plants.

Plankton—free-floating, mostly microscopic aquatic organisms.

Planktonic—algae that are suspended in water as opposed to attached and living on the bottom (benthic).

Prototothecosis—a disease in humans and animals caused by the green algae, *Prototheca moriformis*.

Thallus—the vegetative body of algae.

Algae—Description

Algae are autotrophic, contain the green pigment chlorophyll, and are a form of aquatic plant. Algae differ from bacteria and fungi in their ability to carry out photosynthesis—the biochemical process requiring sunlight, carbon dioxide, and raw mineral nutrients. Photosynthesis takes place in the chloroplasts, which are usually distinct and visible. They vary in size, shape, distribution, and number. In some algal types, the chloroplast may occupy most of the cell space. They usually grow near the surface of water because light cannot penetrate very far through water. Although in mass they are easily seen by the unaided eye (multicellular forms like marine kelp), many of them are microscopic. Algal cells may be nonmotile, motile by one or more flagella, or exhibit gliding motility as in diatoms. They occur most commonly in water (fresh and polluted water, as well as in salt water) in which they may be suspended (planktonic) phytoplanktons or attached and living on the bottom (benthic). A few algae live at the water-atmosphere interface and are termed neustonic. Within the fresh and saltwater environments, they are important primary producers (the start of the food chain for other organisms). During their growth phase, they are important oxygen-generating organisms and constitute a significant portion of the *plankton* in water.

Characteristics Used in Classifying Algae

According to the five-kingdom system of Whittaker, the algae belong to seven divisions distributed between two different kingdoms. Although there are seven divisions of algae, only five are discussed in this text and are listed as follows:

> Chlorophyta—Green algae
>
> Euglenophyta—Euglenids
>
> Chrysophyta—Golden-brown algae, diatoms
>
> Phaeophyta—Brown algae
>
> Pyrrophyta—Dinoflagellates

The primary classification of algae is based on cellular properties. Several characteristics are used to classify algae, including (1) cellular organization and cell wall structure; (2) the nature of chlorophyll(s) present; (3) the type of motility, if any; (4) the carbon polymers that are produced and stored; and (5) the reproductive structures and methods. Table 4.3 summarizes the properties of the five divisions discussed in this text.

Algal Cell Wall

Algae show considerable diversity in the chemistry and structure of their cell walls. Some algal cell walls are thin, rigid structures usually composed of cellulose modified by the addition of other polysaccharides. In other algae, the cell wall is strengthened by the deposition of calcium carbonate. Other forms have chitin present in the cell wall. Complicating the classification of algal organisms are the *Euglenids*, which lack cell walls. In *diatoms* the cell wall is composed of silica. The *frustules* (shells) of diatoms have extreme resistance to decay and remain intact for long periods of time, as the fossil records indicate.

Chlorophyll

The principal feature used to distinguish algae from other microorganisms (e.g., fungi) is the presence of chlorophyll and other photosynthetic pigments in the algae. All algae contain chlorophyll *a*. Some, however, contain other types of chlorophylls. The presence of these additional chlorophylls is characteristic of a particular algal group. In addition to chlorophyll, other pigments encountered in algae include fucoxanthin (brown), xanthophylls (yellow), carotenes (orange), phycocyanin (blue), and phycoerythrin (red).

Motility

Many algae have flagella (threadlike appendages). The flagella are locomotor organelles that may be the single polar or multiple polar types. The Euglena is a simple flagellate form with a single polar flagellum. Chlorophyta, on the other hand, have either two or four polar flagella. Dinoflagellates have two flagella of different lengths. In some cases, algae are nonmotile until they form motile gametes (a haploid cell or nucleus) during sexual reproduction. Diatoms do not have flagella but have gliding motility.

Algal Nutrition

Algae can be either autotrophic or heterotrophic. Most are photoautotrophic; they require only carbon dioxide and light as their principal source of energy and carbon. In the presence of light, algae carry out oxygen-evolving photosynthesis; in the absence of light, algae use oxygen. Chlorophyll and other pigments are used to absorb light energy for photosynthetic cell maintenance and reproduction. One of the key characteristics used in the classification of algal groups is the nature of the *reserve polymer* synthesized as a result of utilizing carbon dioxide present in water.

Algal Reproduction

Algae may reproduce either asexually or sexually. There are three types of asexual reproduction: binary fission, spores, and fragmentation. In some unicellular algae, binary fission occurs where nuclear division is followed by the division of the cytoplasm, forming new individuals like the parent cell. Some algae reproduce through spores. These spores are unicellular and germinate without fusing with other cells. In fragmentation, the thallus breaks up and each fragment grows to form a new thallus.

Sexual reproduction can involve the union of cells where eggs are formed within vegetative cells called oogonia (which function as female structures) and sperm are produced in a male reproductive organ called antheridia. Algal reproduction can also occur through a reduction of chromosome number and/or the union of nuclei.

Characteristics of Algal Divisions

Chlorophyta (Green Algae)

The majority of algae found in ponds belong to this group; they also can be found in salt water and the soil. Several thousand species of green algae are known today. Many are unicellular,

Table 4.3 Comparative Summary of Algal Characteristics

Algal Group	Common Name	Structure	Pigments	Carbon Reserve Materials	Motility	Method of Reproduction
Chlorophyta	Green algae	Unicellular to multicellular	Chlorophylls a and b, carotenes, xanthophylls	Starch, oils	Most are nonmotile	Asexual and sexual
Euglenophyta	Euglenoids	Unicellular	Chlorophylls a and b, carotenes, xanthophylls	Fats	Motile	Asexual
Chrysophyta	Golden-brown algae, diatoms	Unicellular	Chlorophylls a and b, special carotenoids, xanthophylls	Oils	Gliding by diatoms; others by flagella	Asexual and sexual
Phaeophyta	Brown algae	Multicellular	Chlorophylls a and b, carotenoids, xanthophylls	Fats	Motile	Asexual and sexual
Pyrrophyta	Dinoflagellates	Unicellular	Chlorophylls a and b, carotenes, xanthophylls	Starch, oils	Motile	Asexual; sexual rare

while some are multicellular filaments or aggregated colonies. The green algae have chlorophylls *a* and *b* along with specific carotenoids, and they store carbohydrates as starch. Few green algae are found at depths greater than 7-10 meters, largely because sunlight does not penetrate to that depth. Some species have a holdfast structure that anchors them to the bottom of the pond and to other submerged inanimate objects. Green algae reproduce by both sexual and asexual means.

Euglenophyta (Euglenids)

Euglenids are a small group of unicellular microorganisms that have a combination of animal and plant properties. Euglenids lack a cell wall, possess a gullet, have the ability to ingest food, have the ability to assimilate organic substances, and, in some species, are absent of chloroplasts. They occur in fresh, brackish, and salt waters, and on moist soils. A typical *Euglena* cell is elongated and bounded by a plasma membrane; the absence of a cell wall makes them very flexible in movement. Inside the plasma membrane is a structure called the pellicle, which gives the organism a definite form and allows for turning and flexing of the cell. Euglenids that are photosynthetic contain chlorophylls *a* and *b*, and they always have a red eyespot (*stigma*) which is sensitive to light. Some euglenids move about by means of flagellum; others move about by means of contracting and expanding motions. The characteristic food supply for euglenids is a lipopolysaccharide. Reproduction in euglenids is by simple cell division.

Chrysophyta (Golden Brown Algae)

The Chrysophycophyta group is quite large (several thousand diversified members). They differ from green algae and euglenids in that (1) chlorophylls *a* and *c* are present, (2) *fucoxanthin*, a brownish pigment, is present, and (3) they store food in the form of oils and leucosin, a polysaccharide. The combination of yellow pigments, fucoxanthin, and chlorophylls causes most of these algae to appear golden brown. The Chrysophycophyta is also diversified in cell wall chemistry and flagellation. The division is divided into three major classes: golden-brown algae, yellow-brown algae, and diatoms.

Some *Chrysophyta* lack cell walls; others have intricately patterned coverings external to the plasma membrane, such as walls, plates, and scales. The diatoms are the only group that have hard cell walls of pectin, cellulose, or silicon constructed in two halves (the epitheca and the hypotheca), called a frustule. Two anteriorly attached flagella are common among *Chrysophyta*; others have no flagella.

Most *Chrysophyta* are unicellular or colonial. Asexual cell division is the usual method of reproduction in diatoms; other forms of *Chrysophyta* can reproduce sexually.

Diatoms have direct significance for humans. Because they make up most of the phytoplankton of the cooler ocean parts, they are the ultimate source of food for fish. Water and wastewater operators understand the importance of their ability to function as indicators of industrial water pollution. As water quality indicators, their specific tolerances to environmental parameters such as pH, nutrients, nitrogen, concentration of salts, and temperature have been compiled.

Phaeophyta (Brown Algae)

With the exception of a few freshwater species, all algal species of this division exist in marine environments as seaweeds. They are a highly specialized group consisting of multicellular organisms that are sessile (i.e., attached and not free-moving). These algae contain essentially the same pigments seen in the golden-brown algae, but they appear brown because of the predominance and the masking effect of a greater amount of fucoxanthin. Brown algal cells store food as the carbohydrate laminarin and some lipids. The brown algae reproduce asexually.

Rhodophyta (Dionflagellates)

The principal members of this division are the dinoflagellates. The dinoflagellates comprise a diverse group of biflagellated and nonflagellated unicellular, eucaryotic organisms. The dinoflagellates occupy a variety of aquatic environments, with the majority living in marine habitats. Most of these organisms have a heavy cell wall composed of cellulose-containing plates. They store food as starch, fats, and oils. These algae have chlorophylls *a* and *c* and several xanthophylls. The most common form of reproduction in dinoflagellates is by cell division, but sexual reproduction has also been observed.

To this point, the chemical and microbial environmental contaminants we have discussed have been the waterborne types most commonly found in and affecting water and wastewater treatment operations. However, while chemical contaminants are limited, microbes are not; they are ubiquitous. The environmental science practitioner needs a well-rounded knowledge of microbial contaminants that are not only common to water bodies, water treatment systems, and water wastestreams, but also those that inhabit our air and soil. Case Study 4.1 contains information on a few of the microbes common in our air (some of which also have a significant interface with our water and soil), especially indoor air, and discusses some of the environmental health problems they present.

Case Study 4.1
Airborne Particulate Matter

Bacteria, pollen, fungal and plant spores, and viruses are all associated with airborne particles. Air conditioners and humidifiers have been identified as devices where pathogenic organisms may concentrate and later be released as concentrated viable aerosols. A variety of biological contaminants can cause significant illness and health risks. These include infections from airborne exposures to viruses that cause colds and influenza, and bacteria that cause tuberculosis (TB) and Legionnaires' disease. They also include respiratory ailments such as asthma, humidifier fever, hypersensitivity pneumonitis, and chronic allergic rhinitis. Such ailments may be caused by exposures to mold (fungi), fungal glucans (consists of glucose residues), mycotoxins, bacterial endotoxins, microbial volatiles, or organic dust.

A classic, well-known case of biological indoor air contamination was the outbreak of Legionnaires' disease among some of those who attended the Pennsylvania American Legion convention in Philadelphia in 1976. The Centers for Disease Control (CDC) conducted an extensive investigation of this incident and isolated the causal organism, the bacterium *Legionella pneumophila*. CDC also identified the most probable mode of transmission: contaminated air entrained in one of the air handling systems that served the hotel lobby. Legionnaires' disease causes pneumonia-like symptoms. Though it has a low attack rate (about 5%), mortality among those affected is high (15 to 20%).

Legionella pneumophila is widely present in the environment and is commonly isolated from surface waters and soil. Relatively resistant to chlorine, it passes through most water treatment systems. Its optimum growth temperature is 33°C and above. As a result, significant growth can occur in cooling tower systems, evaporative condensers, domestic and institutional water heaters, spas, and hot tubs. It gets entrained in a building's air supply via drift from cooling towers and evaporative condensers. When bacteria levels are high, and when *L. pneumophila* is aerosolized, disease can attack a susceptible population. The various risk factors identified for Legionnaires's disease include middle age, smoking, alcohol consumption, and travel.

PROTOZOA AND OTHER MICROORGANISMS

Protozoa

The *protozoa* ('first animals') are a large group of eukaryotic organisms (more than 50,000 known species that have adapted a form or cell to serve as the entire body). All protozoans are single-celled organisms. Typically, they lack cell walls but have a plasma membrane that is used to take in food and discharge waste. They can exist as solitary or independent organisms, e.g., the stalked ciliates such as *Vorticella* sp., or they can colonize like the sedentary *Carchesium* sp. Protozoa are microscopic and get their name because they employ the same type of feeding strategy as animals. Most are harmless, but some are parasitic. Some forms have two life stages: active *trophozoites* (capable of feeding) and dormant *cysts*.

As unicellular eukaryotes, protozoa cannot be easily defined because they are diverse and, in most cases, only distantly related to each other. As stated earlier, protozoa are distinguished from bacteria by their eucaryotic nature and by their usually larger size. Protozoa are distinguished from algae because protozoa obtain energy and nutrients by taking in organic molecules, detritus, or other protists rather than from photosynthesis. Each protozoan is a complete organism and contains the facilities for performing all the body functions for which vertebrates have many organ systems.

Like bacteria, protozoa depend upon environmental conditions (the protozoan community quickly responds to changing physical and chemical characteristics of the environment), reproduction, and availability of food for their existence. Relatively large microorganisms, protozoans range in size from 4 microns to about 500 microns. They can both consume bacteria (limit growth) and feed on organic matter (degrade waste).

Interest in types of protozoa is high among water treatment specialists because certain types of protozoans can cause disease. In the United States the most important of the pathogenic parasitic protozoans is *Giardia lamblia*, which causes a disease known as *giardiasis* (to be discussed in detail later). Two other parasitic protozoans that are carriers of waterborne disease are *Entamoeba histolytica* (amoebic dysentery) and *Cryptosporida* (Cryptosporidosis).

To address the increasing problem of waterborne diseases, the U.S. Environmental Protection Agency (EPA) implemented its Surface Water Treatment Rule on 29 June 1989, in part because of the occurrence of Giardia and Cryptosporidium spp. in surface water supplies. The Rule requires both filtration and disinfection of all surface water supplies as a means of primarily controlling *Giardia* spp. and enteric viruses. Since implementation of its Surface Water Treatment Rule, the EPA has also recognized that *Cryptosporidium* spp. is an agent of waterborne disease.

Classification

Protozoa are divided into four groups based on the method of motility. The *Mastigophora* are motile by means of one or more flagella; the *Ciliophora* by means of shortened modified flagella called *cilia*; the *Sarcodina* by means of amoeboid movement; and the *Sporozoa*, which are nonmotile. In Table 4.4 all four groups are listed, but, for the purposes of this text, only the first three, *Mastigophora, Ciliates*, and *Sarcodina*, will be discussed in detail.

Table 4.4: Classification of Protozoans

Group	Common name	Movement	Reproduction
Mastigophora	Flagellates	Flagella	Asexual
Ciliophora	Ciliates	Cili	Asexual by transverse fission Sexual by conjugation
Sarcodina	Amoebas	Pseudopodia	Asexual and Sexual
Sporozoa	Sporozoans	nonmotile	Asexual and Sexual

Mastigophora (Flagellates)

These protozoans are mostly unicellular, lack specific shape (have an extremely flexible plasma membrane that allows for the flowing movement of cytoplasm), and possess whip-like structures called flagella. The flagella, which can move in whip-like motion, are used for locomotion, as sense receptors, and to attract food.

These organisms are common in both fresh and marine waters. The group is subdivided into the *Phytomastigophrea*, most of which contain chlorophyll and are thus plant-like. A characteristic species of *Phytomastigophrea* is the *Euglena* sp., often associated with high or increasing levels of nitrogen and phosphate in the treatment process. A second subdivision of *Mastigophora* is the *Zoomastigopherea*, which are animal-like and nonpigmented.

Ciliophora (Ciliates)

The ciliates are the most advanced and structurally complex of all protozoans. Movement and food-getting is accomplished with short hairlike structures called *cilia*, which are present in at least one stage of the organism's life cycle. There are three groups of ciliates: (1) free-swimmers, (2) crawlers, and (3) stalked. The majority are free-living. They are usually solitary, but some are colonial and others are sessile. They are unique among protozoa in having two kinds of nuclei: a micronucleus and a macronucleus. The micronucleus is concerned with sexual reproduction. The macronucleus is involved with metabolism and the production of RNA for cell growth and function.

Ciliates are covered by a pellicle, which may act as a thick armor. In other species, the pellicle may be very thin. The cilia are short and usually arranged in rows. Their structure is comparable to flagella, except that cilia are shorter. Cilia may cover the surface of the animal or may be restricted to banded regions.

Sarcodina

Members of this group have fewer organelles and are simpler in structure than the ciliates and flagellates. *Sarcodina* move about by the formation of flowing protoplasmic projections called *pseudopodia*. The formation of pseudopodia is commonly referred to as *amoeboid movement*. The amoebae are well known for this mode of action (See Figure 4.8). The pseudopodia not only provide a means of locomotion but also serve as a means of feeding; this is accomplished when the organism puts out the pseudopodium to enclose the food. Most amebas feed on algae, bacteria, protozoa, and rotifers.

To gain understanding of a specific biological process and its importance in actual use, we provide a description of the wastewater-activated biosolids process in Case Study 4.2.

Case Study 4.2
Activated Biosolids Process

The activated biosolids (sludge) process originated in England. The name comes from the production of an activated mass of microorganisms capable of aerobically stabilizing the organic content of the waste. Activated biosolids is a biological process where contact occurs with bacteria, protozoa, fungi, and other small organisms such as rotifers and nematodes. The bacteria are the most important group of microorganisms; they are the ones responsible for the structural and functional activity of the activated biosolids floc. In this process, the bacteria and other organisms must be brought in contact with the organic matter in the wastewater. This is usually accomplished through rapid mixing, a process (provided by large mixers) augmented by aeration. Agitation and aeration work hand-in-hand to mix the returned biosolids with effluent from primary treatment, to keep the activated biosolids in suspension, and to supply oxygen to the biochemical reactions necessary for the stabilization of the wastewater.

The activated biosolids process is typified by the successive development of protozoa and mature floc particles. This succession can be indicated by the presence of the type of dominant protozoa present. At the start of the activated biosolids process (or recovery from an upset condition), the amoebae dominate. As the process continues (uninterrupted or without upset) small populations of bacteria begin to grow in logarithmic fashion, which, as the population increases, develop into mixed liquor. When this occurs, the *flagellates* dominate. When the biosolids attain an age of about three days, lightly dispersed floc particles begin to form, and bacteria increase. At this point, the free-swimming ciliates dominate. The process goes on. Floc particles begin to stabilize, taking on irregular shapes, and starting to show filamentous growth. Then the crawling ciliates dominate. Eventually, mature floc particles develop, increase in size, and large numbers of crawling and stalked ciliates are present. When this occurs, the succession process has reached its terminal point.

The succession of protozoan and mature floc particle development just described details the occurrence of phases of development in a step-by-step progression. This is also the case when protozoan succession is based on other factors such as dissolved oxygen and food availability.

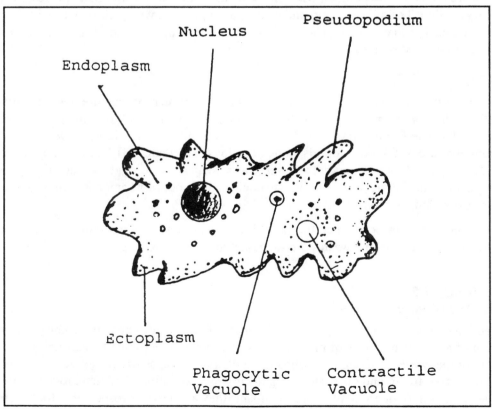

Figure 4.8 Amoeba

Probably the best way in which to understand protozoan succession based on dissolved oxygen and food availability is to view the wastewater treatment plant's aeration basin as a "stream within a container." Using the saprobity system to classify the various phases of the activated sludge process in relation to the self-purification process that takes place in a stream, you can see a clear relationship between the two processes based on available dissolved oxygen and food supply. The following explanation may help you understand the process.

The stream self-purifies and stabilizes over distance. That is, as the polluted (low dissolved oxygen and food supply) stream flows from point of pollution, it stabilizes. In the aeration basin, on the other hand, waste stabilization is based on time (biosolids age), not distance. Each condition described in the saprobity system is experienced in the aeration basin, with the exception of the last phase.

In the stream, a number of distinct zones of pollution can be identified according to their degree of pollution, content of dissolved oxygen, and the types of biotic indicators that are present. The zones used to describe the extent of these conditions are listed as follows:

1) *Polysaprobic Zone*—Point in the stream where pollution occurs and dissolved oxygen declines

2) *alpha Mesoprobic Zone*—Point in the stream where pollution is heavy and dissolved oxygen is low

3) *beta Mesoprobic Zone*—Point in the stream where pollution is moderate and dissolved oxygen is increasing

4) *Oligosaprobic Zone*—Point where pollution is low and dissolved oxygen levels are almost normal

5) *Xenosaprobic Zone*—Point in the stream where pollution is nonexistent and dissolved oxygen is normal

Except for the last zone (the *xenosaprobic zone*), the other zones and their associated conditions of pollution and dissolved oxygen contents are similar to the environment within the activated biosolids aeration basins.

Competition for Food in Activated Sludge Systems

Any change in the relative numbers of bacteria in the activated biosolids process has a corresponding change to microorganism population. Decreases in bacteria increase competition between protozoa and result in succession of dominant groups of protozoans.

The degree of success or failure of protozoa to capture bacteria is dependent upon several factors. For example, those with more advanced locomotion capability are able to capture more bacteria. Individual protozoan feeding mechanisms are also important in the competition for bacteria. At the beginning of the activated sludge process, amoebae and flagellates are the first protozoan groups to appear in large numbers. They are aided in surviving on lower quantities of bacteria because their energy requirements are lower than other protozoan types. Since little bacteria is present, competition for dissolved substrates is low. However, as the bacteria population increases, these protozoans are not able to compete for available food. This is when the next group of protozoans enters the scene: the free-swimming protozoans.

The free-swimming protozoans take advantage of the large populations of bacteria because they are better equipped with food-gathering mechanisms than the amoebae and flagellates. The free-swimmers are not only important because of their insatiable appetites for bacteria, but are also important in floc-formation. By secreting polysaccharides and mucoproteins that are absorbed by bacteria, which make the bacteria "sticky" through biological agglutination (biological gluing together), they stick together, and more importantly, stick to floc. Large quantities of floc are prepared for removal from secondary effluent and are either returned to aeration basins or wasted.

The crawlers and stalked ciliates succeed the free-swimmers. The free-swimmers are replaced in part due to the increasing level of mature floc, which retards their movement. The environment provided by the presence of mature floc is more suited to the needs of the crawlers and stalked ciliates. The crawlers and stalked ciliates also aid in floc formation by adding weight to floc particles, thus enabling removals.

Protozoa are important members of the microorganism population of the activated sludge process in wastewater treatment. Not only do they consume and thus remove bacteria from the activated sludge and secondary effluent, they also help with nitrification. In addition, the protozoans act as parameters of sludge health and effluent quality. By simple examination and identification of the protozoan population in activated biosolids, it is possible to determine whether or not loading is at acceptable or unacceptable levels. The presence of a particular species of protozoa can also indicate whether or not the process is operating correctly. The protozoan varieties can also indicate changes taking place in the strength and composition of the wastewater.

The importance of using protozoans as parameters to indicate biosolids health and effluent quality cannot be over emphasized. To gain a better understanding of how protozoan indicators can be used to determine the quality level of process operation, we provide the following parameters:

- "Healthy biosolids" is indicated when large varieties of crawlers and stalked ciliates are observed. A condition such as just described can indicate to the wastewater specialist that the process is producing a high quality effluent with BOD ranging from 1-10 mg/l.

- "Intermediate biosolids" is indicated by a preponderance of all three ciliated groups. When this occurs, the indication is that the effluent is of satisfactory quality with a BOD ranging 11-30 mg/l.

- "Poor biosolids" is indicated when the population is dominated by free-swimmers and flagellates. The effluent is generally turbid and of low quality with BOD at levels greater than 30 mg/l.

The indicators of effluent quality can be used in other ways. A significant shift from the patterns described may indicate that the sludge age is significantly high and/or may indicate the presence of excessive nutrient levels (nitrogen or phosphate). The absence of (too few) protozoans in activated sludge processes can also be indicative of process problems. For example, when the protozoan population is too low or absent, the F/M ratio may be too high (an overloaded condition).

Environmental Factors Affecting Protozoan Population

The population, activity, and diversity of the protozoa in activated biosolids and other treatment processes are affected by environmental factors. The availability of non-toxic bacteria is important. The dissolved oxygen levels are important even though the protozoans are generally aerobic. (Dissolved oxygen content is indicative of degree of pollution in the system.) Toxicants like chemical surfactants affect the plasma membrane and enzyme systems of protozoans; using surfactants could also lead to the development of bacteria that are harmful to protozoa. pH level is important. Most protozoans have an optimum upper and lower pH range. Shifting pH may favor one variety of protozoa over the other. Significant amounts of rainfall can affect the protozoa; significant decreases in overall population can occur through hydraulic washout.

Rotifers

Rotifers make up a well-defined group of the smallest, simplest multicellular microorganisms and are found in nearly all aquatic habitats. They are strict aerobes that range in size from about 0.1 - 0.8 nm. Often associated with aerobic biological processes in wastewater treatment plants, they are seen either grazing on bacteria or attached to debris by their forked tail or toe (See Figure 4.9). Rotifers promote microfloral activity and decomposition, enhance oxygen penetration in activated sludge and trickling filters, and recycle minerals in each. Most descriptions apply to the female because the male is much smaller and structurally simpler. Rotifers form into various shapes—spherical, sac-shaped, and/or worm-shaped. Their forms are composed of three zones. At their anterior end, rotifers possess actively moving cilia that frequently beat in a circular motion for motility and food gathering. The main body form below the head possesses a thick cuticle that terminates at the foot end. The foot possesses adhesive (cement) glands and toes for attachment to substratum. Rotifers are unique in that they possess the ability to chew their food by using a modified muscular pharynx called a *mastax*. Rotifers require high levels of dissolved oxygen; thus their presence indicates water with a high level of biological purity.

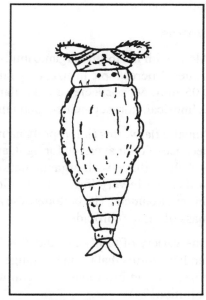

Figure 4.9 Philodina, a common rotifer

Rotifers possess reproductive organs in the form of gonads. They are separated into two different orders according to the number of gonads they possess. For example, in the order *Monogononta* the rotifers possess one gonad; in the order *Digononta* the rotifers possess two gonads.

Movement by rotifers is accomplished either by the free-swimming or crawling mode. Free-swimmers move by the beating action of rings of cilia on the epidermal area of the head. When each ring of cilia beats it gives the impression of a wheel with spokes. The frequency of this beating motion is quite high. Rotifers of this type move in a forward direction at a slow pace.

Rotifers that move using a crawling motion employ an interesting technique to accomplish their movement. While attached by its adhesive glands and toes to an old substratum, the rotifer will extend its body. While extending, the rotifer's head will use its adhesive glands to attach to a new substratum. Then the toes are released from the old substratum. The body contracts so that the foot reaches around and attaches to the substratum close to the head. The head then releases and the body is extended to its normal posture.

Rotifers feed on algae, bacteria, protozoa, and dead organisms.

Crustaceans

Because they are important members of freshwater zooplankton, microscopic crustaceans are of interest to water and wastewater specialists. These microscopic organisms are characterized by having a rigid shell structure. They are multicellular animals that are strict aerobes, and as primary producers, they feed on bacteria and algae. They are important as a source of food for fish. Additionally, microscopic crustaceans have been used to clarify algae-laden effluents from oxidation ponds.

Worms

Along with inhabiting organic muds, worms (Nematodes and Flatworms) also inhabit biological slimes. Microscopic in size, they range in length from 0.5 to 3 mm and in diameter from 0.01 to 0.05 mm. Most species are similar in appearance. They have bodies covered by cuticle, are cylindrical, non-segmented, and taper at both ends.

Aquatic flatworms (improperly named because they are not all flat) feed primarily on algae, and because of their aversion for light, are found in the lower depths of pools. Surface waters that are grossly polluted with organic matter (especially domestic sewage) have a fauna that is capable of thriving in very low concentrations of oxygen. A few species of tubificid worms dominate this environment. The bottoms of severely polluted streams can be literally covered with a writhing mass of these tubificids.

One variety of tubificids, the *Tubifix* (commonly known as sludge worms) are small, slender, reddish worms that normally range in length from 25 to about 50 mm. They are burrowers; their posterior end protrudes to obtain nutrients. When found in streams, *Tubifix* are indicators of pollution.

ENZYMES

In many environmental treatment processes, biological activities are used to degrade organic matter. To accomplish this, the environments in these processes must accommodate the appropriate types of microorganisms that are capable of performing the organic degradation function.

Enzymes, present within the microorganisms and the surrounding environmental media (water, air, and soil), are the essential biological catalysts that enable microorganisms to break down organics. A *catalyst* is defined as a substance that modifies and increases the rate of chemical reaction without being consumed in the process.

Microorganisms must first be acclimated to their environments before they can produce the enzymes they need to break down organics. Different organics are broken down by specific enzymes. In this breaking down process, the enzyme works to speed up the rate of hydrolysis of complex organic compounds and the rate of oxidation of simple compounds by decreasing the activation energy required.

The living cell is the site of tremendous biochemical activity, called *metabolism* (to be discussed in detail later). Metabolism is the process of chemical and physical change, which goes on continually in the living organism. Conversion of food to usable energy, buildup of new tissue, replacement of old tissue, disposal of waste products, reproduction—all are activities that are characteristic of life.

For illustrative purposes, consider the activities that characterize life as being related to the activities that take place on the assembly line in a factory. R. Breslow (1990), in *Enzymes: The Machines of Life*, saw this correlation when he called enzymes the "machines of life." In a sense, Breslow's inference of the machine-like enzyme is fitting, when you take into account how an enzyme can repeat a particular process several times a second, or even faster (like a machine), with similar results. Taking this concept a step further for the purpose of clarity, consider the enzyme as a small machine: (in this case) raw material goes in, finished product comes out. Just like a machine, an enzyme is specialized. Finally, when one considers a living cell as a small factory containing thousands of different types of specialized machines (enzymes), enzyme function becomes clear.

The phenomenon of *catalysis* makes possible biochemical reactions necessary for all life processes. Catalysis is defined as the modification of the rate of a chemical reaction by a catalyst. The catalysts of biochemical reactions are enzymes; they are responsible for bringing about all of the chemical reactions in living organisms. Without enzymes, these reactions take place at a rate far too slow to keep pace with metabolism.

Nature of Enzymes

Enzymes are essentially proteins formed by the polymerization of some or all the amino acids; 20 amino acids are found in proteins. Enzymes are high molecular weight compounds (ranging from 10,000 to 2,000,000) made up of chains of amino acids linked together by peptide bonds. In the overall linking process, a water molecule is removed between the carboxyl group of one amino acid and the amino group of the next one. Several steps are involved in the actual sequence used for synthesis of proteins, including enzymes, so that chemical energy can be supplied from other molecules.

Most enzymes are pure proteins. However, other enzymes require the participation of small non-protein groups, which may be organic or inorganic, before their catalytic activity can be exerted. These non-protein groups are called *cofactor* (the activator). In some cases these cofactors are non-protein metallic ion activators (e.g., ions of iron) that form a functional part of the enzyme. When the cofactor and the protein part (the *apoenzyme*) of the enzyme are present, the entire active complex is called the *holoenzyme*. This relationship can easily be seen in the following:

$$\text{Apoenzyme} + \text{Cofactor} = \text{Holoenzyme} \tag{4.1}$$

The structural nomenclature of enzymes is effected by the way in which the cofactor is attached to the apoenzyme (See Figure 4.10). For example, if the cofactor is firmly attached to the apoenzyme it is called a *prosthetic group*. When the cofactor is loosely attached to the apoenzyme, it is called a *coenzyme*.

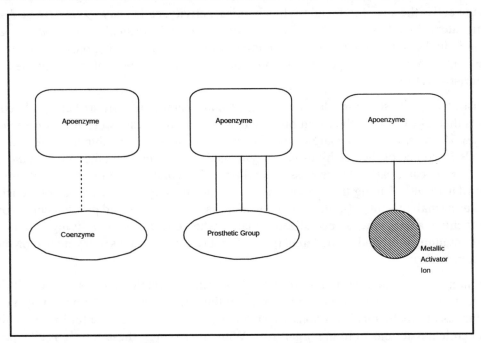

Figure 4.10 Holoenzymes showing apoenzymes and various types of cofactor (adapted from Witkowski and Power, 1975, p. 7)

Action of Enzymes

Keep in mind that enzymes increase the speed of reactions without themselves undergoing any permanent chemical change (i.e., they do not alter their equilibrium constants). They are neither used up in the reaction nor do they appear as products of the reaction. This basic enzymatic reaction process can be seen in the following:

$$\text{Substrate + Enzyme (catalyzes reaction)—Product + Enzyme} \qquad (4.2)$$

In the enzymatic reaction process just shown, note that the end product includes the enzyme, which was not altered or destroyed. The enzyme functions by combining in a highly specific way with its substrate, the substrate being changed without the enzyme itself being changed.

Much effort in research has been expended in trying to determine how enzymes lower activation energy of reactions. What is clear is that enzymes bring substrates together at the enzyme's active site to form an enzyme-substrate complex (See Figure 4.11).

In the enzyme-substrate complex, the substrate is attached by weak bonds to several points in the active site of the enzyme. This bringing together of enzyme and substrate allows for their concentration, which lowers the activation energy required to complete the reaction. Note that most of these reactions take place at relatively low temperatures ranging from 0° to 36°C.

Efficiency, Specificity, and Classification of Enzymes

Enzymes are extremely efficient. Only minute quantities of an enzyme are required to accomplish at low temperatures what normally would require, by ordinary chemical means, high temperatures and powerful reagents. For example, one ounce of pepsin can digest almost two tons of

Figure 4.11 Enzyme function showing the interaction of the substrate and enzyme with the resulting product (adapted from Prescott et al., 1993, p. 141)

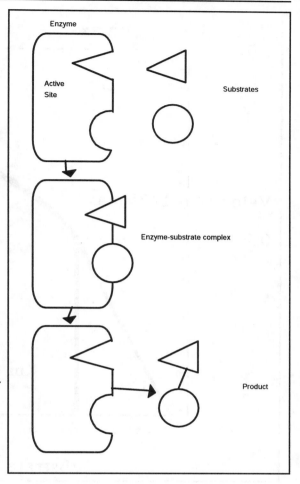

egg whites in a few hours, whereas it would take about 15 tons of strong acid 36 hours at high temperature without the enzyme.

Along with being efficient and extremely re-active, enzymes are characterized by a high degree of specificity. That is, just as a certain key will not fit or unlock each and every lock, enzymes also require an exact molecular fit between the enzyme and the substrate.

By 1956 the number of known enzymes was rapidly increasing. In 1961 the International Union of Biochemistry published an enzyme classification scheme that is universally used today. With the exception of the originally studied enzymes (rennin, pepsin, and trypsin), most enzyme names end in *ase*. Standards of enzyme nomenclature, initiated by the International Union of Biochemistry, recommended that enzymes should be named in terms of both the substrate acted upon and the type of reaction catalyzed.

Effect of Environment on Enzyme Activity

Several factors affect the rate at which enzymatic reactions proceed. These factors include substrate concentration, enzyme concentration, pH, temperature, and the presence of activators or inhibitors.

1) Substrate Concentration

At low substrate concentrations, an enzyme makes product slowly. However, if the amount of enzyme is kept constant and the substrate concentration is gradually increased, the reaction velocity will increase until it reaches a maximum (usually expressed in terms of the rate of product formation). After this point, increases in substrate concentration will not increase the velocity because the available enzyme molecules are binding substrate and converting it to product as rapidly as possible—the enzyme has reached the saturation point and is operating at maximal velocity. To gain an in-depth understanding of this enzyme saturation process, you must study in detail saturation kinetics (Michaelis-Menten kinetics), a study beyond the scope of this text. However, a fundamental appreciation of what is occurring during the enzyme-saturated-with-substrate phenomenon can be gained by studying the graphical representation in Figure 4.12.

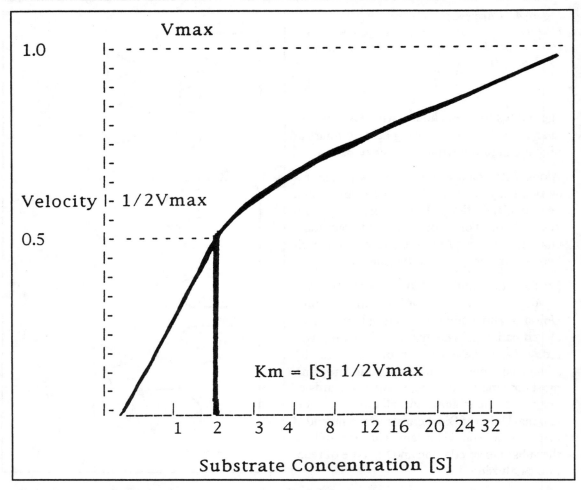

Figure 4.12 The effect of substrate concentration—the dependence of Velocity on Substrate Concentration for a simple One-Substrate Enzyme-catalyzed Reaction. This substrate curve fits the Michaelis equation, which relates reaction velocity (v) to the substrate concentration (S).

From Figure 4.12 you can see that the maximum velocity (V max) is the rate of product formation when the enzyme is saturated with substrate and making product as fast as possible. The *Michaelis constant (Km)* is the substrate concentration required for the enzyme to operate at half its maximal velocity. Theoretically, when the maximum velocity has been reached, all of the available enzyme has been converted to Enzyme-Substrate Complex. This point on the graph is designated *Vmax*. By using this maximum velocity and an equation, Michaelis developed a set of mathematical expressions to calculate enzyme activity in terms of reaction speed from measurable data.

2.   Enzyme Concentration

To study the effect of increasing the enzyme concentration upon the reaction rate, the reaction must be independent of the substrate concentration. Any change in the amount of product formed over a specific timeframe will be dependent upon the level of enzyme present.

3.   pH

Enzymes also change activity with alterations in pH. The most favorable pH value (the point where the enzyme is most active) is known as the optimum pH. When the pH is much higher or lower than the enzyme's optimum value, activity slows and the enzyme is damaged.

4.   Temperature

Enzymes also have temperature optima for maximum activity. Like most chemical reactions, the rate of an enzyme-catalyzed reaction increases with temperature. However, if the temperature rises too much above the optimum, an enzyme's structure will be denatured (disrupted) and its activity lost. The temperature optima of a microorganism's enzymes often reflect the temperature of its habitat. This is demonstrated by bacteria that grow best at high temperatures; they often have enzymes with high temperature optima.

5.   Inhibitors

Enzyme inhibitors are substances that slow down (or in some cases stop) catalysis. An inhibitor competes with the substrate at an enzyme's catalytic site and prevents the enzyme from forming product.

METABOLIC TRANSFORMATIONS

Between initial absorption and final excretion, many substances are chemically converted by the organism. This assembly-line activity that occurs in microorganisms during the processing of raw materials into finished products is called metabolic transformation. Metabolic transformations are mediated by enzymes. Environmental science students and practitioners must have an understanding of these metabolic transformations.

General Metabolism

Metabolism is derived from the Greek word *metabole,* which means to change. Change is what metabolism is all about. In attempting to further characterize a living organism's metabolic process(es) (its metabolism), several descriptions are available. For instance, in an organizational sense, an organism's metabolism (with its associated processes) is its capability to self-organize. Metabolism can also be defined as the total chemical and physical processes by which the functional and nutritional activities of an organism are maintained. In scientific terms, metabolism is generally referred to as the entire set of chemical reactions by which a cell produces and forms the various molecules it needs to maintain itself. In simple terms, metabolism can be characterized as the flow of energy through the organism.

Any definition or explanation of an organism's metabolism must include an explanation of the metabolic processes involved. These processes are well-known and well-documented. For instance, the two general categories of metabolism are *catabolism* and *anabolism*. In *catabolic* reactions, complex compounds are broken down with a release of energy. These reactions are linked to *anabolic* reactions, which result in the formation of important molecules. As a result of chemicals and associated reactions, biological cells are dynamic structures that are continually undergoing change.

During metabolism, the cell takes in nutrients (to be discussed later), converts them into cell components, and excretes waste into the external environment (see Figure 4.13). Microbial cells are made up of chemical substances, and when the cell grows, these chemical constituents increase in amount. The chemical substances that cells need come from the environment; that is, from outside the cell. Once inside the cell, the cell transforms these substances into the basic constituents of which the cell is composed.

Metabolic reactions require energy for the uptake of various nutrients and for locomotion in motile species. Microorganisms are placed into metabolic classes based on the source of energy they use. In describing these classes, the term *troph* is used (from the Greek meaning *to feed*). Thus, microorganisms that use inorganic materials as energy sources are called *lithotrophs* (*litho* is from the Greek for *rock*). Microorganisms that use organic chemicals as energy sources are called *heterotrophs* (which means feeding from sources other than oneself). Microorganisms that use light as an energy source are called *phototrophs* (photo is from the Greek for *light*). Most bacteria obtain energy from chemicals taken from the environment and are called *chemotrophs*.

Although an in-depth discussion of the metabolic processes of all microorganisms is beyond the scope of this text, environmental science practitioners need to be well-grounded in the fundamental concepts that we briefly cover in the following discussion. In particular, we provide information so that a familiarity with cell metabolism and the basics of biochemistry of microbial growth can be understood.

The chemical reactions occurring in cells are accompanied by changes in energy. A chemical reaction can occur with the release of free energy (called *exergonic*), or with the consumption of free energy (called *endergonic*). The free energy of these reactions can be expressed quantitatively.

Before a chemical reaction can take place, the reactants in a chemical reaction must first be activated. This activation requires energy. The amount of activation energy required can be decreased by the use of a catalyst, and the catalysts of living cells are enzymes. As stated in the preceding section, enzymes are proteins that are highly specific in the reaction that they catalyze.

The utilization of chemical energy in living organisms involves *oxidation-reduction* reactions, which involve the transfer of electrons from one reactant to another. *Oxidation* is defined as the removal of an electron or electrons from a substance. A *reduction* is defined as an addition of an electron (or electrons) to a substance. In the oxidation-reduction reaction, a transfer of electrons from one reactant to another takes place. The energy source (the *electron donor*) moves up one or more electrons, which are transferred to an *electron acceptor*. In this process, the electron donor is oxidized and the electron acceptor is reduced. One of the most common electron acceptors of living organisms is molecular oxygen. The tendency of a compound to accept or release electrons is expressed quantitatively by its *reduction potential*.

The transfer of electrons from donor to acceptor in a cell involves one or more intermediates, referred to as *electron carriers*. Some electron carriers are freely diffusible, transferring electrons from one place to another in the cell; others are firmly attached to enzymes in the cell membrane.

Two of the most common electron carriers are the coenzymes *NAD* and *NADP*. NAD+ (mincotinamide-adenine dinucleotide) and NADP+ (NAD-Phosphate) are freely diffusible carriers of hydrogen atoms, and always transfer two hydrogen atoms to the next carrier in the chain.

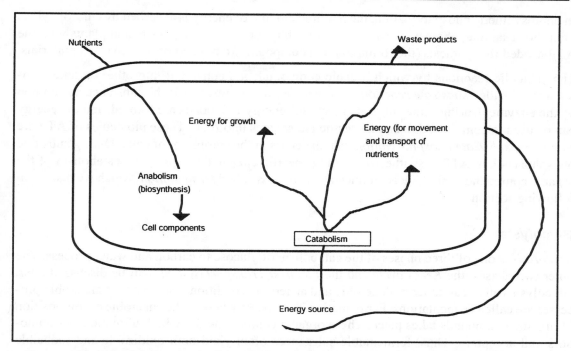

Figure 4.13 A simplified view of cell metabolism (adapted from Brock and Madigan, 1991, p. 93)

In most cases, biological reactions are catalyzed by specific enzymes, which can react only with a limited range of substrates. Oxidation-reduction reactions may be considered to proceed in three stages: (1) removal of electrons from the primary donor, (2) transfer of electrons through a series of electron carriers, and (3) addition of electrons to the terminal acceptor. Each step in the reaction is catalyzed by a different enzyme, each of which binds to its substrate and to its specific coenzyme. After a coenzyme has performed its chemical functions in one reaction, it can diffuse through the cytoplasm until it attaches to another enzyme that requires the coenzyme return to its original form; then the process can be repeated again.

Neither chemicals from the environment nor sunlight can be used directly to fuel a cell's energy-requiring processes. Therefore, the cell must have ways of converting sources of energy into a usable form of energy. In the presence of sunlight and certain chemicals, cells can make specific high-energy compounds with which they can satisfy their energy demands; one of these important compounds is *adenosine triphosphate (ATP)*.

The process of making ATP involves combining adenosine diphosphate (ADP) and inorganic phosphate (*Pi*) as shown in the following:

$$ADP + Pi + ENERGY \rightarrow ATP \qquad (4.3)$$

The energy required for this reaction can be obtained in three different ways: *Photosynthetic phosphorylation* (the changing of an organic substance into an organic phosphate), *substrate phosphorylation*, or *oxidative phosphorylation* (occurs on the membranes of mesosomes and related structures of procaryotes), depending upon the source of energy.

In photosynthetic phosphorylation, the required amount of energy is absorbed by chlorophyll as light. For example, photosynthesis supplies the blue-green bacteria, algae, and plants with the ATP needed for synthesis (the formation of a compound from its constituents) of all materials.

The catabolic reactions by which organic compounds are converted into other organic compounds are called *substrate reactions*. As stated earlier, a substrate is the substance acted upon by the enzyme. During some substrate reactions, energy-rich bonds are formed, and the energy can be used to combine ADP and inorganic phosphates into ATP. These molecules of ATP are formed by *substrate phosphorylation*, which occurs in the cytoplasm of cells. During substrate phosphorylation, ATP is synthesized during specific enzymatic steps in the catabolism of the organic compound. This ATP is produced by a process called *fermentation*, which we discuss in following section.

Glycolysis

Glycolysis is one of three phases of the catabolism of glucose to carbon and water process. The other two phases, the *Krebs cycle* and the *Electron Transport System*, will be discussed later. Glycolysis can occur under both aerobic and anaerobic conditions. Some of the anaerobic processes are called fermentations. Fermentation is a process whereby the anaerobic decomposition of organic compounds takes place. These organic compounds serve both ultimate electron donors and acceptors. Thus, fermentable substances often yield both oxidizable and reducible metabolites (organic compounds produced by metabolism).

The energy-converting metabolism (fermentation) in which the substrate is metabolized without the involvement of an external oxidizing agent is more easily understood by looking at a *metabolic pathway*. For example, in some bacteria, the fermentation of glucose begins with a pathway called glycolysis.

Glycolysis (sometimes referred to as the *Embden-Meyerhof-Parnas (EMP) pathway*) involves the breakdown or splitting of glucose (sugar) in a catabolic reaction that converts 1 molecule of glucose into 2 molecules of the end product, *pyruvic acid*. In this pathway, energy from energy-yielding (exergonic) reactions is used to phosphorylate ADT—that is, ATP is synthesized from ADP. This is an example of substrate phosphorylation where energy from a chemical reaction is used directly for the synthesis of ATP from ADP.

The end product in the energy-yielding glycolysis process is the release of a small amount of energy used for various cell functions, and the loss of larger amounts of energy in the form of fermentation products. Common fermentation products of glycolysis include ethanol, lactic acid, alcohols, and gaseous substances that are produced, for example, by certain bacteria.

Respiration

The process by which a compound is oxidized using oxygen as an *external* electron acceptor is called *respiration*. Using an external electron acceptor is important because in the fermentation process little energy is yielded, mainly because only a partial oxidation of the starting compounds occurs in this process. However, if some external terminal acceptor (oxygen for example) is present, all substrate molecules can be oxidized completely to a by-product (carbon dioxide). When this occurs, a far *higher* yield of ATP is possible.

Because an external oxidizing substance is used, the substrate undergoes a *net* oxidation (See Figure 4.14). The oxidation of a substrate provides more energy than that obtainable from the same substrate as fermentation.

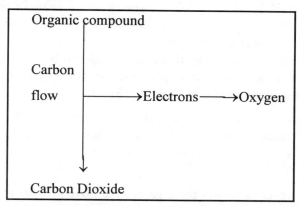

Figure 4.14 Aerobic respiration: the process by which a compound is oxidized using oxygen as an external electron acceptor.

Krebs Cycle

The Krebs cycle is sometimes called the *Citric Acid Cycle* or Tricarboxylic Acid Cycle (TCA), which is commonly called the "energy wheel" of cellular metabolism (See Figure 4.15) because it is a cyclical sequence of reactions crucial to supplying the energy needs of cells.

```
KREBS CYCLE

Overall Reaction:      Pyruvate + 4NAD + FAD→3CO2 + 4NADH + FADH

                       GDP + Pi      →      GTP
                       GTP + ADP     →      GDP + ATP
                                                              15 ATP
Electron-transport     4 NADH =      12 ATP
phosphorylation        FADH =        2 ATP
```

Figure 4.15 A summary of the overall reaction of Krebs cycle.

When oxygen is available to the cell, the energy in pyruvic acid is released through aerobic respiration. In the TCA cycle, pyruvate is first decarboxylated (the removal of a carboxyl group from a chemical compound), leading to the production of one molecule of NADH and an acetyl coupled to coenzyme A (Acetyl-CoA). The addition of the activated 2-carbon derivative acetyl CoA to the 4-carbon compound oxaloacetic acid forms citric acid, a six-carbon organic acid. The energy of the high-energy acetyl-CoA bond is used to drive the synthesis. After undergoing dehydration, decarboxylation, and oxidation, two additional carbon dioxide molecules are released. Eventually, oxalacetate is regenerated and serves again as an acetyl acceptor, thus completing the cycle. In the course of the cycle, three NADHs, one FADH, and one ATP are produced by substrate phosphorylation. The presence of an electron acceptor in respiration allows for the complete oxidation of glucose to carbon dioxide, with a greater yield of energy.

Electron Transport System (ETS)

The *Electron Transport System* (ETS) is a common pathway for the use of electrons formed during a variety of metabolic reactions. Most molecules are prevented from going into and out of an organism's cells by the cytoplasmic membrane. This is the case, of course, so that the cell can control its internal environment. During metabolism, however, the cell must be able to take in various substrates and get rid of waste; this is accomplished by transport systems. In some organisms—Gram-negative bacteria, for example—the transport system is located in membranes other than the cytoplasmic membrane (see Figure 4.16).

A typical ETS is composed of electron carriers. In a bacterium, the ETS involved with respiration occurs in the cytoplasmic membrane. ETS has two functions: (1) to accept electrons from electron donors and transfer them to electron acceptors, and (2) to save energy during electron transfer by synthesis of ATP.

The two protein components that form the ETS are the *flavoproteins* and *cytochromes*. *Flavoproteins* are proteins (enzymes) containing riboflavin, which act as dehydrogenation catalysts or hydrogen-carriers in a number of biological reactions. The flavin portion, which is bound to a protein, is alternately reduced as it accepts hydrogen atoms and oxidized when electrons are passed on. Riboflavin (also called vitamin B-2) is a required organic growth factor for some organisms.

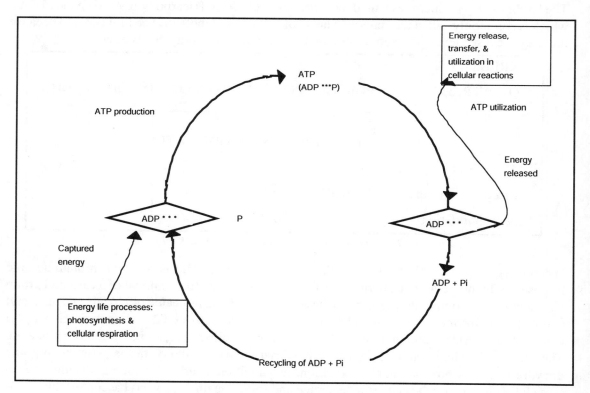

Figure 4.16 The formation of ATP, a substance that fuels all living organisms. By means of phosphorylation, energy-rich bonds (***) are formed and used to combine ADP and Pi into ATP, which is used to fuel the life processes. Then Pi and ADP are used again in a continuous cycle. (Adapted from Wistreich and Lechtman, 1980, p. 273)

Cytochromes are iron-containing proteins that receive and transfer electrons by the alternate reduction and oxidation of the iron atoms; they are important in cell metabolism. Cytochromes in the ETS are known for (among other things) their reduction potentials. One cytochrome can transfer electrons to another that has a more positive reduction potential and can itself accept electrons from cytochromes with a less positive reduction potential.

Autotrophic and Heterotrophic Metabolism

Autotrophs can use carbon dioxide as their major carbon source for the formation of essential biochemical compounds. Photosynthetic autotrophic bacteria combine carbon dioxide with ribulose diphosphate to form other macromolecules, which can be used for energy. The nonsynthetic chemosynthetic autotrophs rely on oxidation of inorganic compounds, including hydrogen, for the energy to fix carbon dioxide.

In heterotrophic metabolism, carbon dioxide cannot be used as a major carbon source. Chemosynthetic heterotrophs perform metabolic reactions involving proteins, lipids, and carbohydrates similar to those performed by other organisms. Heterotrophic organisms that are phototrophic can adjust to varying amounts of oxygen.

Microbial Nutrition

In earlier chapters, we presented various aspects of the chemical makeup of cell constituents. It may be useful at this point to review a summary of data that is known about the chemical composition of a bacterial cell (Table 4.5).

From Table 4.5 you can see that cells contain large numbers of water, inorganic, and organic molecules, but consist primarily of macromolecules such as proteins and nucleic acids. The cell is capable of obtaining most of the water it needs from the environment in usable form, whereas macromolecules are synthesized inside the cell.

Table 4.5: Chemical Composition of a Bacterial Cell

Molecule	Percent Wet Weight	Percent Dry Weight
Water	70	—
Total Macromolecules	26	96
Proteins	15	55
Polysaccharide	3	5
Lipid	2	9
DNA	1	3
RNA	5	20
Total Monomers	3	0.5
Amino Acids	0.5	0.5
Sugar	2	2
Nucleotides	0.5	0.5
Inorganic ions	1	1
Totals	100%	100%

Data taken from Neidhart, F. C. (ed.) Escherichia coli and Salmonella typhimurium—Cellular and Molecular Biology. American Society of Microbiology, Washington, D.C., 1987.

The mass of the cell primarily consists of four types of atoms: carbon, oxygen, nitrogen, and hydrogen. A number of other atoms are functionally important to the cell, but are less apparent. These include calcium, magnesium, iron, zinc, and phosphorus, all present in microbial cells, but in lesser amounts than carbon, hydrogen, oxygen, and nitrogen.

Nutrition

Nutrients used by organisms and obtained from the environment can be divided into two classes: (1) *macronutrients*, which are required in large quantities; and (2) *micronutrients*, which are required in lesser quantities.

Macronutrients

Most procaryotes require an organic compound to obtain their source of carbon. Bacteria have demonstrated that they can assimilate a wide variety of different organic carbon compounds to make new cell material. Major macronutrients such as amino acids, fatty acids, organic acids, sugars, and others have been shown to be used by a variety of bacteria. The major macronutrients in the cell, after carbon, are nitrogen, sulfur, phosphorous, potassium, magnesium, calcium, sodium, and iron. Table 4.6 shows some of the common forms of these major elements needed for biosynthesis of cell components.

Table 4.6: Macronutrients

Elemental forms found in the Environment	Element
Carbon dioxide Organic compounds	Carbon
Water Organic compounds	Hydrogen
Water Oxygen gas	Oxygen
Ammonia Nitrate Amino acids	Nitrogen
Phosphate	Phosphate
Hydrogen sulfide Sulfate Organic compounds	Sulfur

Micronutrients

Micronutrients (trace elements) are also required, and are just as critical to the overall nutrition of a microorganism as macronutrients. For example, cobalt is needed for the formation of vitamin B-2, zinc plays a role in the structure of many enzymes, molybdenum is important for nitrate reduction, and copper is important in enzymes involved in respiration.

Bacterial Growth

In microbiology, growth may be defined as an increase in the number of cells or in cellular constituents. If the microorganism is a multinucleate (*coenocytic*) organism in which nuclear division is not accompanied by actual cell division (as with bacteria), growth results only in an increase in cell size and not in cell number. In bacteria, as a general rule, growth leads to a rise in cell number because reproduction is by binary fission, where two cells enlarge and divide into two progeny of about equal size. Other bacterial species increase their cell numbers asexually by budding, e.g., as with the mycoplasma.

Population Growth

For the environmental scientist, investigating the growth of individual microorganisms is not usually convenient or practical because of their small size. They normally follow changes in the total population number when studying growth.

Growth (defined earlier as an increase in the number of microbial cells in a population) is measured as an increase in microbial mass. The change in cell number or mass per unit time is the growth rate.

When bacterial cells are introduced into a suitable medium and held at the optimal growth temperature, at set intervals a small volume of medium is withdrawn and cultured (bacteria growing in or on a medium) and a count can be made of the cells contained (counting methods will be discussed later). In this way the development of a population (i.e., the increase in cell numbers with time) can be observed and followed. By plotting the number of cells against time, a *growth curve* can be obtained. The actual shape of each portion of the curve and the actual numbers of organisms obtained varies between species and different types of media used.

Bacterial Growth Curve

The growth of bacteria can be plotted as the logarithm of cell number versus the incubation time. The resulting curve has four distinct phases. Note that these phases are reflections of the events in a population of bacteria or other microorganisms, not of individual cells. The terms *lag*, *log*, *stationary*, and *death phase* do not apply to individual cells, but only to populations of cells (See Figure 4.17).

Lag Phase of Growth

When bacteria are inoculated into a culture medium, growth usually does not begin immediately. Apparently this lag in growth, which may be brief or extended, represents a transition period on the part of bacteria transferred to new conditions. During this transition period, time is required for acclimation and for performing various functions such as synthesis of new enzymes. When conditions are suitable, binary fission begins, and after an acceleration in the rate of growth, the cells enter the logarithmic (log) phase.

The Logarithmic Phase

During the log phase, the number of bacteria is increasing at the maximal rate possible. The bacteria increase in a geometric progression—one splits to make two, two splits to make four, four to eight, and so on. Because each individual divides at a slightly different moment, the

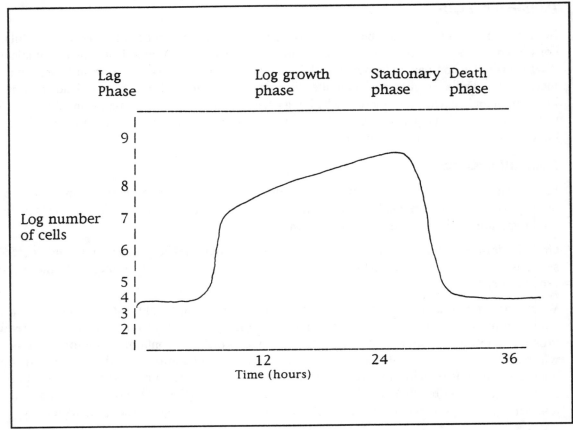

Figure 4.17 Bacterial Growth Curve. The four phases of the growth curve are identified on the curve and discussed in the text. (Adapted from McKinney, R. E. Microbiology for Sanitary Engineers, 1962, p. 118)

growth curve rises smoothly (in a straight line) rather in discrete jumps. From this logarithmic increase in cell number, the average time for a cell to divide, the *generation time*, can be calculated.

Stationary Growth Phase

Eventually population growth ceases and the growth curve becomes horizontal (Figure 4.17). When this occurs, the growth and death rates are more nearly identical, and a fairly constant population of bacteria is achieved. This uniformity in population number is reached primarily because either an essential nutrient of the culture medium is used up, or (for aerobic organisms) oxygen is limited to an inhibitory level and logarithmic growth ceases. The stationary phase leads eventually to the death phase, in which the number of living cells in the population decreases.

Death Phase

In the limited environment of the batch culture, conditions develop that accelerate the rate of death. When this occurs, the population is said to be in the death phase. The death phase is brought about because of environmental changes such as nutrient deprivation and the buildup of toxic wastes. For a while, the cells will persist. Some will tolerate the ever-increasing accumulation of wastes and will survive on the lysed cellular contents of the dead cells. At some point, however, further degradation of conditions will cause even the hardiest organism to die.

The Effect of Environmental Factors on Growth

The growth of microorganisms is greatly affected by the chemical and physical conditions of their environments. An understanding of environmental influences helps in the control of microbial growth and in the understanding of the ecological distribution of microorganisms.

Temperature

Temperature is one of the most important environmental factors affecting the growth and survival of microorganisms. In turn, one of the most important factors influencing the *effect* of temperature upon growth is *temperature sensitivity* of enzyme-catalyst reactions. As temperature rises, enzyme reactions in the cell proceed at more rapid rates (along with increased metabolic activity) and the microorganism grows faster. However, above a certain temperature, growth slows. Eventually, as the temperature continues to increase, enzymes and other proteins are denatured and the microbial membrane is disrupted. The microorganism is damaged or killed off. Usually, as the temperature is increased, functional enzymes operate more rapidly up to a point where the microorganisms may be damaged and inactivation reactions set in.

Because of these opposing temperature influences, for every organism, there is a minimum temperature below which growth no longer occurs, an optimum temperature at which growth is most rapid, and a maximum growth temperature above which growth is not possible (See Figure 4.18). The temperature optimum is always nearer the maximum than the minimum. Although these three temperatures (called the *cardinal temperatures*) are generally characteristic for each type organism, they are not rigidly fixed, but often depend to some extent on other environmental factors such as available nutrients and pH.

The cardinal temperatures of different microorganisms differ widely (Table 4.7). Some microbes have temperature optimums ranging from 0°C to as high as 75°C. The temperature range through which growth occurs is even wider, from below freezing to greater than boiling. Generally, the growth temperature range for a particular microbe is about 30 to 40 °C.

Table 4.7: Approximate Temperature Ranges for Microbial Growth

Microorganism	Temperature Ranges (Degrees C)		
	Minimum	Optimum	Maximum
Bacteria (nonphotosynthetic)	-10 to 85	10 to 105	25 to 110
Bacteria (photosynthetic)	70	30 to 80	45 to 85
Eucaryotic Algae	-40 to 35	0 to 50	5 to 57
Fungi	0 to 25	5 to 50	15 to 60
Protozoa	2 to 29	20 to 45	31 to 49

Adapted from Prescott et al. *Microbiology*, (1993), p. 126

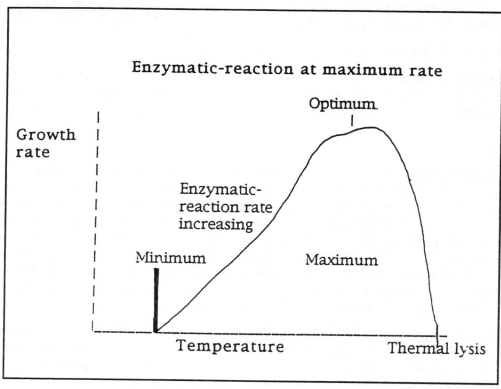

Figure 4.18 Effect of temperature on growth rate and the enzymatic-reaction activity that occurs as the temperature increases.

Microorganisms occupying those groups listed in Table 4.7 can be placed in one of four other groups based on their temperature ranges for growth.

1) *Psychrophiles*—(with low-temperature optima) grow optimally at or below 15°C, do not grow above 20°C, and have a lower limit for growth of 0°C or below. These micro-organisms are readily isolated from polar seas. Psychrophilic microorganisms have adapted to their cold environments in several ways. For example, their transport systems, protein synesthetic mechanisms, and enzymes function at low temperatures. Because psychrophiles are found in environments that are constantly cold and they are rapidly killed by warming to room temperature, their laboratory study is difficult. Psychrophilic microorganisms include a number of Gram-negative bacteria, certain fungi and algae, and a few Gram-positive bacteria.

2) *Psychotrophic*—(with low enzyme and low temperature optima) can grow at low temperatures ranging from 0° to 5°C, but they grow optimally above 15°C and have an upper limit for growth ranging between 25° and 30°C, with maximum about 35°C. Psychrotrophic fungi and bacteria are major contributors to spoilage of refrigerated food such as meat, milk, vegetables, and fruits; only when these foods are frozen is microbial activity (growth) not possible.

3) *Mesophilic*—(with mid-range-temperature optima) grow optimally at temperatures between 20° and 45°C. Most microorganisms probably fall within this category, and they

include those bacteria pathogenic to humans and other animals.

4)   *Thermophilic*—(with high-temperature optima) can grow at temperatures of 55°C or higher. Their growth minimum is around 45°C, and they often have optima between 55° and 65°C. A few thermophiles have maxima above 100°C. These thermophiles occur in composts, hydrothermal vents on the ocean floor, and in hot springs.

pH

When attempting to control microbial growth, controlling pH is one of the best methods. Most bacteria grow best at or near neutral pH 7, and the majority cannot grow under either acidic or alkaline conditions. Acidity or alkalinity of a solution is expressed by its *pH value*. pH value can be defined as the measure of the hydrogen ion activity of a solution. pH can be defined as the negative logarithm of the hydrogen ion concentration. This hydrogen concentration is important because it affects the equilibrium relationship of many biological systems that function only in a very narrow pH range.

As previously stated, pH dramatically affects microbial growth. Each species has a definite pH growth range and pH growth optimum that can be clearly seen in Table 4.8.

Table 4.8: Approximate Effect of pH on Microbial Growth

Microorganism type	Lower Limit Range	Upper Limit Range
Bacteria	0.5	9.5
Algae	0.0	9.9
Fungi	0.0	7.0
Protozoa	3.2	9.0

Adapted from Prescott, et al., Microbiology, *(1993), p.125*

In culturing microorganisms, pH adjustment of the culture medium is a common practice. For example, if the pH is too acidic, an alkaline substance such as sodium hydroxide (caustic solution) can be added. If the culture medium pH is too alkaline, an acidic substance may be added to adjust the pH. In general, different groups of microorganisms have pH preferences. Most bacteria and protozoa, for example, prefer a pH between 5.0 and 8.0. Most fungi and some algae prefer slightly more acidic surroundings, ranging from a pH of about 3 to 6.

Sometimes adding a pH buffer to the culture medium to compensate for changes (to keep the pH relatively constant) in pH fluctuations caused by microorganisms as they grow is desirable. Often indicator dyes are added to culture media and can provide a dual visual indication of the initial pH and changes (in the color of the dye) in pH resulting from growth activity of microbes.

Each species has a definite pH growth range and pH growth optimum. This important characteristic allows the water and wastewater specialist significant control of various microbial populations. Most natural environments have pH values between 5 and 9; organisms in this optima range are most common. Microorganisms that live at low pH are called *acidophiles*. Those microorganisms with a high optima (ranging from 8.5 to 11.5) are called *alkalophiles*. Note that acidophiles and alkalophiles may not grow at all or only very slowly at pH 7 (neutral pH).

Water Availability

The availability of water is another environmental factor that can affect the growth of microbes. In some microbes (e.g., bacteria), approximately 80% or more of their mass is water. In the growth phase, nutrients and water products enter and leave the cell, respectively, in solution. For these microbes to grow, they must be in or on an environment that has adequate available water or ions in solution.

All microorganisms (like all other organisms) need water for life. Indeed, availability of water is critical. Water availability does not depend only on water content of the environment, but also on the substances that are present; that is, some substances can absorb water and do not readily give it up.

Water availability is generally expressed in physical terms as *water activity (a*w*)*, the amount of free or available water in a given substance. The water activity of a solution is 1/100 the relative humidity of the solution when expressed as a percent. Water availability is expressed as a ratio of the vapor pressure of the air over the solution divided by the vapor pressure at the same temperature of pure water, as demonstrated in the following:

$$aw = \frac{P \text{ solution}}{P \text{ water}} \qquad (4.4)$$

Values of water activity vary between 0 and 1. Some representative values are given in Table 4.9.

Water diffuses from a region of high water—low solute concentration to a region of lower water—higher solute concentration. If pure water and a salt solution are separated by a semipermeable membrane, water will diffuse *from* the pure water *into* the salt solution by *osmosis*. The cytoplasm of most cells has a higher solute concentration than the environment; thus water diffuses into the cell. If the cell is in an environment of low water activity, water will flow out of the cell. An environment such as salt or sugar in solution has low water activity and causes the cell to give up water. When this occurs, the plasma membrane shrinks away from the wall (plasmolysis), the cell dehydrates, damaging the membrane, and the cell ceases to grow.

Table 4.9: Water Activity of Various Materials

Material	Water Activity Value
Pure Water	1.000
Human Blood	0.995
Bread	0.950
Ham	0.900
Jams	0.800
Candy	0.700

Adapted from Brock and Madigan, Biology of Microorganisms, *(6th ed.), (1991), p. 329.*

Oxygen

Microorganisms vary in their need for, or tolerance of, oxygen. Some bacteria, for example, need oxygen for growth. Others need the *absence* of oxygen for growth. Still others can grow regardless of the presence or absence of oxygen. Microorganisms can be grouped depending on the effect of oxygen. A microorganism that is able to grow in the presence of atmospheric oxygen is an *aerobe*, whereas one that can grow in its absence is an *anaerobe*. Organisms that are dependent upon atmospheric oxygen for growth are called *obligate aerobes*.

Microorganisms that do not require oxygen for growth, but grow better in its presence are called *facultative anaerobes*. The *strict* or *obligate anaerobes*, on the other hand, do not tolerate oxygen at all and will die in its presence.

Some aerobes must be aerated to grow, because oxygen is poorly soluble in water. The oxygen used up by microorganisms during growth is diffused from air too slowly. To compensate for this shortage of oxygen in cultured aerobes, forced-aeration is desirable. This can be accomplished by forcing sterilized air into the medium, or by simply shaking the tube or flask vigorously.

The anaerobic culture requires the *exclusion* of oxygen. This is a difficult task; oxygen, obviously, is readily available in the air. To vacate air from the culture medium in which anaerobes are to be grown, it is necessary to completely fill tubes to the top with culture medium and seal with tight fitting stoppers. This procedure is useful for providing anaerobic conditions for organisms not too sensitive to small amounts of oxygen. A *reducing agent* could also be added to the medium that reacts with oxygen and excludes it from the culture medium. A common reducing agent used in this procedure is *thioglycollate*. The nature of bacterial oxygen responses can be readily determined by growing bacteria in culture tubes filled with a solid culture medium or media treated with a reducing agent. To easily detect the presence of oxygen in the medium, an indicator dye is usually added because the dye will change color and indicate the penetration level of oxygen in the medium.

PATHOGENICITY

> *I can think of a few microorganisms, possibly the tubercle bacillus, the syphilis spirochete, and malarial parasite, and a few others, that have a selective advantage in their ability to infect human beings, but there is nothing to be gained, in an evolutionary sense, by the capacity to cause illness or death. Pathogenicity may be something of a disadvantage for most microbes, carrying lethal risks more frightening to them than to us. The man who catches a meningococcus is in considerably less danger for his life, even without chemotherapy, than meningococci with the bad luck to catch a man. Most meningococci have the sense to stay out on the surface, in the rhinopharynx. During epidemics this is where they are to be found in the majority of the host population, and it generally goes well. It is only in the unaccountable minority, the "cases," that the line is crossed, and then there is the devil to pay on both sides, but most of all for the meningococci. —Thomas*

Thomas's account of the dilemma meningococci face upon entering the body notwithstanding, the prevention of an invasion by water-air-soil borne pathogens into human beings and other forms of life is what environmental scientists are most concerned with. For example, the most frequent cause of waterborne disease in public water supplies in the United States is inadequate water treatment, whether treatment is nonexistent or is ineffective due to treatment process breakdown, especially in disinfection. When inadequate water treatment or treatment system breakdown occur, most diseases that are transmitted via the untreated water are caused by contaminated fecal material. Table 4.10 shows the public water system breakdowns in the United States reported by the Centers for Disease Control (CDC) during the 1986-88 timeframe.

Table 4.10: Water System Breakdowns Leading to Waterborne Disease Outbreaks, 1986-88

Type of breakdown or deficiency	Community Water Systems	Noncommunity Water Systems	Total
Untreated surface water	1	1	2
Untreated ground water	3	9	12
Treatment	11	12	23
Distribution system	6	0	6
Miscellaneous	1	2	3
Total	22	24	46

Source: Centers for Disease Control, Atlanta

Even before Mary Mallon (a.k.a. Typhoid Mary) was "cooking up" her daily food preparations (she was an American cook in the late 1800s and early 1900s) and passing on typhoid to her unsuspecting victims, a relationship had been suspected between microorganisms and disease. As a matter of historical record, even before the discovery of waterborne, disease-causing microorganisms, a relationship between water and the spread of disease had been suspected. Mankind's quest for the truth about disease and disease-causing microorganisms began almost 2600 years ago. This was the time frame when Hippocrates, the father of medicine, suspected that "different" waters caused several different diseases. Proof took several more centuries, until Leeuwenhoek invented the microscope in 1675. Leeuwenhoek's microscope, with several later refinements, opened up the microscopic world to man. *Giardia lamblia* was the first microorganism studied and described by Leeuwenhoek. Eventually, in the nineteenth century, other great men of science such as Koch and Pasteur disproved the ancient theory of poisonous vapors arising from decaying filth as the cause of disease; instead, they developed the germ theory.

At the beginning of this section, Lewis Thomas discussed the plight of the meningococci that might have the misfortune of entering the human body. After having read Thomas' discussion, the reader may have developed the misperception that pathogenic microorganisms are defenseless and not at home within our bodies. This is not the case. Microorganisms can adapt. As a case in point, consider the following descriptive account of the adaptive challenges faced by waterborne pathogenic bacteria ingested by humans.

> *In water, this bacterium was experiencing an environment where the temperature was well below 37 degrees C, nutrient concentrations and osmotic strength were low, and pH was near neutral. At least some oxygen was available. When the bacterium is ingested, it suddenly encounters a higher temperature, higher osmotic strength, a transient exposure to low pH in the stomach, followed by a rise in pH in the intestine and high concentrations of membrane-disrupting bile salts. Also, the environment of the small intestine, and to a greater extent the colon, is anaerobic.*

The bacterium will find abundant sources of carbon and energy in the intestine, but the forms of these compounds will be different from those it may have been using in water. Keep in mind that the incoming bacterium does not have much time to adapt to its new environment because it takes only a few hours to transit the small intestine and reach the colon, where it also will encounter stiff competition from the resident microflora.—Salyers and Whitt

Given the adaptive challenges faced by the waterborne bacteria and their ability to make these adaptations, the fact that pathogenic bacteria have evolved defense mechanisms that make their destruction increasingly difficult is not surprising. This is the ultimate challenge confronting water and wastewater treatment specialists of the future. Given the fact that pathogenic microorganisms have the remarkable ability to adapt and survive, it follows that advanced biological treatment methodologies will have to evolve and adapt as well.

Causal Factors for Transmission of Disease

Certain factors must exist for the transmission of disease. These factors are related to the diseased individual (the *host*), the microorganism (the *agent*), and the environment. Among the many diseases of microbial origin, some are caused by viruses, some by fungi, some by bacteria, and others by protozoa. Note that disease does not necessarily follow exposure to a given pathogen. For disease to develop, several factors must *all* be present. These necessary factors are described in the following:

1)   *Pathogen* (causative agent). Any microorganism that can cause disease is called a pathogen. Normally this pathogen is parasitic and lives on or in another organism. In some diseases, the link between pathogen and disease is very specific; that is, the pathogen may only infect certain species. Along with being selectively specific, for a pathogen to cause the occurrence of disease is dependent on various factors. These factors include the extent or degree of resistance of the host, and the capacity to produce disease (the *virulence*) of the pathogen.

2)   *Pathogen's reservoir*. The pathogen lives and reproduces (multiplies) in a reservoir and is unable to reproduce or grow outside the reservoir. Humans, animals, plants, organic matter and/or soils can serve as the pathogen's reservoir. The human body is a significant reservoir of microbial pathogens; these organisms are a normal part of humanity's microflora and usually do not behave as pathogens unless disturbed in surgery or in an injury. When this occurs, the body's specific or nonspecific defenses have been impaired; the type of pathogenic contamination that then takes place is known as *opportunistic*.

3)   *Movement of pathogens from reservoir*. Movement of pathogens in humans may be from body openings in his/her respiratory, urinary, and intestinal systems, from open infections, or by mechanical means delivered by disruptions or damage caused by injuries.

4)   *Transmission of disease to another host*. Transmission can be accomplished directly or indirectly. In direct transmission, disease passes (via an animal bite, for example) immediately to a new host. Indirect transmission is accomplished through vectors or ve-

hicles. These vectors can be mosquitoes, ticks, fleas, and other invertebrates. The vehicles are substances such as milk, food, air, and other nonliving substances.

5)   *Virulence of the pathogen.* The occurrence of disease depends upon the pathogen's virulence (its capacity to cause disease).

6)   *Degree of resistance to transmission.* Humans possess mechanisms of defense against disease. For example, human skin provides the primary defense against infectious disease. For most diseases to pass or be transmitted through the skin's resistance barrier, the skin must be broken by wounding or bites of insects (as explained in 4 above).

For disease to spread, some vehicle, medium, or opportunity for spread must exist. This is where water and wastewater treatment specialists play such a vital role. If untreated water is ingested by an individual, the individual may become a carrier of waterborne disease. Typhoid Mary infected several people with typhoid fever, but she never demonstrated obvious symptoms of this deadly disease; thus, she was a carrier.

Pathogenic organisms found in water and wastewater may be from carriers of disease. The principal categories of pathogenic organisms found in water and wastewater are bacteria, viruses, protozoa, and helminths. Such highly infectious organisms have been responsible for countless numbers of deaths, and currently continue to cause the death of large numbers of people and animals in underdeveloped parts of the world.

Problems related to waterborne diseases in underdeveloped parts of the world are well-documented. In one instance, for example, a report filed by Ries et al. (1992) in Peru in 1991, in Piura, a Peruvian city of more than 350,000, recorded that within a 2 month period more than 7922 cases and 17 deaths were attributed to a cholera epidemic. During the investigation, a hospital-based culture survey showed that approximately 80% of diarrhea cases were cholera. A study of 50 case-patients and 100 matched controls demonstrated that cholera was associated with drinking unboiled water, drinking beverages from street vendors, and eating food from vendors. Note that in a second study, patients were more likely than controls to consume beverages with ice. Ice was produced from municipal water. Testing of municipal water supplies revealed no or insufficient chlorination, and fecal coliform bacteria were detected in most samples tested. With epidemic cholera spreading throughout Latin America, these findings emphasize the importance of safe municipal drinking water. More specifically, the results of this study indicate that even though we have come a long way in the fight against waterborne disease and have been successful in most instances, parts of the world still struggle with this life-threatening problem.

Parasites and Pathogens

One form of pathogenicity is demonstrated when parasitic microorganisms derive their nutrients for growth and reproduction from living on or in a viable host. Another form of pathogenicity occurs when toxic substances are produced by pathogens.

Microbial parasites include viruses, protozoans, and helminths, while bacteria, fungi, and actinomycetes are pathogens. *All* viruses are parasitic. Protozoans are parasites that commonly enter and live in the gastrointestinal system of humans and animals. The helminths (worms) have two parasitic forms: roundworms and flatworms (tapeworms).

Like their ability to adapt and survive in the human body, parasites and pathogens are not defenseless against the efforts of humans to destroy or remove them. This ability of pathogenic microorganisms to adjust or adapt should come as no surprise, since humans, through evolution, have also been able to develop ways to protect themselves (build up resistance) from parasites and bacteria.

Remember that during man's evolutionary process and even with his ability to fight off disease, pathogenic microorganisms have not remained static. Pathogenic microorganisms have also evolved. Consider the effect that certain antibiotics have today on the same pathogens they were used successfully against only a few years ago. Today, these antibiotics may not be as effective because pathogens have evolved ways to circumvent their effects to the point that they may no longer provide a defense against disease. Salyers and Whitt (1994) point out that it is most striking in the way in which pathogenic bacteria have been able to develop strategies that enable them to survive against the various defensive actions implemented by man. For example, bacteria cellular structures such as capsule formation, thickened cell walls, and spore formation are defensive mechanisms that bacteria have developed with time and constant exposure to various antibiotics. "Bacteria are clearly Machiavellian" (p.2).

Bacteria are not alone in their resistance to preventing their own destruction. Take for example the viruses; they are protected from most chemical treatment processes because of the very nature of their own chemical composition. Protozoans provide other examples of adaptation and adjustment. For instance, they have developed a specialized hyaline wall structure that protects its cyst forms. The helminths are often able to resist destruction because of the tough shell that surrounds their eggs.

Parasitic protozoans can survive outside the intestinal tract in feces as a cyst with a thick hyaline wall that provides resistance to treatment.

The highest prevalence of giardiasis in the United States is in communities using surface water supplies, where potable water treatment consists primarily of disinfection. Giardiasis can also be contracted from contaminated food, mountain streams, and wastewater contaminated with infected feces. The best control that can be used to protect against protozoan waterborne infection is to use proper sanitation procedures (i.e., boil water and wash hands with soap and water).

SUMMARY

Microbiology is basic to the groundwork of environmental science, to the day-to-day functioning of what working environmental scientists use every day. This technical information and the hands-on skills of identification and classification, testing, enumeration, and assessment tie in directly with organic chemistry and with the toxicology concepts presented in the next chapter.

Cited References

Bergey's Manual of Systematic Bacteriology, 8th edition. Buchanan, R. E. and Gibbons, N. E. (eds.), Williams and Wilkins, (1974).

Breslow, R., *Enzymes: The Machines of Life*, Burlington, NC: Carolina Biological Supply Co. (1990).

McKinney, R. E., *Microbiology for Sanitary Engineers*, New York: McGraw-Hill, (1962).

Salyers, A. A. and Whitt, D. D., *Bacterial Pathogenesis: A Molecular Approach*, Washington, DC: American Society for Microbiology (1994).

Thomas, L., *The Lives of a Cell*, New York: Viking Press (1974).

Suggested Readings

Brock, T. D. and Madigan, M. T., *Biology of Microorganisms*. Englewood Cliffs, NJ: Prentice-Hall (1991)

Lynn, L., *Environmental Biology*, Northport, NY: Kendall-Hunt, (1995).

Metcalf and Eddy (revised by Tchobanoglous, G. and Burton F. L.) *Wastewater Engineering, Treatment, Disposal, and Reuse*, 3rd ed., New York: McGraw-Hill, Inc. (1991).

Ries, A. A., Vugia, D. J., Beingolea, L., Palacios, A. M., Vasquez, E., Wells, J. G., Garcia, N., Swerdlow, D. L., Pollack, M. and Bean, N. H., Cholera in Piura, Peru: A Modern Urban Epidemic, *Journal of Infectious Disease*, 166(6): 1429-1433, (1992)

Riddihough, G., Picture of an Enzyme at Work, *Nature*, 362, 793, (1993).

Singleton, P., *Introduction to Bacteria*, 2nd ed., New York: John Wiley and Sons, (1992).

Singleton, P. and Sainsbury, D., *Dictionary of Microbiology and Molecular Biology*, 2nd ed., New York: John Wiley and Sons, (1994).

Spellman, F. R., *Microbiology for Water/Wastewater Operators*, Lancaster, PA: Technomic Publishing Co., (1997).

Wistreich, G. A., and Lechtman, M.D., *Microbiology*, 3rd ed., New York: Macmillan Publishing Co., (1980).

CHAPTER 5

ENVIRONMENTAL TOXICOLOGY

MIGRANT FARMWORKER POISONED

A 41-year old migrant farmworker was brought to Rachel's Creek Hospital emergency department by his co-workers. All spoke only Portuguese, and no interpreter was available. However, the co-workers indicated that the patient had suffered from severe abdominal pain, nausea, vomiting, weakness, diarrhea, and increased salivation for several hours.

The patient was a well-developed, well-nourished Portuguese woman who seemed anxious and restless. Her speech slurred, and despite profuse sweating she appeared somewhat dehydrated. Blood pressure was 165/110, pulse 94, and respiration 25 and labored. Physical findings included a watery nasal discharge, bilateral pinpoint pupils, profuse salivation, and marked respiratory wheezing. Occasional, diffusely distributed, uncoordinated, uncontrolled twitching of muscles (fasciculation) were noted. The patient was oriented to day and place, but mental status could not be evaluated further due to the language barrier.

Arterial blood gases showed mild respiratory acidosis, serum electrolytes were normal, and glucose was 195 mg per deciliter. An appropriate drug was administered without improvement.

During the hour following her arrival the patient became progressional, dyspneic and finally required intubation. When an interpreter arrived, he learned from the patient's co-workers that the patient had been accidentally sprayed by a pesticide applicator a few hours earlier at the cabbage field where they all worked. She had sustained a concentrated exposure, both respiratory and percutaneous, to the pesticide. Her co-workers were unable to identify the pesticide.

Within an hour of the exposure, the patient had begun to complain of chest tightness, nausea, and difficulty with swallowing (dysphagia). She vomited several times, passed three loose stools, and developed generalized muscular weakness over several hours. It was then that she had been brought to the hospital.

NOTE: The text above is a fictional account of an incident—not unlike similar incidents that actually occur on a continuing basis in agricultural settings throughout the world—involving the misuse of pesticides. These common incidents demonstrate that the misapplication of toxic chemicals can impact people. However, as someone studying environmental science, you should look beyond the immediate effects to people. Think about the possible significance to the bigger picture: environmental damage.

Many pesticides are used throughout the world. Those at risk of occupational poisoning include individuals involved in production and application, as well as field workers. Incidents of pesticides carried home on clothing (and affecting the health of young children in particular) are also increasing. Increased use of pesticides at home has been associated with an increased number of childhood poisonings, and pesticides are frequently implicated in suicidal ingestions.

The patient described in this example was acutely exposed to parathion (an organophosphate). After receiving proper medical treatment, the patient went on to complete recovery. Her poisoning was reported to the departments of public health and environmental quality as required by state law. A follow-up study of the farm hands at the cabbage field revealed that most had suffered from nausea, vomiting, diarrhea, and increased secretions during work. Under intense regulatory and legal pressure, the grower soon introduced a series of procedural safeguards for pesticide application.

CHAPTER OUTLINE

- Introduction
- Dose Response
- Environmental Toxicology

INTRODUCTION

Toxicology is the study of the adverse effects of chemicals on living organisms. Often, toxicology is more simply defined as the science that deals with the nature and effects of poisons—or even more simply as the science of poisons. The development of toxicology as an independent science closely follows and parallels the growth pattern of the chemical industry and its production of non-natural toxicants. *Environmental Toxicology* (which blends the principles and practices of environmental science with those of toxicology, and is sometimes referred to as *ecological toxicology* or *ecotoxicology*) is the branch of toxicology that addresses the effect of toxic substances, not only on the human population (population dynamics, community structure, and ecosystems), but also the environment in general, including air, soil, surface water, and ground water. To devise and recommend mitigation procedures for the effects of pollution, environmental toxicologists attempt to understand, monitor, and predict the consequences of a wide variety of toxic pollutants.

A toxicant or toxin is a chemical that can cause serious illness or death. Toxicity is a physiological property of matter that defines the capacity of a chemical to harm or injure a living organism by other than mechanical means. Toxicity entails a definite dimension—amount or quantity. The toxicity of a chemical depends on the degree of exposure.

DOSE-RESPONSE

In the science of toxicology, a phrase that has become well known to most students and practitioners is: *It's the dose that makes the poison*. What this means, of course, is that a person can be exposed to just about any chemical and not receive and/or feel any ill effects—if the exposure dose is small or below the "toxic" level.

Consider arsenic. Arsenic is a poison. Anyone who has been exposed to crime shows on television or has had the opportunity to see or read Joseph Hesselring's *Arsenic and Old Lace* is familiar with the potent effects of arsenic. Arsenic trioxide has been a known poison from at least the time of the Middle Ages.

Most people immediately correlate the word arsenic with another word: poison. Yet most people do not know that arsenic itself does not kill, but that it is the amount of arsenic (the dose) ingested that kills.

Arsenic, like several other heavy metals, tends to accumulate in the body. Thus, ingestion of a small dose of arsenic may seemingly exert no adverse effect. However, ingestion of multiple small doses could cause death.

Toxicologists base all toxicological considerations on the *dose-response relationship*. A dose is administered to test animals, and, depending on the outcome, is increased or decreased until a range is found where at the upper end all animals die and, at the lower end, all animals survive. The data collected are used to prepare a *dose-response curve* relating percent mortality to dose administered.

To understand the dose-response relationship and its importance to environmental scientists and others, let's take a closer look at the dose-response curve and how it is used.

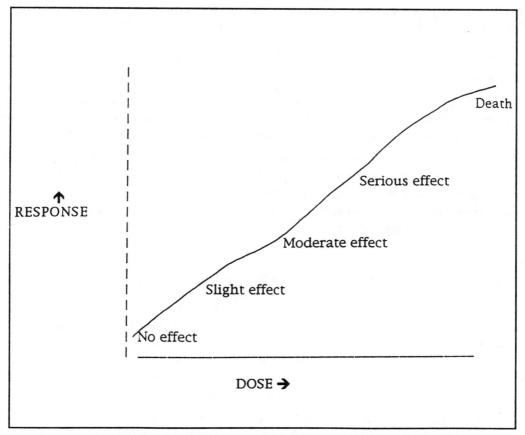

Figure 5.1 Dose-response curve for a typical chemical

To determine what dose of a particular chemical causes which kinds of toxic effects, scientists often administer a wide variety of doses to experimental animals. During this process, the usual response pattern is: a very small dose causes no observable effects, at a higher dose some toxicity is observed, at a still-higher dose there is greater toxicity, and at a high enough dose the animal dies. This gradual increase in toxic effects is known as dose-response and is often presented graphically as shown in Figure 5.1.

If we assume that the dose-response curve shown in Figure 5.1 represents what happens when chemical T is given to a mouse, the next question is: What happens if chemical T is given to a rat? Will the dose-response curve look the same? If not, how will it differ? The answer is that the dose-response curve will probably have a similar shape, but it will start rising at either a lower or higher dose than in the mouse. An entire group of curves that describe what happens when chemical T is given to a number of different animals has been determined (see Figure 5.2), demonstrating that some animals are more sensitive to chemical T than others—important because what this has really shown in this relationship is that a smaller dose is needed to produce the same toxic affect—the "toxic dose" varies from one kind of animal to another. The dose is species-specific.

Remember that a sufficiently small amount of most chemicals is not harmful. These chemicals possess a threshold of effect—a "no effect" level. The most toxic chemical known (if present in small enough amounts) will produce no measurable effect. It might damage a few cells, but no effect (liver damage, for example) will be measurable. As the dose increases, a point when the first measurable effect is noted occurs. The toxic potency of a chemical is defined by the relationship between the dose of the chemical and the response that is produced in a biological system. A high concentration of a toxic substance in a target organ (e.g., the liver), for example, will cause a severe reaction, and a low concentration, a less severe reaction.

The "qualitative" information provided above is a rough estimate of the relative toxicity of various chemicals. In science, however, we are interested in making reproducible "quantitative" determinations of toxicity. To do this we use and follow specific experimental protocols.

The toxicologist is interested in determining whether or not a specific substance is harmful to organisms. He or she understands that this determination depends on the properties of the chemical, the dose, the route by which the substance enters the body (there are four routes by which a substance may enter or act on the body: (1) inhalation, (2) ingestion, (3) injection, and (4) contact with or absorption through the skin), and the susceptibility or resistance of the exposed individual.

ENVIRONMENTAL TOXICOLOGY: PRACTICAL APPLICATIONS

The environmental toxicologist is concerned with determining the concentration of toxic contaminants, and with how they affect all living organisms that inhabit our environment. Taking the lead in making these determinations is the U.S. Environmental Protection Agency (EPA). The EPA is required to characterize and evaluate known hazardous waste sites where toxic releases to the environment have been observed. Under congressional mandate, the EPA must provide a means of responding to toxic substance releases to the environment.

What role does the environmental scientist and/or toxicologist play in the EPA's national monitoring and cleanup efforts of toxic waste sites and chemical spills and releases? They play a very

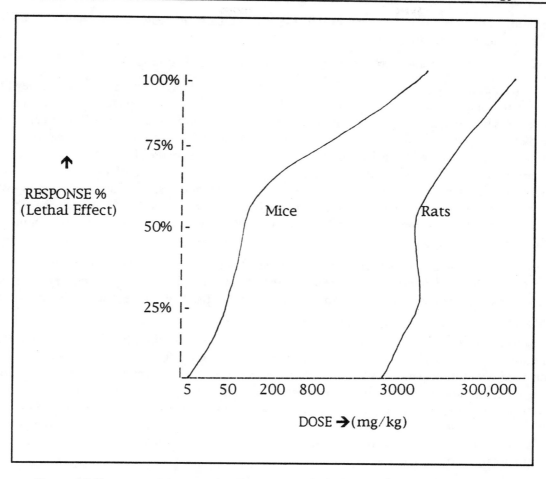

Figure 5.2 Variation of dose-response for the same chemical in two different species.

important role—most of the personnel performing these tasks for the EPA are environmental scientists and toxicologists.

On a local level (in the real, everyday work world), environmental toxicologists do most of their experimentation, studies, and analysis in the laboratory. However, they also do a fair amount of work in the field, even though most toxicological analysis requires extensive work in a fully equipped laboratory. Let's take a look at an example of field level work that the environmental scientist/toxicologist might be involved with in determining water quality without the aid of a laboratory.

Example of Biotic Index Field Level Work

Four indicators of water quality—coliform bacteria count, concentration of dissolved oxygen (DO), biochemical oxygen demand (BOD), and the *Biotic Index*—are of interest to and are often used by environmental scientists. The biota that exist at or near a stream, for example, are direct indicators (a biotic index) of the condition of the water. This biotic index is often more reliable than many of the laboratory chemical tests that environmental scientist/toxicologists use in attempting to determine the pollutant level in a stream. Actually, indicator species help determine when pollutant levels are unsafe.

How does the biotic index actually work? How does it indicate pollution?

Certain common aquatic organisms, by indicating the extent of oxygenation of a stream, may be regarded as indicators of the intensity of pollution from organic waste. The responses of aquatic organisms in streams to large quantities of organic wastes are well documented. They occur in a predictable, cyclical manner. For example, upstream from an industrial waste discharge point, a stream can support a wide variety of algae, fish, and other organisms, but in the section of the stream where oxygen levels are low (below 5 ppm), only a few types of worms survive. As streamflow courses downstream, oxygen levels recover and those species that can tolerate low rates of oxygen (such as gar, catfish, and carp) begin to appear. Eventually, at some further point downstream, a clean water zone reestablishes itself and a more diverse and desirable community of organisms returns.

During this characteristic pattern of alternating levels of dissolved oxygen (in response to the dumping of large amounts of biodegradable organic material), a stream, as stated above, goes through a cycle. This cycle is called an *oxygen sag curve*. Its state can be determined using the biotic index as an indicator of oxygen content.

The biotic index is a systematic survey of invertebrate organisms. Since the diversity of species in a stream is often a good indicator of the presence of pollution, the biotic index can be used to correlate with water quality. A knowledgeable person (an environmental scientist or toxicologist, for example) can easily determine the state of water quality of any stream simply through observation—observation of types of species present or missing is used as an indicator of stream pollution. The biotic index, used in the determination of the types, species, and numbers of biological organisms present in a stream, is commonly used as an auxiliary to BOD determination in determining stream pollution.

The biotic index is based on two principles:

1)   A large dumping of organic waste into a stream tends to restrict the variety of organisms at a certain point in the stream.

2)   As the degree of pollution in a stream increases, key organisms tend to disappear in a predictable order.

The disappearance of particular organisms tends to indicate the water quality of the stream.

There are several different forms of the biotic index. In Great Britain, for example, the Trent Biotic Index (TBI), the Chandler score, the Biological Monitoring Working Party (BMWP) score, and the Lincoln Quality Index (LQI) are widely used. Most forms use a biotic index that ranges from 0 to 10. The most polluted stream, which contains the smallest variety of organisms, is at the lowest end of the scale (0); the clean streams are at the highest end (10). A stream with a biotic index of greater than five will support game fish; a stream with a biotic index of less than four will not support game fish.

Because they are easy to sample, macroinvertebrates have predominated in biological monitoring. In addition, invertebrates can be easily identified by comparing with identification keys, which are portable and conveniently used in field settings. Present knowledge of invertebrate tolerances and responses to stream pollution is well documented. In the United States, for example, the EPA has required states to incorporate narrative biological criteria into their water quality standards since 1993.

Macroinvertebrates are a diverse group. They demonstrate tolerances that vary between species. Discrete differences tend to show up, containing both tolerant and sensitive indicators.

The biotic index provides a valuable measure of pollution, especially for species that are very sensitive to lack of oxygen. Consider the stone fly. Stone fly larvae live underwater and survive best in well-aerated, unpolluted waters with clean gravel bottoms. When the stream deteriorates from organic pollution, stone fly larvae cannot survive. The degradation of stone fly larvae has an exponential effect upon other insects and fish that feed off the larvae; when the stone fly larvae disappear, so in turn do many insects and fish.

Table 5.1 shows a modified version of the BMWP biotic index. Since the BMWP biotic index indicates ideal stream conditions, this index takes into account the sensitivities of different macroinvertebrate species to stream contamination. Aquatic macroinvertebrate species are represented by the diverse populations and are excellent indicators of pollution. These organisms are large enough to be seen by the unaided eye. Most aquatic macroinvertebrates live for at least a year. They are sensitive to stream water quality, on both a short term and long-term basis. Mayflies, stone flies, and caddis flies are aquatic macroinvertebrates considered clean-water organisms; they are generally the first to disappear from a stream if water quality declines, and are therefore given a high score. Tubicid worms (which are tolerant of pollution) are given a low score.

In Table 5.1, a score from 1-10 is given for each family present. A site score is calculated by adding the individual family scores. The site score (total score) is then divided by the number of families recorded to derive the Average Score Per Taxon (ASPT). High ASPT scores result from such taxa as stone flies, mayflies, and caddis flies present in the stream. A low ASPT score is obtained from heavily polluted streams dominated by tubicid worms and other pollution-tolerant organisms.

Table 5.1: The BMWP Score System (modified for illustrative purposes)

Families	Common-Name Examples	Score
Hepatagenidae	Mayflies	
Leuctridae	Stone flies	10
Aeshnidae	Dragonflies	8
Polycentropidae	Caddis flies	7
Hydrometridae	Water Strider	
Gyrinidae	Whirligig beetle	5
Chironomidae	Mosquitoes	2
Oligochaera	Worms	1

Organisms having high scores, especially mayflies and stone flies (the most sensitive), and others (dragonflies and caddis flies) are very sensitive to any pollution (deoxygenation) of their aquatic environment.

The biotic index example above shows that environmental scientists/toxicologists make use of the fact that unpolluted streams normally support a wide variety of macroinvertebrates and other aquatic organisms with relatively few of one kind in making determinations about water quality in the field. While some aquatic species, such as mayflies and stone flies, are more sensitive than others to certain pollutants and succumb more readily to the effects of pollution, other species, such as mussels and clams, accumulate toxic materials in their tissues at sub-lethal levels. These species can be monitored (*must be* monitored to protect public health) to track pollution movement and buildup in aquatic systems.

Using a biotic index to determine the level of pollution in a water body demonstrates only one application. B. M. Levine et al. (1989) point out that similar determinations regarding soil quality can be made by observing and analyzing organisms (such as earthworms) in the soil. Studies conducted to assess the impact of sewage biosolids (sludge) treatments on old-field communities revealed that earthworms concentrate cadmium, copper, and zinc in their tissues at levels that exceed those found in the soil. Cadmium levels even exceed the concentrations found in the biosolids. Thus, earthworms may provide an "index" to monitor the effects of biosolids disposal on terrestrial communities.

In addition to determining the level of pollution in water and soil, environmental scientists/toxicologists also monitor the air we breathe. A few horrendous incidents involving acute exposure to dangerous airborne chemicals have been well documented. The sudden accidental release of methylisocyanate (MIC) at Bhopal, India is one infamous example. Such acute exposures like Bhopal make headlines in the press, but chronic (long-term) exposures to sub-lethal quantities of toxic materials present a much greater hazard to public health, and are of major concern to environmental toxicologists. Millions of urban residents are continually exposed to low levels of a wide variety of pollutants. Many deaths attributed to heart failure or diseases like emphysema may actually be brought on by a long-term exposure to sub-lethal amounts of pollutants in the air.

SUMMARY

All organisms are exposed to toxic substances. Some of these substances (lead and mercury, for example) have always been in the environment in trace amounts. However, the industrial processes (technological advancements) of the present time concentrate substances like lead and mercury to dangerous levels—and then release them into the environment. Many of the toxic chemical products and by-products produced today that enter the environment were unknown a few decades ago. There is an increasing concern that the environment we live in today—the air that we breathe, the water we drink, and the food that we eat—endanger our health. This is where environmental scientists/toxicologists come in. Their concern is with the study, detection, and mitigation of all such toxicants and their potential impact on our environment—and on our lives and life around us.

Cited References

Levine, M. B., Hall, A. T., Barret, G. W., and Taylor, D. H., Heavy-Metal Concentration During Ten Years of Sludge Treatment to an Old-Field Community, *Journal of Environmental Quality*, 18, No. 4: 411-418 (1989).

Suggested Readings

Environ, *Elements of Toxicology and Chemical Risk Assessment*, Washington, DC: Environ Corporation, (1988).

Huff, W. R., Biological indices define water quality standards. *Water Environment and Technology*, 5, 21-22, (1993).

Jefferies, M and Mills, D., *Freshwater Ecology: Principles and Applications*, London: Belhaven Press, (1990).

Kamrin, M. A., *Toxicology*, Chelsea, Michigan: Lewis Publishers, (1989).

Mason, C. F., Biological aspects of freshwater pollution. *Pollution: Causes, Effects, & Control*. (Ed.) R. M. Harrision. Cambridge, Great Britain: The Royal Society of Chemistry, (1990).

Meyer, E., *Chemistry of Hazardous Materials*, 2nd ed., Englewood Cliffs, NJ: Prentice-Hall, (1989).

O'Toole, C (Ed.). *The Encyclopedia of Insects*, New York: Facts on File, Inc., (1986).

Spellman, F. R., *Stream ecology and Self-Purification: An Introduction for Wastewater and Water Specialists*, Lancaster, PA: Technomic Publishing Company, (1996).

U.S. Environmental Protection Agency, *Superfund Public Health Evaluation Manual*, Washington, DC: Office of Emergency and Remedial Response, (1986).

Wooten, A. *Insects of the World*. New York: Facts on File, Inc. (1984).

CHAPTER 6

ENVIRONMENTAL GEOLOGY AND GROUNDWATER HYDROLOGY

Plants absorb energy from the sun. This energy flows through a circuit called the biota, which may be represented by a pyramid consisting of layers. The bottom layer is the soil. A plant layer rests on the soil, an insect layer on the plants, and so on up through various animal groups to the apex layer, which consists of the larger carnivores. —Aldo Leopold

CHAPTER OUTLINE

- Introduction
- What is Geology?
- Formation and Types of Rocks
- Formation of Soil
- Soil Characteristics
- Soil Profile
- Soil Function
- Groundwater Hydrology

INTRODUCTION

As with biology, chemistry, ecology, and other related sciences, a fundamental knowledge of geology is a prerequisite to meeting and understanding the many environmental science challenges we face as we enter the 21st century. Geology is a wide field, but in this text we focus on only one division of geological science, soil science, because of its natural interface with the other environmental media: air and water. The study of soil not only can be fascinating and intellectually stimulating, it is also the ideal medium in which to observe practical applications for basic principles of biology, chemistry, other related sciences, and water hydrology.

But to discuss soil without discussing the natural interface that is soil plus water hydrology would be like launching a hot air balloon without the hot air needed to lift it skyward. More specifically, there is a natural interface of air, soil (minerals), living organisms, and water. This natural interface can be seen in any good-quality loam surface soil where these four mediums are mixed in complex patterns. Figure 6.1 shows the proportions of soil volume occupied by each medium.

To better understand and appreciate the message presented in Figure 6.1, consider a handful of dirt, a clump you turn over and over in your hand. You ask yourself, "Is this clump of dirt solid?" It appears to be—but it is not. That clump is really a mixture of components. Only about half of the soil volume contains solid material (mineral and organic). The other half consists of pore spaces filled with air and water.

In water hydrology (especially groundwater hydrology), the environmental scientist must have a thorough understanding of the machinations of the hydrologic cycle, the water-soil surface

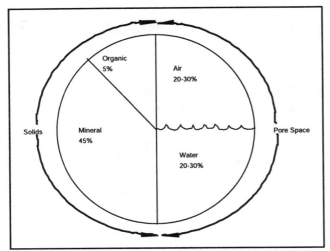

Figure 6.1 Typical volume concentration of a loam surface soil. The curved line between water and air indicates that the proportion of these two componenets fluctuates as soil becomes wetter or drier.

interface, and the role soils play in sub-surface water supplies, because the focus of environmental legislation has been directed toward cleaning up our waters—both on the surface and below ground. The environmental practitioner must have a fundamental understanding of geology and water hydrology to be able to effect cleanup of the problems of pollution, use the technology used to prevent contamination, or decontaminate subsurface problems. The following sections will serve to raise your understanding of the basic science involved, and in turn will help you understand the more complex environmental issues related to soil and groundwater problems that are discussed in detail later in this text.

WHAT IS GEOLOGY?

In a traditional sense, we can say that geology is the science of the earth, its origin, composition, structure, and history. On a much simpler plane, we can say that the earth is comprised of three main areas, or spheres: the *lithosphere* (rock and soil), the *hydrosphere* (water), and the *atmosphere* (air). In this chapter we are concerned with the study of geologic material (primarily rocks and soils—the lithosphere) through which groundwater moves.

FORMATION AND TYPES OF ROCKS

There are three types of rocks: igneous, sedimentary, and metamorphic. *Igneous* rocks, which at one time were in a molten or liquid state, are the many types of volcanic material that then hardened into rock form. Granite, for example, is an igneous rock found in the core of many mountain areas.

Sedimentary rocks have one thing in common: they are all composed of small units, ranging in size from molecules up through dust particles, to pebbles and large boulders, all brought together and deposited on the surface of the earth's crust. All of the mineral matter comprising these rocks was once part of other rocks.

The third and final category of rocks is *metamorphic*. The term metamorphic means simply "changed in form." The name concentrates attention on the processes by which the rock evolves. All of the rocks in this class were once either igneous or sedimentary, but have been changed by pressure, heat, or chemical action of liquids or gases so that their original nature has been significantly altered. Pressure is caused by burial of rock types beneath the surface, and heat is applied either as a function of depth or of proximity to molten rock from *magma* beneath the surface of igneous intrusions into the upper rock layers.

Soil is derived from *parent material* rock.

FORMATION OF SOIL

Soil is a combination of minerals, organic material, living organisms, air, and water, and is the thin covering over the surface of the Earth. Soil is initially formed from rock, the parent material, through a combination of physical and biological events. Soil building begins with the physical fragmentation of ancient layers of rock or more recent geologic deposits from lava flows and/or glacial activity. The kind of parent material and climate determines the kind of soil formed. Physical factors that bring about the fragmentation or chemical change of the parent material are known as *weathering*. Abrasion and temperature changes are two primary agents of weathering (see Figure 6.2).

Because rock does not expand evenly, heat can cause a large rock to fracture. Pieces fall off, resulting in smaller chunks, which are eventually reduced in size by other processes. One of these processes is the repeated freezing and thawing of water. Water finds its way into cracks and crevices, then expands as it freezes, causing the rock to be broken into smaller pieces over time. In the seasonal cycles of freeze and thaw, the cracks become progressively wider and wider, causing the rock to eventually fracture, crack, and break into smaller pieces.

Abrasion is also an important weathering process that eventually reduces large, solid rock masses to soil. This physical breakdown of rock is commonly caused by glacial action where rock fragments grind against each other. This results in small fragments and smoother surfaces. Carried by the glaciers, these fragments and particles are deposited when the ice melts. Weathering by abrasion also occurs when wind and moving water cause small particles to collide. Moving water in streams and rivers also causes rock fragments to

*Figure 6.2 Parent rock exposed to weathering
(photo by John Goeke)*

collide and rub together, causing their surfaces to become smooth. Similarly, wind picks up particles and causes them to collide with objects like rock formations, resulting in fragmentation of both the rock formation and the wind-driven particles.

Moving water and wind also cause smaller particles to be pushed along to some other location, where new surfaces are exposed to further weathering. Many of the landscapes throughout the world were created by a combination of moving water and wind that removed easily transported fragments and particles, while more resistant rock formations remained. These remaining formations are still undergoing the weathering process; someday, they too will be reduced to small fragments and particles that will end up as soil (see Figure 6.3).

In addition to the forces of changing temperatures, glaciers, moving water, and wind, certain chemical activities also alter the size and composition of parent material and participate in the soil-making process. These chemical activities include the action of atmospheric air working to chemically oxidize small rock fragments to different compounds, and hydrolysis, through which water combines with water molecules—oxidized and hydrolyzed molecules are more readily soluble in water and may be removed by rain or moving water. Acid rain also assists in dissolving rocks.

This rock fragmentation process continues as the first organisms gain a foothold in broken-down or modified parent material. *Lichens* often form a *pioneer community*—the first successful integration of plants (including animals and decomposers) into a bare rock community. The lichen-formed pioneer community traps small particles and chemically alters the underlying rock (see Figure 6.4). When plants and animals become established, they contribute increasing amounts of organic matter, which is incorporated with small rock fragments to form soil. *Humus*, dead organic matter, becomes mixed with the top layers of rock particles—constituting a critical soil ingredient. It supplies some of the nutrients needed by plants, and also increases soil acidity so that inorganic nutrients become more available to plants. These inorganic soil constituents are more soluble under acidic conditions. Some crops (corn and wheat, for example) grow best in soils with a pH between 5.0 to 7.0. When humus is added to a soil with a pH above 7.0, the soil becomes more productive because the humus increases the soil acidity. Along with affecting the soil's nutrient availability, humus modifies soil texture. A soil with a loose, crumbly texture allows water to soak in and permits air to be incorporated into the soil. A compact soil allows water to run off and prevents adequate aeration.

Burrowing animals, fungi, soil bacteria, and the roots of plants are also important in the biological process of soil formation. The earthworm is one of the most important burrowing animals. These animals literally eat their way through soil, mixing organic and inorganic material. This mixing action obviously results in an increase in the amount of nutrients available for plant use. Earthworms increase the fertility of the soil by bringing nutrients from the deeper layers of the soil up into the area where the roots of plants are concentrated. Burrowing earthworms also improve soil aeration and drainage. When earthworms collect dead organic matter from the surface and transport it into burrows, this adds organic matter into the soil.

Earthworms are not the only organisms that work to improve soil quality. Bacteria and fungi are decomposers. They, along with other organisms, reduce organic matter to smaller particles, improving soil quality.

*Figure 6.3
Smoothing
results of
running water
on rock
formations*

*Figure 6.4 Rock
with colonies of
lichens forming*

Over time (from as little as a few weeks to hundreds of years, depending on climate conditions) the physical, chemical, and biological processes have formed the soils we have today. Soil formation is a slow but continuous process.

SOIL CHARACTERISTICS

Soil characteristics include soil texture, structure, atmosphere, moisture, biotic content, and chemical composition. Soil texture (or feel) is determined by the mineral present in the soil. It determines the oxygen (i.e., proportions of sand, silt, and clay) present in the soil and determines the oxygen-holding and water-holding capacity of a soil. Gravel and sand, the larger particles, allow water and air to penetrate the soil because their shapes permit many tiny spaces between individual particles. Not only does water and air flow through these spaces, but water also drains very rapidly, and as it drains, it carries valuable nutrients to lower soil layers where they are normally beyond the reach of the roots of plants. Clay particles are generally packed close together to form waterproof layers, so soils with large amounts of clay content do not drain well, are poorly aerated, and retain nutrients.

Soil rarely consists of one type or size of particle. Soil is a mixture of various particle sizes and types, resulting in soils being classified into several different types. *Loam* is the ideal soil because it combines the good aeration and drainage properties of large particles with the nutrient retention ability of clay particles.

Although soil texture helps to determine soil structure, remember that soil structure and texture are different. Soil texture, or *tilth*, refers to the way various soil particles clump together, or how the soil particles are arranged. The structure of a particular soil is strongly influenced by the amount of clay and organic matter it contains. Sand particles do not clump; therefore sandy soils lack structure. Clay soils tend to stick together in large clumps. Clay and organic particles, because of their chemical and physical properties, are able to link with other particles, forming larger soil aggregates. Being small and numerous, clay particles create a large surface area in a given amount of soil, which provides a surface for water and nutrients to cling to. In contrast, soils lacking clay or organic matter have an unstable structure and are likely to form dusts or loose sand, which can easily blow away.

A good soil, composed of small clumps, crumbles easily when squeezed by hand. This ability to crumble is known as *friability*. Sandy soils are friable; clays are not. Soil that will crumble has adequate spaces to allow air and water to mix with soil. The air in these spaces provides a source of oxygen for plant root cells. Water for the roots occupies the remaining soil space.

SOIL PROFILE

The *soil profile* is a series of horizontal layers of different chemical composition, particle sizes, and amounts of organic matter that differ greatly from region to region. Each recognizable layer is known as a *horizon* (see Figure 6.5). Each soil horizon has a characteristic color, texture, structure, acidity, and composition. Road cuts and other excavations can expose soil profile. Horizons within a soil may vary in thickness and have somewhat irregular boundaries, but generally they parallel the land surface.

The uppermost layer of the soil contains more nutrients and organic matter (e.g., decomposed plant leaves and roots) than the deeper layers. This layer is known as the *A horizon* or *topsoil*. The A horizon may vary in thickness from a few centimeters to over a meter in some areas. The majority of the living organisms and nutrients are found near the top of the A horizon. The lower portion of the A horizon often contains few nutrients because they are leached by flowing water to the B horizon.

The layer underlying the A horizon, the *B horizon* (also known as the subsoil), contains comparatively less organic matter than the horizon nearer the surface. The B horizon also contains less organic material and fewer organisms, but it accumulates nutrients leached from higher levels.

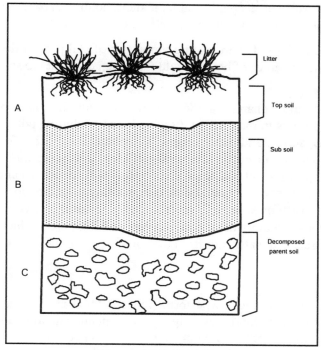

Litter

Top soil

Sub soil

Decomposed
parent soil

A

B

C

Figure 6.5 Soil profile

The area below the B horizon is known as the *C horizon* and consists of weathered, unconsolidated parent material. The C horizon is outside the zones of major biological activities, lies above the impenetrable layer of bedrock, and is generally little affected by the processes that formed the horizons above it.

Note that soil profiles and the factors that contribute to soil development are extremely varied.

SOIL FUNCTION

The soil's primary function is to support the growth of plants by providing a medium for plant roots and supplying nutrient elements that are essential for the plant. The various properties of the soil determine the nature of the vegetation present, and to a lesser degree, the number and types of animals that the vegetation can support. A secondary function of soil is to act as a recycling system for nature (and for us). Within the soil, dead bodies of animals, plants, people, and their waste products are assimilated, and their basic elements (nutrients) are made available for reuse. Soil also functions as an engineering medium. Soil not only provides the foundation for virtually every type of construction project, but is also an important building material (earth fill and bricks). Obviously, one of the most important functions soil provides is a habitat for a host of living organisms. Finally, soil functions to control the fate of water in the hydrologic system. Soil affects water loss, utilization, purification, and contamination.

Soil Provides a Medium for Plant Growth

That soil provides physical support and anchors the root systems of plants so they won't fall over is obvious. Less obvious is what else soil provides to plants. We all know soil is the primary medium for plant growth—but what is involved in this process?

Plant roots are the primary residents of subsurface soil. Roots depend on the process of respiration to obtain energy. Since root respiration, like our own respiration, produces carbon dioxide (CO_2) and uses oxygen (O_2), an important function of soil is "gas transfer"—allowing CO_2 to escape and fresh O_2 to enter the root zone. This gas transfer is accomplished via the network of soil pore spaces.

An equally important function of soil is to absorb rainwater and hold it where it can be used by plant roots. When exposed to sunlight, plants require a continuous stream of water to use in nutrient transport, cooling, turgor maintenance (swelling or distention from water), and photosynthesis. Whether it's raining or not, plants use water continuously. Thus the water-holding capacity of soils is essential for plant survival. The deeper the soil, the more holding capacity it normally has for storing water to allow plants to survive long periods without rain.

The soil also moderates temperature fluctuations. The insulating properties of soil protect the deeper portion of the root system from the extremes of hot and cold that often occur at the soil surface.

As well as moderating moisture and temperature changes in the root environment, the soil also supplies plants with inorganic mineral nutrients in the form of dissolved ions, including potassium, iron, copper, nitrogen, phosphorus, sulfur, and many others. The plant takes these elements out of the soil solution (the thin aqueous film surrounding soil particles) and incorporates most of them into the various organic compounds that constitute plant tissue. Soil supports plant growth by providing a continuing supply of dissolved mineral nutrients in amounts and relative proportions appropriate for plant growth. Enzymes, organic metabolites, and structural compounds making up a plant's dry matter consist mainly of carbon, oxygen, and hydrogen, which the plant obtains via photosynthesis from air and water, not from the soil.

Soil Recycles Raw Materials

How important is the raw-material recycling continuously performed by soil? So essential that without the constant reuse of nutrients that this natural recycling provides, plants and animals would have run out of nutrients eons ago, and in fact, would not have been able to establish life on any large scale. Instead of layers of plants, the world would be covered with a layer of dead plants, animals, and waste. Soil also plays a key role in the geochemical cycles. Soils assimilate large quantities of organic waste, turning this waste into humus, converting the nutrients in wastes to forms that can be used by plants and animals, and returning the carbon to the atmosphere as gaseous carbon dioxide, where it will be reused via plant photosynthesis. Large accumulations of carbon in soil organic matter can have a major impact on the greenhouse effect (to be discussed in detail later).

Soil Is an Engineering Medium

Generally, soil is a firm, solid base on which build all kinds of contructions, from roads to airports. However, some soils are not as stable as others and are unsuitable to be built on. Whether a soil is or is not suitable for a particular type of construction or excavation is a decision that can only be accurately made by professional soil engineers.

Soil Provides a Medium for Organisms

For those who consider soil as nothing more than a pile of organic debris and broken rocks, consider, for a moment, picking up a handful of dirt. When you do, what you really have in your hand is an entire ecosystem. A handful of certain types of soil may be the home of several 100-million organisms, belonging to several thousand species. This one handful of soil could contain a full range of organisms, including predators, prey, producers, consumers, decomposers, and parasites. When someone questions the validity of placing soil in the same category of importance as the other two environmental media, water and air, he or she only need study the microworld contained in one handful of soil.

Soil Regulates Our Water Supplies

Excluding the relatively small quantity of precipitation that falls directly into bodies of fresh water, nearly every drop of water in our lakes, rivers, streams, estuaries, and aquifers has either traveled through the soil or flowed over its surface. Consider a steep hillside, joined by a gully housing a slow-moving stream. Some of the rain falling on the steep hillside may soak into the ground, where the water may be stored in the soil and used by the trees, bushes, flowers, and grass that cover it, while some may seep slowly down through the soil layers to the groundwater, eventually entering the stream over a period of a few months or years. As the water flows through the soil, it is purified by soil processes that remove many impurities and kill disease organisms.

Soil is not only a natural storehouse for water, it is also a filter—probably the best filter there is. Nature has a way of perfecting things and processes that we attempt to imitate but never really duplicate.

Environmental scientists are concerned with the quality and quantity of the water in our lakes, rivers, streams, and underground aquifers. A primary mission in environmental science is to prevent, mitigate, or abate the pollution that threatens the quality of our waters for fishing, drinking, and swimming. To better fulfill this obligation, environmental science practitioners must have a fundamental knowledge of soil science and groundwater hydrology.

GROUNDWATER HYDROLOGY

Hydrology is the science concerned with the occurrence and circulation of water in all its phases and modes. Hydrology begins with the hydrologic cycle, the means by which water is circulated in the biosphere (See Figure 6.6).

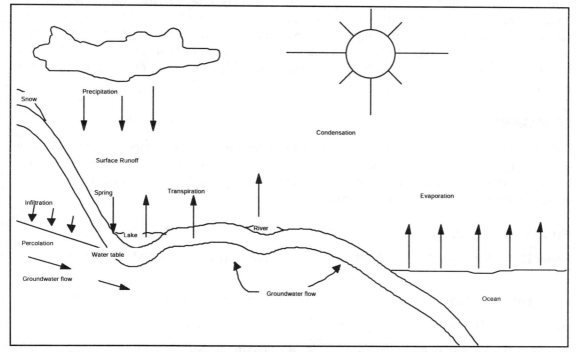

Figure 6.6 Hydrologic cycle (water cycle) (Adapted from Blackman, W. C., Jr., Basic Hazardous Waste Management, *1993, p. 53)*

Evapotranspiration (the combination of evaporation and transpiration of liquid water in plant tissue, and in the soil to water vapor in the atmosphere) from the land mass plus evaporation from the oceans is counterbalanced by cooling in the atmosphere and precipitation over both land and oceans (see Figure 6.6). The hydrological cycle requires that on a worldwide basis the evaporation and precipitation be equal. However, oceanic evaporation is greater than oceanic precipitation, thus excess precipitation is given to the land. Eventually this land precipitation ends up in lakes, rivers, and streams and thus eventually returns to the sea, completing the cycle.

Most of the earth's water (about 97%) is salty; only a small portion (less than 3%) is fresh. Three-quarters of all fresh water is found in polar ice caps and glaciers, and nearly one-quarter is found underground in water-bearing porous rock or sand or gravel formations. 96% of the United State's water supply is groundwater.

Groundwater percolates downward through the soil after a rain or snow or seeps downward from surface water and is stored in an *aquifer*. An aquifer is a water-bearing geologic formation composed of layers of sedimentary material such as sand, gravel and porous rock. Water fills the cracks and crevices of the rock and the pores between the particles of sand and gravel. The depth at which the aquifer begins is known as the *water table*. Before it reaches the aquifer, water passes through an unsaturated zone, where pores contain both water and air. Plants remove some of the water; the rest continues to move downward to the saturated zone. Thus groundwater is that part of underground water below the water table, and soil moisture is that part of underground water above the water table. Shallow, unconfined aquifers are recharged by water percolating downward from soils and materials directly above the aquifer.

Historically, groundwater has been considered so safe to drink that many water companies deliver it untreated to their customers. We have learned quickly, however, that some of our groundwater is becoming contaminated with hazardous substances from landfills, septic systems, and surface impoundments. The challenge for environmental practitioners is to prevent groundwater contamination, and to restore quality and quantity when contamination has occurred. However, remember that once groundwater becomes contaminated, restoration is difficult, if not impossible

SUMMARY

For most of us, soil is easier to ignore than is our water or air. Most people don't realize how directly our water supply is linked to healthy soil. The study of soil science and groundwater hydrology has never been more important to environmental scientists, practitioners, and students—and to all other organisms on earth.

Suggested Readings

Bohn, H. L., McNeal, B. L., and O'Connor, G. A., *Soil Chemistry*, 2nd ed., New York: Wiley, 1985.

Bouwman, A. F. (ed.) *Soils and the Greenhouse Effect*, New York: Wiley, 1990.

Brady, N. C., *The Nature and Properties of Soils*, New York: Macmillan, 1974.

Courtney, F. M. and Trudgill, S. T., *Soil: An Introduction to Soil Study*, Baltimore: E. Arnold, 1984.

Ellis, B. G. and Foth, H. D., *Soil Fertility*, 2nd ed., Boca Raton, FL: CRC Press/Lewis Publishers, 1997.

Freeze, R. A. and Cherry, J. A., *Groundwater*, Englewood Cliffs, NJ: Prentice-Hall, 1979.

Gupta, R. S., *Hydrology and hydraulic Systems*, Englewood Cliffs, NJ: Prentice-Hall, 1989.

Heath, R. C., *Basic Ground-Water Hydrology*, U.S. Government Printing Office: U.S. Geological Survey Water-Supply Paper 2220, 1983.

Jenny, H., *Factor of Soil Formation: A System of Quantitative Pedology*, New York: McGraw-Hill, 1941.

Linsley, R. K., Hohler, M. A. and Paulhus, J. L., *Hydrology for Engineers*, 2nd ed., New York: McGraw-Hill, 1975.

Paddock, J., Paddock, N. and Bly, C., *Soil and Survival: Land Stewardship and the Future of American Agriculture*, San Francisco: Sierra Club Books, 1986.

Petersen, G. W., Cummingham, R. L. and Matelski, R. P., Moisture Characteristics of Pennsylvania Soils: III. Parent Material and Drainage Relationships, *Soil Science Society Amer. Proc.*, 35:115-119, 1971.

Soil Survey Division Staff, *Soil Survey Manual*, Agric. Handbook 18, Washington, DC: U.S. Govt Printing Office, 1993.

Spellman, F. R., *The Science of Water: Concepts and Applications*, Lancaster, PA: Technomic Publishing Company, 1998.

U.S. Environmental Protection Agency, Drinking Water in America: An Overview, *EPA Journal*, Sept. 1986.

Willis, R. and Yek, W. W-G., *Groundwater System Planning & Management*, Englewood Cliffs, NJ: Prentice-Hall, 1987.

CHAPTER 7

ENVIRONMENTAL SAMPLING AND ANALYSES

THE ENVIRONMENTAL PRACTITIONER'S TOOL BOX

In a serious mechanic's toolbox, you would expect to find a complete set of tools. These tools would include sockets, wrenches, drive tools, screwdrivers, clamps, tap and die sets, micrometers, calipers, combination squares, tape measures, and other specialty tools. But the serious mechanic has other tools that are not as obvious. These tools include knowledge and skill.

The environmental technician, scientist, and/or practitioner also has a toolbox. This toolbox includes the tools of chemistry, biology, ecology, toxicology, geology, groundwater hydrology, soil science, and other related sciences. Like the mechanic, the environmental practitioner must know how to use these tools. He or she needs a well-rounded education in the basic sciences in order to properly evaluate environmental contamination.

But this is not all the environmental scientist needs to know. The mechanic and the environmental practitioner not only must all have their set of basic tools, they must also know how to use them.

From the environmental practitioner's standpoint, possession of scientific knowledge puts the tools in your hand—but more is required. The environmental practitioner must take those tools and apply them. The principal way in which they are applied is through sampling and analyzing the media (air, water, or soil) with which they are concerned.

CHAPTER OUTLINE
- Introduction
- Environmental Sampling and Analysis
- General Considerations for a Sampling Program
- General Evaluation Methods for Environmental Mediums

INTRODUCTION

Through improper use, storage, and/or disposal, contaminants find their way into air, groundwater, and soil. Once a hazardous material has been spilled, discharged, or has leaked, its impact and movement depends on several factors. In air, quality is impacted by those things we can see, either with the naked eye or under the microscope (pollen, dust, etc.), and those substances we can't see (ozone, sulfur dioxide, carbon dioxide, etc.). In soil and groundwater contamination, when a contaminated material has leaked or been discharged to the surface, its movement depends on the nature of the material and the soil horizon. In this chapter, we discuss various means of sampling air, soil, and groundwater. This chapter provides the environmental practitioner with those additional "tools" he or she needs to perform the critical task of evaluating environmental contamination problems.

ENVIRONMENTAL SAMPLING AND ANALYSIS

Obtaining reliable environmental samples is a difficult process. The key word here is "reliable." In general, for the purpose of reliability, the sampling effort should be directed toward taking and obtaining *representative samples*. According to the U.S. Environmental Protection Agency's (EPA) Office of Research and Development, a representative sample (representativeness) expresses the degree to which data accurately and precisely represent a characteristic of a population, parameter variations at a sampling point, a process condition, or an environmental condition. Taking representative samples is critically important.

However, sampling is more involved than just ensuring representativeness. Sampling operations involve more than that. You must first start with a sampling plan, and then execute it in a reliable fashion. Specific needs, of course, will dictate which techniques are actually incorporated into a sampling plan and which ones are rejected. Techniques not selected should be rejected only because they don't meet sampling goals, not because they were overlooked.

In the sampling plan, the main objective is to take representative samples of a heterogeneous and dynamic piece of our environment in order to analyze for components that constitute a very small fraction of the samples [often at or below the parts-per-billion (ppb) range]. Working with tiny fractions of samples is only one of many problems involved in sampling. Other complicating factors are that the matrix is usually very complex, opening the door for analytical interference, such as false positive and masking. Other types of interference can be introduced during transport and preservation. Further complicating the sampling and analytic process is the reactivity (instability) of some analytes.

Sampling is therefore and without a doubt the weakest link in the planning-sampling-analysis-reporting process. In short, we can say that the reliability of the overall data obtained cannot be greater than that of the reliability of the weakest part of the chain of events constituting an environmental sampling and analysis effort. The bottom line is that any analytical report is only reliable if the samples are representative of their source.

GENERAL CONSIDERATIONS FOR A SAMPLING PROGRAM

NOTE: Much of the general sampling information provided in the following section is based on information contained in the U.S. Department of Commerce's *Handbook for Sampling and Sample Preservation of Water and Wastewater*, 1982.

No single sampling program can apply to all types of environmental media. Nevertheless, each sampling program should consider:

- Objectives of Sampling Program
- Location of Sampling Points
- Types of Samples
- Sample Collection Methods
- Measurements
- Field Procedures

Objectives of Sampling Programs

We use sampling and analyses programs for four major reasons: planning, research or design, process control, and regulation. These objectives in an overall environmental media quality program are interrelated and cover different stages from planning to enforcement. Based on these objectives, the different sampling programs are compared in general terms in the following.

An environmental planner monitors to:

1) Establish representative baseline media quality conditions

2) Determine assimilative capacities of specific media

3) Follow the effect of a particular project or activity

4) Identify pollutant source

5) Assess long-term trends

6) Allocate waste load

7) Project future media characteristics

Sampling Locations

Usually, the sampling program objectives define the approximate locations for sampling (for example, influent and effluent to a treatment plant or water supply intake). Often, however, the sampling program objectives give only a general indication (the effect of surface runoff on stream quality) when assessing the quality of drinking water supplies for a community.

Most sampling surveys and subsequent analyses are performed to meet the requirements of federal, state, or local regulations. An example of regulatory monitoring is the National Pollutant Discharge Elimination System (NPDES) established in accordance with the Federal Water Pollution Control Act Amendments and the EPA's 503 regulations dealing with the disposal and incineration of wastewater biosolids in accordance with the Clean Air Act (CAA). Specific objectives in collecting regulatory data vary considerably and often overlap, but generally are performed to:

1) Verify self-monitoring data,

2) Verify compliance with a regulatory permit,

3) Support enforcement action,

4) Support permit reissuance and/or revision, or

5) Support other program elements such as media quality standards requiring various data.

Types of Samples

The type of sample collected depends on the variability of flow, variability of media (water, air, soil) quality, accuracy required, and the availability of funds for conducting the sampling and analytical programs.

We are concerned with two types of sampling techniques: *Grab* and *Composite* Samples.

Grab samples are individual discrete samples collected over a period of time not exceeding 15 minutes. A grab sample can be taken manually, using a pump, scoop, vacuum, or other suitable device. Collecting a grab sample is appropriate when it is needed to:

1) Characterize media quality at a particular time

2) Provide information about minimum and maximum concentrations

3) Allow collection of variable sample volume

4) Collaborate composite samples

5) Meet a requirement of a discharge permit

A grab sample should be used when (1) the media being sampled does not flow continuously; (2) the media or waste characteristics are relatively constant; (3) the parameters to be analyzed are likely to change with storage, such as with dissolved gases, residual chemical, soluble chemical, oil and grease, microbiological parameters, organisms, and pH; (4) information on maximum and minimum air variability is desired; (5) the history of media quality is to be established based on relatively short time intervals; and (6) the spatial parameter variability is to be determined (for example, the parameter variability throughout the cross section and/or depth of a stream or large body of water).

A composite sample is a sample formed by mixing discrete samples taken at periodic points in time or a continuous proportion of the flow. The number of discrete samples that make up the composite depends upon the variability of pollutant concentration and flow. A sequential composite is defined as a series of periodic grab samples, each of which is held in an individual container, then composited to cover a longer time period.

A composite sample is used when (1) determining average concentrations, and (2) when calculating mass/unit time loading.

Sample Collection Methods

Samples can be collected manually or with automatic samplers. Whichever technique is employed, the success of the sampling program is directly related to the care exercised in sample collection. Optimum performance is obtained by using trained technicians.

Manual sampling involves a minimal initial cost. The human element is the key to the success or failure of manual sampling programs. Well suited to a small number of samples, it is costly and time-consuming for routine and large sampling programs. The advantages of manual sampling include low capital cost, no maintenance, and the fact that extra samples can be collected in a short time. Some of the disadvantages of manual sampling include the probability of increased variability from sample handling, inconsistency in collection, and high cost of labor. Additionally, sampling is a repetitious and monotonous task for personnel.

Automatic samplers are cost effective, versatile, and reliable. They also increase and improve capabilities, allowing greater sampling frequency and increased sampling needs because of regulatory requirements. They are available with widely varying levels of sophistication, performance, mechanical reliability, and cost. Additional advantages of using automatic samplers include the consistency of the samples, minimal labor requirements, and the capability of collecting multiple samples for visual estimate of variability and analysis of individual samples.

The disadvantages include considerable maintenance, the fact that size is restricted to general specifications, inflexibility, and the ever-present potential of sample contamination.

Measurements

Inaccurate measurements for samplings such as flow measurements lead to inaccurate proportional composite samples, which in turn lead to inaccurate results. Care must be exercised in selecting a measurement site. The ideal site provides the desired measurement to meet program objectives, ease of operation and accessibility, and personnel and equipment safety.

A media flow measurement system usually consists of a primary device that has some type of interaction with the media, and a secondary device that translates this interaction into a desired readout or recording.

Flow measurement methods, depending on the type of media to be sampled, can be broadly grouped into four categories:

1) Closed circuit flow measurement

2) Flow measurement for pipes discharging to atmosphere

3) Open channel flow measurement

4) Miscellaneous methods of measurement

Field Procedures

In environmental sampling, field operations are critical. If proper precautions are not exercised in field procedures, the entire sampling program becomes meaningless—despite adequate planning, analytical facilities, and personnel. The key to the success of a field sampling program lies in good housekeeping, collection of representative samples, proper handling and preservation of samples, and appropriate chain of custody procedures.

GENERAL EVALUATION METHODS FOR ENVIRONMENTAL MEDIUMS

Ambient Air Quality Evaluation

To evaluate ambient air quality, some fundamental knowledge of how air is contaminated by pollutants is needed. This air contamination process begins with a source, either stationary or mobile, which releases contaminants to the atmosphere. When contaminants are released, they are subject to atmospheric *dispersion*, *transformation*, and *depletion* mechanisms.

The dispersion of contaminants (pollutants) in the atmosphere is determined by wind conditions and atmospheric turbulence. Horizontal winds play a significant role in the transport and dilution of pollutants. As the wind speed increases, the column of air moving by a source in a given period of time also increases. The concentration of pollutants is an inverse function of wind speed; that is, if the emission rate is relatively constant, a doubling of the wind speed will halve the pollutant concentration. Atmospheric turbulence is the result of two factors: (1) Air does not flow smoothly near the earth's surface because of friction created by the earth's surface and by man-made structures such as buildings, and (2) atmospheric heating also causes turbulence.

Transformation refers to the chemical transformations that take place in the atmosphere (such as the conversion of an original pollutant to a secondary pollutant, i.e. ozone).

Depletion refers to the fact that pollutants emitted into the atmosphere do not remain there forever. Weather affects air pollution in several ways. Precipitation helps cleanse the air of pollutants, although those pollutants are then transferred to the other media (soil and water). Winds and storms may deplete (by dilution) air pollutants, making them less troublesome in the area of their release. Air heated by the sun rises, carrying pollution with it. When the rising air mass is accompanied by wind, the pollution is carried aloft and diluted with clean air.

In evaluating ambient air quality, the most widely used methods are monitoring and modeling. Monitoring involves the use of measuring devices to determine the concentration of a specific contaminant, at a specific location, at a specific time. The main advantage of monitoring is that an exact level of contaminant can be determined, subject only to the accuracy and level of detection limitations of measurement method. The main disadvantages are (1) monitoring is expensive, (2) the actual contribution of specific emission sources to the total measured concentration can be difficult to determine, and (3) the impact of proposed sources not yet in operation cannot be assessed with monitoring methods.

Modeling refers to the use of mathematical representations of contaminant dispersion and transformation to estimate ambient pollutant concentrations. The main advantage of modeling is that it can be used to estimate concentrations at several hundred locations at relatively low cost. The main disadvantage of modeling lies in the fact that a mathematical model cannot exactly replicate the complexities of atmospheric dispersion and transformation. To be effective, a model must be tested with actual data over an appropriate period of time.

Soil and Groundwater Quality Evaluation

The soil is a primary recipient (a sink or depository) of the waste products and chemicals used in modern society. Once these pollutants enter the soil, they become part of a cycle that affects all forms of life. When soils are polluted, the degradation goes well beyond the soil itself. In many cases the damage also extends to water, air, and living organisms. In this section we are primarily concerned with evaluating the quality of soil and groundwater. To do so, we must have an understanding of the mechanics of soil/groundwater pollution.

Once contaminants such as organic chemicals (hydrocarbons or pesticides) reach the soil, they move in one or more directions (depending in part on their chemical structures). They may (1) vaporize into the atmosphere; (2) be absorbed by the soil; (3) move downward through the soil and be leached from the soil; (4) undergo chemical reactions on the surface or within the soil; (5) be broken down by soil microorganisms; (6) wash into streams, rivers, or lakes in surface run-off; or (7) be taken up by plant or soil animals and move up the food chain.

Before the environmental practitioner can evaluate the horizontal or vertical extent of contaminated horizons, he or she must understand the general principles of how contaminants are transported. The permeability of the soil and the viscosity of the contaminants influence the speed at which contaminants move through the soil. Gravity's force (the main cause of vertical movement of a contaminant through the soil) is dependent on the volume of contaminant, the depth of the water table, and the density and viscosity of the contaminants. Not every contaminant reaches the water table. On their journey downward through soils, contaminants often are stopped naturally whenever they encounter an impermeable soil layer. Movement is also stopped whenever the surrounding soils absorb the contaminant.

If a certain contaminant is discharged directly onto soil in a large enough quantity (a large spill), chances are good that it will eventually reach the groundwater. Once a contaminant reaches the groundwater, its movement is dependent on the physical properties (specific gravity and solubility) of the contaminant. The specific gravity of the contaminant determines the extent the contaminant will move through the water, while the solubility will help determine how fast the contaminant will be transported through the groundwater.

Contaminants less dense than water and with high solubility will often be found moving horizontally in the upper sections of the aquifer. Contaminants more dense than water and with low solubility will generally be found moving vertically through the aquifer and at higher concentrations at depth.

Soil and Groundwater Sampling

Several tools and methods are available and commonly used in soil and groundwater sampling. One of the most common methods used for large spill areas is *geophysical testing*, which involves using resistivity and conductivity meters (which can evaluate site conditions up to 200 feet below land surface) to evaluate the subsurface layers, locate the water table, and map contaminate contours. These surveys can be conducted easily and quickly by traversing the suspect contaminated area to locate and evaluate the horizontal perimeter of the contaminated area.

Probably the easiest and most commonly used method to obtain accurate soil samples *is soil boring*. Soil boring normally involves the use of augers. A hand-operated *soil auger* is a pole-like device with a T handle that usually features a 3-inch diameter high-strength-barrel sample collection device on one end. The device is hand operated by rotating the T bar and turning the device vertically through the soil up to a depth of about 10 feet. General-purpose augers are used in wet sands and gravels, dry or damp soils, and loams. Open-sided augers are used in clay soils. Augers are also available for use in sand and muddy soils. The key advantage in using soil augers is that they permit direct evaluation of soil samples collected from various depths. Augered soil samples can be immediately examined visually for contamination, examined with field equipment during soil boring, or can be preserved and evaluated in the laboratory. When samples need to be taken at a depth greater than 10 feet, usually a *split spoon* type sampler is attached to a drilling rig.

An even easier method of soil sampling for shallow excavations can be accomplished by using a shovel, spade, or scoop. The shovel breaks the soil surface (surficial soil) to the required depth. A normal lawn-and-garden spade is used to remove the soil. After the topsoil is removed, a steel scoop is used to remove a sample for field or laboratory analysis.

When a large amount of contaminant is spilled, trenching and excavation is used. Trenching and excavation normally involve the use of large earth-moving machines to remove large layers of soil. The obvious advantage gained in this operation is that it allows for direct observation of site-specific conditions and field and lab analysis.

Normally, soil borings are conducted where the resulting data will provide conclusions as to the potential migration of contaminants onto a particular property. In some cases, groundwater monitoring well installation may be favored over random soil borings since the cost-effective characterization of groundwater quality is the main objective of many investigations.

Groundwater monitor wells are installed after the contaminated area has been identified. These wells provide access to groundwater for sampling purposes. Samples are evaluated for dissolved contaminate and free product.

Several common drilling methods are available for well construction, and usually the "one up three down" layout rule is followed in their installation. This consists of placing one well upgradient from the spill area and placing three downgradient. The upgradient well provides data on groundwater that are not influenced by contamination from the spill. The downgradient wells are strategically placed to intercept any waste migrating from the spill area.

After the monitor wells are installed, groundwater samples are collected. These samples are collected for laboratory analysis to confirm the type of contaminant and concentration of dissolved contaminants in the groundwater.

Laboratory Analysis

To accurately evaluate the extent and concentrations of contamination at a spill site, samples are taken to an environmental laboratory for analysis. All laboratory analyses must be performed at an approved analytical laboratory, using EPA-approved protocol and quality control measures. All groundwater samples undergo analyses for temperature, pH, organic vapor concentration, and conductivity. Additional parameters may also be chosen based upon the results of subsequent investigations and regulatory requirements.

SUMMARY

The basic techniques for air, soil, and water sampling are relatively simple. Learning the techniques presents little problem, but to be able to properly collect samples in the field is harder than the descriptions of the techniques suggest. Sampling and analyses require practice—practical experience in using the environmentalist's available tools.

Suggested Readings

American Conference of Governmental Industrial Hygienists, *Air Sampling for Evaluation of Atmospheric Contaminants*. 8th ed., Cincinnati, OH: ACGIH, 1995.

Black, H. H., Procedure for Sampling and Measuring Industrial Wastes. *Sewage & Industrial Wastes.*, 24, pp. 45-65, January 1952.

Boulding, J. R., *Description and Sampling of Contaminated Soils: A Field Guide*. 2nd ed., Boca Raton, FL: Lewis Publishers, 1994.

Cahill, L. and Kane, R., *Environmental Audits*, 6th ed., Rockville, MD: Government Institutes, 1989.

Handbook for Sampling & Sample Preservation of Water and Wastewater, Springfield, VA: U.S. Dept of Commerce, 1982.

Metcalf and Eddy, Inc. *Wastewater Engineering: Treatment, Disposal, Reuse*. 3rd ed., New York: McGraw-Hill, 1991.

Pasquill, F. and Smith, F. B., *Atmospheric Diffusion*. New York: Ellis Horwood, 1990.

Tan, K. H., *Environmental Soil Science*. New York: Marcel Decker, Inc., 1994

Testa, S. M., *The Reuse & Recycling of Contaminated Soil*, Boca Raton, FL: CRC/Lewis Publishers, 1997.

Testa, S. M. and Patton, D., "Don't dig clean soils—selective excavation can cut project costs," *Soils*, pp. 31-33, December, 1993.

The NALCO Water Handbook. F. N. Kemmer (ed.), 2nd ed., New York: McGraw-Hill, 1988.

Turner, D. B., Atmospheric Dispersion Modeling: A Critical Review, *JAPCA*, 29:502-519, 1979.

Water Measurement Manual. U.S. Bureau of Reclamation. Washington, DC: U.S. Govt. Printing Office, p. 16, 1967.

CHAPTER 8

TECHNOLOGY AND THE ENVIRONMENT

We are just outside a large natural cave set deep under a solid outcropping that formed a fairly significant mountain meadow before the last glacial ice-sheet gouged, gorged, ground, and pulverized it down to its present size and shape. By our current calendar it's 15,543 BC.

The colossal sheet of ice is in retreat. When it was at its full width, depth, and length, it extended several hundred miles beyond the cave site to a V-shaped valley, where glacial melt fed a youthful, raging river that ran through the valley's bottomland.

A small, steady stream of melt-water courses almost in a straight line past one side of the cave, down toward that valley, where it will join and feed the river.

On the other side is a sloping field of young grass, brush, and flowers. Up close, we see the stark remnants of the terminal moraine that formed this abrupt slope with its fresh cover of grass and blossoms. A closer look reveals a dark heap at the base of the slope—a heap of trash: skin, sinew, bone, decaying corpses, burnt remnants of past hunts and feasts. We know only too well the refuse, filth, and discards that people leave behind as their foulest signature; someone must live close by. Perhaps in the cave. Let's take a look inside.

Walking up to the mouth of the cave, we tread carefully, not wanting to disturb the occupants. Remember that we're talking about caveman here—no language in common, no culture in common. Could we have anything in common with such primitive people?

But all is quiet, and with no overt threat present, our curiosity overcomes our caution, and we walk into the opening chamber.

And something reaches out and grabs us—not a caveman, but a stench, a horrible stench, too horrible to describe. With our fingers clamped tightly to our noses, we move on, too interested to retreat.

We can see fairly well in this chamber, because daylight pours in through the entrance. We take a few steps and stop to look around.

The walls of the cave are covered with black soot. A pit near the cave wall to the right, under attack by millions of flies and other insects, provides much of the stench: the latrine. A heap of detritus similar to the dump outside provides the rest of the reek.

It dawns on us that this cave is abandoned. We have no doubt as to why. The largest by-product manufactured by mankind has taken over—the cave is a garbage dump.

Perhaps, back in other chambers, deeper in the cave, exist cave paintings and remnants of ceremony. But we don't have the tools to explore them with us today, and we retreat, grateful for a breath of fresh air.

Outside, a few hundred feet from the cave and within sight of the garbage heap, we stop to contemplate what we've seen. Fifteen thousand years from now, archaeologists will find this cave and explore it thoroughly, learning information that will give us insights into the world of the people whose former home we have just visited. But the remains the archaeologists will find will be altered by fifteen thousand years of history. The picture they see will be incomplete, scattered by the natural interference life causes, giving mystery to the short and brutal lives of our ancestors.

But here, right now, we see similarities. We foul our environment in the same ways, and in more. But the caveman had a huge advantage over modern man, in that respect. When his living quarters became too foul for comfort, he could pick up whatever he considered of value that he and his tribe could carry and move on. A fresh site was always just around the next bend in the river. The pollution he created was completely biodegradable. In a few years, this cave could house humans again.

Although we have our similarities with those far-off ancestors, one stark difference is plain: we cannot destroy and pollute our environment with impunity. We can no longer simply pull up stakes and move on. What we do to our environment has ramifications on a scale that we cannot ignore—or avoid.

CHAPTER OUTLINE

- Introduction
- The Impact of Technology on Air Quality
- Sources of Water Pollution
- Sources of Soil Pollution

INTRODUCTION

In prehistoric times, humans had many of the same bad habits of environmental abuse that we have today. However, cavemen did have a distinct advantage over modern man: When prehistoric humans contaminated their environment, the damage was local and on a much smaller scale. Prehistoric man simply pulled up stakes and moved on to another cave. His environmental abuse was on a much smaller impact-of-scale for two reasons: (1) There were fewer human inhabitants on earth, and (2) technology had not progressed beyond the use of fire for cooking/ heating and the manufacturing of a few rudimentary tools for hunting and gathering.

Today's technology would drive prehistoric man to seek shelter in the depths of his cave in terror. We take for granted great advances in technology—advances that have given us the potential to completely destroy life on earth as we know it. We also possess the chronic potential to destroy all life through many practices that pollute the environment.

Advances in technology have allowed us to do many things. We can clear-cut entire forests in short order. We can blow megatons of pollutants into the atmosphere. We can pour endless streams of toxic wastes into our oceans, rivers, streams, and lakes. We can carve out huge gaping holes in our soil and dump huge quantities of waste into them. Because of these great advances in technology, we can literally change the landscape of earth, the quality of the air we breathe, the quality of the water we drink, and the quality of our soil that all life depends on. We can do all these things—and we do them.

The consequences of mankind's technological progress have taken us from caves to "lifestyles" with amenities most people in modern societies have come to expect today. But with these changes come tradeoffs. Some people question whether these technological advances are advantageous in the long run. Not all technologies should be automatically labeled bad. A few would argue that the technological advances in medicine, agriculture, weather forecasting, and many other important areas are all bad. Technology has a plus side; it can be put to work for the betterment of mankind. We can only hope that in the coming years we will use technology to our benefit—to make our lives better and to sustain and protect our environment at the same time.

In this chapter we briefly discuss technology and its impact on our environment—its impact on our air, water, and soil quality. Later, in media-specific chapters, we discuss how technology can be used to neutralize the adverse effect we have on our environment. We stress the positive: how advanced technology can be used in an earth-friendly manner.

THE IMPACT OF TECHNOLOGY ON AIR QUALITY

When a volcano erupts and spews millions of tons of ash into the atmosphere, or when a large forest area is struck by lightning and a small fire turns into a rampaging blazing swath of death and destruction, producing and pouring millions of tons of smoke, ash, and cinders into the atmosphere, or when, at the local level, a neighborhood house catches on fire and the smoke and ash contaminate the local air, you have no difficulty in recognizing what air pollution is all about. It is about having to draw air into your lungs that makes you cough, that can make you sick, that smells foul, tastes foul, that fills your lungs with particulate matter—that is contaminated.

These natural and accidental man-made air contaminant producers are, if not common, at least well known. However, several other events—man-made processes, chemical reactions, and machines—also contaminate our atmosphere—the air we all must breathe.

In terms of effect on the environment, we've made a huge leap forward from the smoke-producing fire of the prehistoric cave dweller to the plume of a 380-meter-tall "super-smokestack" venting pollutants from a modern copper smelter. This huge leap forward is the direct result of the technological advances that developed slowly from about 4,000 BC to the rapid, ever-increasing, mind-boggling pace in modern times.

Air pollution presents one of the greatest risks to our health and environment. The water most of us drink comes clean and pure (because of technology) from the taps in our houses. Most of the food we eat comes ready-to-cook, wrapped in plastic at a supermarket. But we must breathe the air that surrounds us, constantly. The health problems caused or aggravated by air pollution are extensive, and include lung diseases (cancer, chronic bronchitis, pulmonary emphysema) and other health problems (eye irritation, bronchial asthma, and neural disorders). Environmental problems range from vegetation and crop damage to increased lake and pond acidity that makes these bodies of water uninhabitable for aquatic life.

As you know, air pollution is not limited to the actions of humans (*anthropogenic* causes) but is also the result of natural phenomena—vulcanism, lightning storms, earthquakes, and others. However, we focus on the anthropogenic sources of air contaminants. We have no control over and cannot accurately predict when natural disasters will occur. But the sources of air pollution related to technological advances we must be concerned with. These technological sources of pollution can be classified into two categories: mobile and stationary sources.

Mobile sources of air pollution account for the production of about 50% of the air pollution in the United States. Mobile sources are classified as any non-stationary source of air pollution: aircraft, boats, trucks, trains, motorcycles, and passenger cars. Exhaust vapors from such sources contain carbon monoxide, volatile organic compounds (VOCs), nitrogen oxide, lead, and particulate matter. VOCs, along with nitrogen oxides, contribute to the formation of *smog*.

Stationary sources of air pollution are those emanating from any point that is fixed or stationary. These sources range from large chemical processing plants to neighborhood gas stations and dry cleaners. Some of the common sources are power plants, printers, steel plants, and coke plants.

Stationary sources generate air pollutants primarily by combusting fuel for energy, and, to a lesser degree, as by-products of industrial processes. Factories, utilities, and commercial and residential buildings that burn oil, wood, coal, natural gas, and other fuels are principal sources of pollutants such as sulfur dioxide, nitrogen oxides, carbon monoxide, VOCs, lead, and particulate matter.

Toxic air pollutants also come from a variety of manufacturing and industrial processes. Hazardous waste disposal facilities, municipal incinerators, utilities, landfills, and fuel oil contaminated with hazardous materials are potential sources of toxic air pollution.

The major pollutants generated by mobile and stationary sources are carbon monoxide, ozone, sulfur dioxide, lead, air toxics, airborne particulates, and acid deposition. A brief description of each of these major pollutants is provided in the following sections. The technological advances that have been made to mitigate the effects of these pollutants is discussed later in those sections dealing with each specific environmental media type: air, water, and soil.

Carbon Monoxide

Carbon monoxide (CO) is a colorless, odorless, and tasteless product of the incomplete combustion of fossil fuels and biomass. Although the dominant production of CO is through natural sources, large emissions to the atmosphere result from a number of anthropogenic sources. When inhaled, CO replaces the oxygen in the bloodstream and can impair alertness, vision, and other physical and mental capacities. For those suffering with lung and heart ailments, inhalation of CO can bring about severe health effects.

If you were to attempt to locate the maximum concentrations of CO at any given time in a particular part of a country (e.g., the United States), all you would have to do is to find areas with high concentrations of people and motor vehicles. Motor vehicles are the main source of carbon monoxide, especially when their engines are burning fuel inefficiently. Industrial processes, incinerators, and wood stoves are other sources of CO.

Ozone

Ozone is a variation of oxygen that possesses a set of physical and chemical properties significantly different from the "normal" form of oxygen. Ozone possesses three instead of the usual two atoms of oxygen per molecule; thus, its chemical formula is represented by O_3. At ambient room temperature, ozone is a pale blue gas with a pungent odor. It is heavier (more dense) than oxygen (vapor density of 1.7) and considerably more soluble in water. Ozone is particularly hazardous for two reasons: (1) it is extremely reactive, and (2) it is very toxic.

Individuals are often familiar with ozone in regard to the protective ozone layer in the earth's *stratosphere* (extends roughly 7 to 30 miles above earth's surface), which serves us in a critical way. It acts as a protective shield, preventing (by absorption) harmful amounts of ultraviolet radiation from penetrating the *troposphere* (the bottom layer of the earth's atmosphere extending to about seven miles above the surface) and reaching the earth's surface.

The benefits of ozone in the upper atmosphere notwithstanding, the presence of ozone in the lower atmosphere is one of the most widespread environmental problems: it can be hazardous to—poisonous to—most living organisms. The presence of ozone in the lower atmosphere is what we are concerned with.

Ozone is produced naturally in the atmosphere when sunlight triggers chemical reactions between atmospheric gases and pollutants such as VOCs and nitrogen oxides. The main source of VOCs and nitrogen oxides is internal combustion engines used in automobiles, buses, trucks, etc. During heavy commuter traffic, ozone levels are usually highest, because large amounts of VOCs and nitrogen oxides are being produced.

Two problems with ozone are related to advancements in technology. The first problem is that the over-production of ozone by anthropogenic emission works to disturb the natural equilibria among stratospheric ozone reactions, consequently decreasing ozone concentration. Because stratospheric ozone absorbs much of the incoming solar ultraviolet radiation, it serves as a UV shield, protecting organisms on the earth's surface. Reducing the ozone concentration in the stratospheric regions strips away that protective shield.

The second problem with ozone normally occurs under two conditions. Turbulence in the upper atmosphere sometimes causes stratospheric ozone to enter the troposphere. On these rare occasions, ozone usually enters for short durations only. But on those occasions, endogenous photochemical reactions take place in the lower troposphere—a primary cause of oxidants in Los Angles-type smog.

Sulfur Dioxide

Sulfur dioxide (SO_2) is a colorless gas possessing the sharp, pungent odor of burning rubber. It is the product of combustion resulting from the burning of sulfur-containing materials (coal and other fossil fuels, for example); sources include refineries, pulp and paper mills, steel and chemical plants, smelters, and energy facilities related to oil and gas production. Residential neighborhoods are directly affected by SO_2 emitted from home furnaces and wood-burning stoves. Near major industrialized areas, sulfur dioxide is a common air pollutant. Excessive levels of sulfur dioxide in ambient air are associated with significant increases in acute and respiratory diseases.

When emitted to the atmosphere, sulfur dioxide can be transported long distances because it bonds to particles of dust or aerosols. With time, sulfur dioxide oxidizes to sulfur trioxide, which dissolves in atmospheric water vapor, forming highly corrosive sulfuric acid. In highly industrialized regions, two major environmental problems develop: acid rain and sulfurous smog.

Acid rain is precipitation contaminated with dissolved acids like sulfuric acid. Acid rain is a threat to the environment because of the damage that it causes to aquatic life when it falls in freshwater lakes. The second problem, *sulfurous smog*, is the haze that develops in the atmosphere when droplets of sulfuric acid accumulate, growing in size until they become sufficiently large to serve as light scatterers.

Lead

Lead is a heavy metal that can cause serious physical and mental impairment. Children are particularly vulnerable to high lead levels. Stationary sources of lead used to consist of nonferrous smelters and battery manufacturers, but their level of lead emissions have been substantially reduced due to EPA regulation. The automobile used to be another significant contributor of atmospheric lead contamination, especially in densely populated areas. Up until the 1970s, tetraethyl lead was a component of gasoline. When combusted (in internal combustion engines) tetraethyl lead was transformed into lead oxide. Tiny particles of lead oxide were emitted as components of vehicular exhaust, and were subsequently inhaled or consumed, directly or indirectly.

The adverse health effects caused by overexposure to lead prompted the EPA to take action to reduce the lead content of all gasoline over time. In addition to phasing down lead in gasoline, the EPA instituted an automotive emission control program in 1975 that required the use of unleaded gas in any car. At the present time, about 70% of the gas sold in the U.S. is unleaded.

Toxic Chemicals

A category of toxic pollutants overlooked in the past (or not understood until recently) is toxic chemical pollution, found in all environmental media. Emission of toxic chemicals into the air by human activities has both acute and chronic effects on human health and the environment. Toxic chemicals are emitted into the atmosphere by industrial and manufacturing processes, solvent uses, hazardous waste handling and disposal sites, incinerators, motor vehicles, and sewage treatment plants. Toxic heavy metals (such as cadmium, chromium, mercury, arsenic, and beryllium) are emitted by smelters, manufacturing processes, and metal refiners. Plastic and chemical manufacturing plants emit toxic organics such as benzene and vinyl chloride. In processes where plastics are burned at high temperatures in incinerators, chlorinated dioxins are emitted.

For people, the most common exposure of these contaminants is by inhalation in the industrial workplace, but most of these toxins are emitted from smoke stacks and tailpipes. More problems occur when these toxins leave the air and fall to the earth, where they are consumed or absorbed by animals, fish, or crops consumed by humans. These airborne toxics also contaminate water sources, including those that supply drinking water to communities. When these toxics enter the body, they accumulate and become highly concentrated in human tissues.

Airborne Particulate Matter

Airborne particulate matter such as dust, smoke, and aerosols may have both long-term and short-term health and environmental effects. These effects range from irritating the eyes and respiratory tract and reducing the body's resistance to infection, to causing chronic pulmonary diseases. Particulates, especially smaller particles (under 0.2 microns) that are able to reach the lower regions of the respiratory tract, affect breathing, aggravate existing respiratory and cardiovascular diseases, alter the body's defense systems against foreign materials, and damage lung tissue. Particulates may also be carcinogenic (such as particulates emitted from diesel engines) and can absorb gaseous pollutants (such as sulfur dioxide) and deliver them directly to the lungs. Particulates that end up in wind-blown dusts can be toxic [such as pesticides and polychlorinated biphenyls (PCBs)].

Major sources of particulates include diesel engines, residential wood combustion, coal-fired power plants, agricultural tilling, construction, and unpaved roads. In addition, particulates are also released into the atmosphere from steel mills, power plants, cotton gins, construction work, demolition, cement plants, smelters, and grain storage elevators.

Particulates are also responsible for soiling and corrosion of building materials, severe reduction in visibility (atmospheric haze), and damage to vegetation.

Acidic Deposition

Acidic deposition (usually referred to somewhat misleadingly as acid rain—misleading because snow and all other types of precipitation are affected, causing acid fog, acid clouds, acid dew, and acid frost as well) not only has caused a lot of confusion, but also controversy. Much scientific evidence points to acidic deposition as an environmental problem. Acidic deposition problems have been of concern not just recently, but for hundreds of years.

Acidic deposition is the phenomenon whereby pollutants affect the chemical nature of precipitation. Precipitation is by nature somewhat acidic. Remember the pH scale? The range is 0-14; 7.0 is neutral, and the more pH drops below 7.0, the more acidity increases. Because the pH scale is logarithmic, there is a tenfold difference between one number and the next. A drop in pH from 6.0 to 5.0 represents a tenfold increase in acidity, and a drop from 6.0 to 4.0 represents a hundred fold increase. Although all rain is slightly acidic, only rain with a pH below 5.6 is considered "acid rain." Changes in precipitation chemistry have been reported for more than thirty years. During that period, national governments in North America and Europe have come to recognize the seriousness of precipitation acidification by sulfur and nitrogen pollutants of man-made origin.

How Significant Is the "Acid Rain" Phenomenon?

In some locations, the acidity of rainfall has fallen well below 5.6. For example, in the northeastern United States, the average pH of rainfall is 4.5, and it is not unusual to have a rainfall with a pH close to 4.0, which is 1000 times more acidic than distilled water. The increased acidity appears to be due to sulfuric (65%) and nitric (30%) acids. The apparent major sources of sulfur and nitrogen oxides for these strong acids include fossil fuel-fired power plants, metal smelters, industrial boilers, and automobiles.

The middle Ohio Valley and the states immediately adjacent to it are the major source region for sulfur and nitrogen oxides (precursors of "acid rain"). Ohio emits twice the levels of sulfur oxide than do all the New England states combined. Significant sulfur oxide emissions also occur in Indiana, Kentucky, Illinois, West Virginia, Tennessee, Missouri, and Michigan. Data collected by different monitoring networks show that the areas of the U.S. receiving the most acid rainfall are downwind and northeast of those states with the highest acid precursor emissions.

Effects of Acid Deposition

The effects of acid rain are normally measured in economic and environmental terms. In terms of economic costs, the effect of acid deposition is difficult to estimate, since it would include damage to agriculture, fisheries, lakes, vegetation, human and animal health, and tourism. However, one fact is certain and easier to determine. The effect of acid deposition can be seen throughout the ecosystem. The ecological effects brought about by acid deposition include damage to leaf surfaces, release of harmful chemicals from the soil that damage root systems, and acid-catalyzed releases of toxic metals such as aluminum, which can filter into lakes and streams, threatening public water supplies and contaminating fish. Metals usually remain inert in the soil until acid rain moves through the ground (scientists call this phenomenon *mobilization*). The acidity of precipitation is capable of dissolving and mobilizing metals including mercury, manganese, and (as indicated earlier) aluminum. Transported by the movement of this acidic water through the ecosystem, these toxic metals accumulate in lakes and streams, where they may threaten aquatic organisms (see Figure 8.1). Nitrates from acid deposition in salt-water estuaries create algal blooms that cause oxygen depletion and suffocate fish and aquatic plants. In combination with ground-level ozone, acid deposition can impair plant growth and damage forests.

Acid rain can also damage material used in construction and sculpture. Building materials such as limestone, marble, carbonate-based paints, and galvanized steel all can be eroded and weakened by the dilute acids found in acid deposition.

SOURCES OF WATER POLLUTION

Protecting drinking water supplies, coastal-zone waters, and surface water is made complex by the variety of sources that affect them. Groundwater is being contaminated by pollution from animal wastes, leaking underground storage tanks, silage liquor, leachate from landfill sites, spoil heaps, solvent discharges to sewers or to land, fertilizers and pesticides, septic tanks, drainage wells, and inadequate sewage treatment plants, threatening a large percentage of the world's drinking water supplies.

Water pollution is both diverse and complex. This combination is clearly evident in the number of different pollutant categories and the way in which they overlap. Water pollutant categories include: oxygen-demanding wastes, disease-causing wastes, toxic and hazardous substances, sediments, thermal pollution, plant nutrients, persistent substances, and radioactive substances. The overlapping of these categories can be seen in improperly treated sewage waste, which may contain organic wastes, disease-causing wastes, plant nutrients, toxic substances, and persistent substances. Pollutants may also fit into more than one category. For example, PCBs are both hazardous and persistent, and one type of pollutant may enter water attached to another type. Organic chemical pollutants often adhere to sediments. Water systems can be contaminated by pollutants from all categories. Various contaminants may also act synergistically to form deadly pollutants.

Figure 8.1 A 20-acre lake in the Appalachian mountains. The lake's pristine appearance is deceptive. It is not visited by fishermen because it has no fish. As a result of acidic deposition, the lake's pH level is 5.9—too low (too acidic) for many fish species.

Oxygen-Demanding Wastes

Oxygen-demanding or organic wastes are small particles of once-living plant or animal matter. Usually suspended in the water column, they can also accumulate in thick layers of sediments on the bottom of lakes or rivers. Human and animal wastes and/or plant residues make up most of the suspended matter. Other sources of oxygen-demanding wastes are from natural runoff from land; industrial wastes from oil refineries, paper mills, and food processing plants; and urban runoff. In a process called *decomposition*, aerobic bacteria use the organic matter as an energy source. As they decompose or consume the organic matter, the bacteria use *dissolved oxygen* (DO) from the water to the detriment of other aquatic organisms, such as fish and shellfish. Bacteria residing in sediments (*anaerobic bacteria*) do not require oxygen to decompose organics; however, they may emit noxious gases such as methane and hydrogen sulfide as a by-product of decomposition.

Biological oxygen demand (BOD) is a measure of the amount of dissolved oxygen needed by decomposers to break down organic materials in a given volume of water. A natural water body (lake or stream) with a high BOD will have a low concentration of dissolved oxygen (DO) because oxygen is being used up by bacteria to decompose organic matter. (NOTE: BOD levels in pure water and typical fresh waters run 0 to 2-5 mg/l, respectively). Dissolved oxygen is a major limiting factor in the aquatic habitat, so oxygen depletion is a serious factor affecting the water quality of a lake or stream.

Disease-Causing Wastes

Disease-causing wastes enter water bodies from sources of untreated human and animal wastes and increase the chance that infectious organisms—*waterborne pathogens*—will occur, causing the outbreak of diseases, including typhoid, infectious hepatitis, cholera, and dysentery. Waterborne pathogens enter the water mainly through the feces and urine of infected persons and animals. Many other diseases are also transmitted by organisms in water. For example, mosquitoes transmit the protozoan that causes malaria, and snails transmit the fluke that causes schistosomiasis.

Toxic and Hazardous Substances

Toxic or hazardous substances are those materials injurious to the health of individual organisms. Sometimes fatal, they disrupt the metabolism of organisms as a result of ingestion or contact. Toxic substances include oils, gasoline, greases, solvents, cleaning agents, biocides, and synthetics. Many rivers, streams, lakes, and bays have thousands of toxic and hazardous chemicals in their sediments. The increasing number of discoveries of hazardous chemicals in dumps and landfills is one of the most serious environmental problems in many industrialized countries. As hazardous contaminants leach from dumps and landfills, they make their way into lakes, streams, rivers, and groundwater supplies, leading to contamination of drinking water supplies.

Several thousand organic chemicals enter aquatic ecosystems every day. Most are by-products of industrial processes, or are present in hundreds of thousands of commonly used products. Some of them are carcinogenic [dioxin and polychlorinated biphenyls (PCBs)]. Inorganic substances (acids, salts, brine, and metals) are also contaminating our water systems. These contaminants are produced by processes such as mining and manufacturing, which produce and release acids into the environment. Salt from road salting and irrigating causes damage, oil and natural gas wells release brine, and manufacturing processes also release metals (chromium, copper, zinc, lead, mercury, etc.) into aquatic ecosystems.

Sediments

Soil particles dislodged by raindrops travel via runoff into steams, rivers, lakes, or oceans and are deposited there as sediments. Although rivers and streams have always transported enormous quantities of sediment to the sea, their sediment loads today are greater than ever (by weight, sediments are the most abundant water pollutant). Soils stripped of vegetation by crop cultivation, timber cutting, strip mining, overgrazing, road building, and other construction activities are subject to high rates of erosion. When eroded, sediments by the millions of tons are deposited into aquatic systems, muddying streams and rivers.

The obvious result of soil erosion is the loss of valuable agricultural soils, but other problems are associated with the wearing down of soil as well. Eroded soil particles eventually fill lakes, ponds, reservoirs, harbors, navigation channels, and river channels. As a result, the accumulation of sediments greatly reduces the attractiveness of lakes and reservoirs, which causes them to lose their recreational value. Sedimentation also impedes navigation, covers bottom-dwelling organisms, eliminates valuable fish-spawning areas, and reduces the light penetration necessary for photosynthesis. There is another problem: soils that are eroded from farmlands sweep nutrients in the form of nitrogen and phosphorus into surface waters. In small quantities these nutrients are not a problem. However, a dramatic increase in the sediment load can cause problems—ecological changes.

The ecological changes that have occurred in the Chesapeake Bay, which were caused partly by the influx of sediments and nutrients, are described in Case Study 8.1.

Case Study 8.1
What's Wrong with the Chesapeake Bay?

Not all that long ago (maybe less than 30 years) inhabitants of towns on the Eastern Shore of the Chesapeake Bay often went wading in clear, knee-deep waters to catch crabs. The crabbers made their way through the lush, waving grasses on the bay's bottom, carrying crabpot-nets attached to long poles, dragging along a container tied by a string to their waists.

They would wait until the crabs scampered out of the grasses and then flip them into their containers, never breaking stride as they continued scooping them up. The water was so clear the crabbers could see their own feet.

Today, the clear water of the past has been replaced with water that is brown and turbid. The crabs have moved on to "greener pastures" and cleaner waters. The lush, thick grasses that tickled the crabber's feet are gone, and so are the crabs.

In less than thirty years, submerged grasses have vanished from many parts of the upper and middle bay. Scientists and environmentalists believe that they are beginning to understand why. The answers lie in assaults on the Chesapeake, many of them caused by the heavy hand of humans. The ecology of the bay has changed. Some scientists, ecologists, and other environmental specialists suspect the bay is dying.

The answer to what is going on in the Chesapeake Bay is complex. Actually, over the last thirty years there have been many answers presented by many different groups and individuals. Many of these answers provided much speculation as to what was causing the Chesapeake Bay problem. Here's an example of one of these answers—one that isn't really the solution to the Chesapeake Bay problem.

Environmental policy-makers in the Commonwealth of Virginia came up with what is called the Lower James River Tributary Strategy on the subject of nitrogen from the lower James River and other tributaries as the possible culprit contaminating the Lower Chesapeake Bay Region. Nitrogen is a nutrient. When in excess, nitrogen is a pollutant. Some theorists jumped on nitrogen as being the cause of a decrease in the oyster and other aquatic organism populations in the Lower Chesapeake Bay Region. Oysters, like crabs, are important to the lower region for both economic and environmental reasons, as well as others. From an environmental point of view, oysters are important to the Lower Chesapeake Bay Region because they have worked to maintain relatively clean water in the past. Oysters are filter-feeders. They suck in water and its accompanying nutrients and other substances. The oyster sorts out the ingredients in the water and uses those nutrients it needs to sustain life. Impurities (including pollutants) are aggregated and excreted by the oyster back into the James River.

In the past (maybe 45 years ago) when oysters thrived in the Lower Chesapeake Bay, they were able to take in turbid water and turn it almost clear in a matter of about three days.

This is not the case today. The oysters are almost all gone. They are no longer colonizing the Lower Chesapeake Bay Region in the numbers they did in the past. They are no longer providing economic stability to watermen; they are no longer cleaning the Bay.

A regional sanitation authority and a local university in the Lower Chesapeake Bay region formed a study group to formally, professionally, and scientifically study this problem. Over a five year period, using *Biological Nutrient Removal* (BNR) techniques at a local wastewater treatment facility, they determined that the effluent leaving the treatment plant and entering the Lower

James River consistently contained to below 8 mg/l nitrogen for five consecutive years.

The first question is: Has the water in the Chesapeake Bay become cleaner, clearer?

The second question is: Have the oysters returned?

The answer to both of these questions, respectively, is no, not really.

Wait a minute—some environmentalists, regulators, and other well-meaning interlopers stated that the problem was nitrogen. If nitrogen levels have been reduced in the Lower James River, shouldn't the oysters start thriving, colonizing, and cleaning the Lower Chesapeake Bay again?

You might think so, but they are not. While it is true that the nitrogen level in the wastewater effluent was significantly lowered through treatment, it is also true that a major point source contributor of nitrogen was reduced in the Lower Chesapeake Bay.

If the nitrogen level has decreased, then where are the oysters?

A more important question is: What is the real problem?

The truth is that no one at this point and time can give a definitive answer to this question.

However, a number of questions need to be answered before another theory on how to clean up the Lower Chesapeake Bay is acted upon: (1) Is nitrogen from the Lower James River and other tributaries feeding the Chesapeake Bay having an impact on the Bay? (2) Is there evidence of low dissolved oxygen in the Lower James River? (3) Although concentrations of nitrogen in the Lower James River exist, are there corresponding high levels of plankton (chlorophyll "a")? (4) Is it true that removing nitrogen for the sake of removing nitrogen would produce no environmental benefits, be very expensive, and divert valuable resources from other significant environmental issues?

Back to the problem with the decrease in the oyster (and crab) population in the Lower James River/Chesapeake Bay Region.

Why have the oyster and crab populations decreased?

One theory states that because the tributaries feeding the Lower Chesapeake Bay (including the James River) carry megatons of sediments into the Bay, they add to the Bay's turbidity problem. When waters are highly turbid, oysters do the best they can to filter out the sediments but eventually decrease in numbers and fade away.

A similar fate may await the crab. Highly turbid waters do not allow the sunlight to penetrate the murky water. Without sunlight, the seagrasses will not flourish. Without the seagrasses, the crab population diminishes.

Is this the answer? Is the problem with the Lower Chesapeake Bay and its oyster population, and the Eastern Shore and its crab population related to turbidity?

Only solid, legitimate, careful scientific analysis, using scientific methods, can provide the answer. There is no question that we need to stop the pollution of our surface water bodies. But shouldn't we use good common-sense and legitimate science instead of jumping to conclusions?

> [As a footnote to Case Study 8.1 we point out that recently, from 1984 through 1996, because of reductions made in nutrient and sediment deposition in the Lower Chesapeake Bay Regions, Bay grasses have rebounded. Hopefully the oyster and crab populations will also rebound.]

Thermal Pollution

Simply stated: *Thermal pollution* occurs when industry returns heated water to its source. Large-scale generation of electricity requires enormous quantities of water for cooling—water that is sometimes drawn from lakes, rivers or streams. Power plants that burn fossil fuels or that use nuclear fuel generate great amounts of waste heat, some of which is removed by circulating cool water around and through hot power-generating equipment. This transfers heat to the water, raising its temperature. When the heated water is discharged, it can have an adverse effect on aquatic ecosystems. For example, thermally polluted water can decrease solubility of oxygen in water, can kill some kinds of fish, can increase the susceptibility of some aquatic organisms to parasites, disease, and chemical toxins. Thermal pollution in general tends to disrupt aquatic ecosystems, and accelerate changes in their composition. Still waters—lakes and bays—are particularly vulnerable to thermal pollution.

Plant Nutrients

The amount of nitrogen and phosphorous in a water system normally limits aquatic plant and algal growth. These plant nutrients are usually found in large amounts in sewage; phosphates are found in some detergents and agricultural and urban runoff. When present in large amounts in still water systems (lakes, bays and ponds), these nutrients stimulate massive, rapid reproduction and growth in algal blooms. Algae imparts a green color to water and forms a green scum on the surface and on rocks near shore. When algae die and decompose, additional nutrients are added to the water system, increasing the BOD.

In Case Study 8.1, we discussed part of the problem created when too much sediment-bearing nutrient contaminant (such as phosphorus or nitrate) finds its way into our water systems. Note that excessive plant nutrients in a water system not only affect the habitats of oysters and crabs, but also impact entire aquatic systems—and public safety. Too much phosphorus in a water system is generally not a human health problem, but excess nitrogen, in the form of nitrates, is. Nitrates are found in fertilizers and organic waste from livestock feedlots. Soluble in water, nitrates do not bind to soil particles, making them highly mobile. Because of their mobility, nitrates wash into surface water and percolate into groundwater. Nitrates in drinking water pose a significant health threat to human populations. In particular, nitrates are a serious health threat to infants. When they reach infant intestinal tracts, nitrates oxidize the hemoglobin in the blood, rendering it unable to carry oxygen. This condition, called methemoglobinemia, can result in brain damage or death.

Persistent Substances

A water pollution threat that has often been ignored is *persistent substance* contamination. In this modern age, some materials (products of scientific and engineering wizardry) persist; they are not normally changed or degraded to harmless substances. These persistent substances include the pesticides DDT and chlordane, metals (mercury, for example), and organic chemicals like PCBs. These substances do not break down easily and tend to magnify throughout the food chain. Consequently, organisms at higher trophic levels (peregrine falcons, bald eagles, and other raptors, for example) suffer the most serious effects of these nonbiodegradeable, persistent substances.

In addition to the persistent chemical substances we've just discussed, another category of persistent substances is playing havoc with our water systems and wildlife, and is quickly gaining in notoriety as a world-class environmental problem: plastics. Plastic six-pack rings, plastic bags, and monofilament line have life expectancies of hundreds of years—they literally take hundreds of years to degrade. In marine systems, where plastics are discarded in record amounts, the problem is particularly troublesome. Balloons and bags are frequently mistaken for food and are ingested by marine wildlife, and become entwined in the stomachs and intestines—killing slowly. Sea creatures become entangled in monofilament line and six-pack rings, causing constriction and loss of movement that can eventually lead to death. On a larger scale, the fishing industry follows the deadly practice of cutting free entangled drift nets, sometimes allowing hundreds of yards of fishing nets to remain in the ocean. Called ghost nets, these abandoned nets entrap and drown countless sea mammals and other creatures on a much larger scale.

Radioactive Substances

Radioactive substances are another category of pollutant affecting aquatic ecosystems. Natural radioactivity occurs in the environment; however, radiation pollution arises from the use of radioactive materials. Three main sources of radiation pollution are nuclear power plants, coal-fired power plants, and nuclear explosions. Radiation pollution can cause genetic defects and cancer. To control radiation pollution, the use of nuclear power plants must be strictly regulated and the processing and shipping of nuclear fuels and wastes must be strictly controlled.

SOURCES OF SOIL POLLUTION

Historically, in solid waste disposal, a common saying and practice was: "I don't want it anymore. Take it to the river and throw it away." Is there any wonder why our river and stream systems became polluted over the years? Look to the historical record of the Thames River in London and the population along its banks to gain understanding of the compound problem of solid waste disposal and river pollution.

Although most early human settlements were located along river courses, not all camps, settlements, and communities were located near rivers or some other water system. Eventually, as populations grew and as improvements were made in transportation, population centers formed in areas remote from water systems. One thing did not change, however. People continued to produce waste. The problem was that there was no convenient body of water to dump the waste in. The waste had to be dumped somewhere. That somewhere was the land.

Land became the convenient dumping ground for wastes, including those removed from air and water. Historically, and not surprisingly, activities such as improper storage and disposal of chemicals also became—and remain—a serious problem.

When land is used for dumping waste, where do the waste products actually end up? What is affected by such dumping practices? The answer is the soil.

With almost every country in the world experiencing a continuing increase in the amount and type of materials being discarded in landfills and other disposal sites, soil is being contaminated at an ever increasing rate. Because of the increased rate of waste disposal on land, soil itself has become a waste product to be disposed of properly. This situation is exacerbated by the limitations associated with the technologies currently available for the remediation of contaminated soil. This is an important issue and will be developed more fully later in this text.

Note that disposal of solid waste materials is not the only casual factor related to soil contamination. Soil contamination is also the result of certain practices conducted worldwide, including agricultural and industrial practices, as well as exploration and production, mining, and nuclear industrial practices. Petroleum-contaminants affect the largest number of sites and the largest total volume of impacted material.

Although generally small in unit size, another large volume contaminator of soil can be directly linked to underground storage tanks (USTs). In fact, petroleum contamination is commonly linked to USTs. Between 1950 and 1970, several million USTs were installed. The EPA estimates that out of an estimated 2.5+ million USTs throughout the United States, more than 400,000 have leaked or are leaking petroleum hydrocarbons. Virtually thousands of USTs remain unrecorded, and their individual impacts on soil and the subsurface remain unknown.

Along with petroleum hydrocarbons, several other chemicals are responsible for soil contamination. The 1979 Eckhardt Committee Survey of the largest chemical manufacturing companies in the U.S. reported 16+ million tons of organic generated waste disposed up to that point. Of this total, almost 10 million tons were untreated (i.e., in ponds, injection wells, lagoons, and landfills). Almost 0.5 million tons were incinerated, and a little more than 0.5 million tons were either reused or recycled. Manufacturing accounts for the largest percentage (85%) of the hazardous waste generated.

Oil field sites account for a large volume of hydrocarbon-contaminated soil. Sources of soil contamination include oil wells, pits and dumps, sumps, leakage from above-ground storage containers, reservoirs, pumping stations, piping ratholes, underground storage tanks, transformers, well cellars, and random leakage or spillage.

However, oil field sites aren't the only culprits. Several high-volume petroleum-handling facilities, including refineries, terminals, and pipeline corridors, are situated close to production fields and residential areas. All of these sites, in the past and in the present, contribute to the overall volume of contaminated soil.

Another operation where a close relationship between site usage and the potential for adverse environmental damage exists is in geothermal power operations. Geothermal power operations involve the use of energy (in the form of superheated steam or water) from the earth's interior conducted to the surface in areas where igneous rocks are in a molten or partly molten state. Typically, geothermal plants consist of a power plant, brine storage holding tanks/ponds, drill sumps, and leach fields. Brine and lead-mine scale are the two hazardous constituents associated with geothermal operations.

For several years goal gasification processes have operated in several locations throughout the United States. Using various gasification processes, tars containing a wide variety of organic and inorganic compounds are commonly produced. These tars have contributed to a large volume of contaminated soil, groundwater, and surface waters.

Mining waste (wastes from mining, milling, smelting, and refining ores and minerals) is a considerable contributor to soil contamination. Environmental problems are primarily associated with disposal of mining waste: the overburden plus tailings (residue of ore processing). If piled into heaps and left at the site, rainwater seeps through mine wastes, which are rich in heavy metals and chemicals, and harmful byproducts like sulfuric acid may be produced. This acid runoff then drains into rivers, streams, soils, and then groundwater.

SUMMARY

Technological advances have allowed our modern culture to achieve incredible goals—and along the way, create incredible problems we must address. But the technology that we've created also affords us ways to solve the problems we have caused. While we paint a bleak picture of the state of our environment, environmental engineers, scientists, and technologists, as well as concerned "civilian" environmentalists (and even some politicians), recognize what needs to be done and work to improve our environmental situation.

Cited References

PEDCO, *PEDCO Analysis of Eckhardt Committee Survey for Chemical Manufacturer's Association*, Washington, DC: PEDCO Environmental Inc., 1979.

Suggested Readings

Adams, D. D. and Page W. P. (Eds.). *Acidic Deposition—Environmental, Economic and Policy Issues*, New York: Plenum Publishers, 1985.

Alexander, M., *Biodegradation and Bioremediation*, San Diego: Academic Press, 1994.

Arms, K., *Environmental Science,* 2nd ed., Saddle Brook, NJ: HBJ College and School Division, 1994.

Beaty, C. B., The Causes of Glaciation, *American Science*, 66:452-459, 1978.

Blumberg, L. and Gottieb, R., *War on Waste—Can America Win Its Battle with Garbage?*, Island Press, Washington, DC: p 301, 1989.

Bridgman, H. A., *Global Air Pollution*, New York: John Wiley and Sons, 1994.

Bowne, N. E., Atmospheric Dispersion, S. Calvert and H Englund (Eds). *Handbook of Air Pollution Technology*, pp. 859-893, New York: John Wiley and Sons, 1984.

Conner, H. R., *Chemical Fixation and Solidification of Hazardous Waste*, New York: Van Nostrand Reinhold, p 692, 1990.

Elsom, D. M., *Atmospheric Pollution—A Global Problem*. 2nd ed. Oxford: Blackwell Publishers, 1992.

Gates, D. M., *Climate Change and Its Biological Consequences*, Sunderland, MA: Sinuer Associates, 1993.

Kerr, R. A., Global Pollution: Is Arctic Haze Actually Industrial Smog?, *Science* 205:290-293, 1979.

Harrision, R. M. (Ed.). *Pollution—Causes, Effects and Control*. Cambridge, UK: Royal Society of Chemistry, Thomas Graham House, 1990.

Jackson, A. R. and Jackson, J. M., *Environmental Science: The Natural Environment and Human Impact*, New York: Longman, 1996.

Lamb, B., Gaseous Pollutant Characteristics, S. Cavert and H. Englund (Eds.). *Handbook of Air Pollution Technology*, 3rd ed. New York: John Wiley & Sons, pp. 65-96, 1984.

Moore, J. W. and Ramamoorthy, S., *Heavy Metals in Natural Waters*, New York: Springer-Verlag, 1984.

Perkins, H. C., *Air Pollution*, New York: McGraw-Hill, 1974.

Rizzo, J. A (ed.), *Underground Storage Tank Management: A Practical Guide*, 4th ed., Rockville, MD: Government Institutes, 1990.

Rodriguez, J. M., Probing Atmospheric Ozone, *Science* 261:1128-1129, 1989.

Seinfeld, J. H., *Atmospheric Chemistry and Physics of Air Pollution,* New York: Wiley-Inter-Science, 1986.

Stern, A. C. and Wohlers, H. C., *Fundamentals of Air Pollution*, New York: Academic Press, 1984.

Testa, S. M., *Geological Aspects of Hazardous Waste Management,* Boca Raton, FL: CRC/Lewis Publishers, p. 537, 1994.

U.S. Environmental Protection Agency, *Air Quality Criteria for Particulate Matter and Sulfur Oxides, Vol. II.* EPA/600/8-82-029a-c, 1982.

U.S. Environmental Protection Agency, *Environmental Progress and Challenges*, August 1988.

U.S. Environmental Protection Agency, *EPA Journal*, 17:1, January/February 1991.

U.S. Environmental Protection Agency, *National Air Pollutant Emission Trends, 1900-1993.* EPA/454/R-94-027, 1994.

U.S. Environmental Protection Agency, *Air Quality Criteria for Particulate Matter. Vol. I,* EPA/600/AP-95/001a, 1995.

U.S. Environmental Protection Agency, *Air Quality Criteria for Particulate Matter, Vol. III*, EPA/600/AP-95/001c, 1995.

PART II

AIR QUALITY

CHAPTER 9

THE ATMOSPHERE: BASIC AIR QUALITY

Whether we characterize it as a caress or a light touch against soft skin, as a gentle breeze, a warm wind, or a blustery gale, as tempest, typhoon, tornado, or hurricane, air is vital.

Air encapsulates us. It surrounds us. We take it in with every breath, our bodies thrive on it, and we fail immediately without it. Literally awash with air, on earth, all life we know of depends on it. It occurs naturally everywhere on earth—the sky begins where the ground ends.

Air is scientifically unique. The combination of common and rare gases we breathe has made life possible. Air, as with water, is a chemical compound found naturally that affects most living organisms in a manner of ways.

We associate air with all the good on earth. We cannot imagine life without breathing—we must constantly quench our thirst for air. Air sustains growth. It creates the subtle and blatant movements that provide us with changing weather patterns. But can we really say emphatically, definitively, that air is only good? No we cannot. It is also true that nothing is safe from air.

Air is odorless, colorless, tasteless. We rarely stop to think about it, unless it brings something to us as a reminder. But it covers the earth completely. Nothing can escape its touch.

Life and air are inseparable. We sometimes call air the breath of life—a fitting name, especially when you consider that air can be the boon or bane of all life, capable in time of sustaining or destroying all life as we define it.

Whether it pushes the blade of a windmill, a billowy cloud, a dust mote, or a feather, whether it sets water lapping against some distant shore, drives a gritty wind that sculpts mountains to sand, or hammers like a fist to flatten whatever stands in its path, air is essential. Air is life. Air is vital.

Air gives us the blessing of communication. From our first cry to our dying breath, our voices travel on a current of air. Air carries sound. Air carries warmth. Air carries cold.

Our very existence depends on air, but we have created a paradox within our vital line to life. Why would we abuse something so vital—something we need to survive—something we cannot live without? Why do we foul the very essence of our lives?

Let us hope that we are not destroying the very air we breathe. Let us hope that technology will aid us in our efforts to retain the quality of the air we need to survive.

We need air as it should be: pure and wholesome, in the perfect mixture of elements we were evolved to inhale, vital to our existence.

CHAPTER OUTLINE

- Introduction
- The Atmosphere
- Basic Air Quality

INTRODUCTION

When we undertake a comprehensive discussion of air, the discussion begins and ends with the earth's atmosphere. The earth's atmosphere is unique. Often described as a thin veil, skin, shroud, envelope, blanket, or an invisible sea of gases that surround the planet, it is probably more correct to state that the atmosphere is an ocean of air. No matter how it is described, the atmosphere is shared by a huge variety of living things—it is vital to life itself. Composed of invisible gases and condensed water vapor, the atmosphere is maintained in place by the earth's gravitational pull.

In this chapter, we cover concepts important to gaining a basic understanding of the earth's atmosphere, which in turn will enable us to better understand our anthropogenic impact on the atmosphere. We also discuss key parameters used to measure air quality. Obviously, air quality is the important parameter that concerns us most. Could we have a concern greater than that of maintaining air quality to sustain life? A full understanding of environmental science requires knowledge of the atmosphere.

THE ATMOSPHERE

We live at the bottom reaches of a virtual ocean of air. Extending upwards nearly 1,000 miles, this massive, restless ocean is dynamic and far different from the watery oceans that cover most of earth's surface. At the bottom of our atmosphere (Greek: *atmos*—vapor, and *sphaira*—sphere), humans and other creatures live.

The Atmosphere: Chemical Composition

Atmospheric air is a mixture of many gases, and also holds many suspended liquid droplets and solid particles. Two gases comprise about 99% of the volume of air near the earth. From Figure 9.1 we see that air is primarily composed of a relatively constant mixture of nitrogen (78%) and oxygen (21%), with nitrogen being about four times as abundant as oxygen. The other main constituents are argon and carbon dioxide.

Many other gases of minute quantities are found in the atmosphere, along with dust, pollen, salt particles, etc. Some of these, especially water vapor and carbon monoxide, vary in concentration, depending on conditions and locality. The amount of water vapor in the air is dependent to a great extent on the temperature (see Chapter 10). Carbon monoxide and carbon dioxide, both by-products of incomplete combustion, are both present in abnormally high concentrations in populated areas.

From Figure 9.1 we see that nitrogen (N_2) is the most abundant gas in the atmosphere, but it has a relatively limited direct role in atmospheric or life processes. It serves as a precursor molecule for the formation of nitrate nitrogen, which is required by plants to make proteins, amino acids, chlorophyll, and nucleic acids, which are essential to all living things. The conversion of nitrogen to nitrate nitrogen is the result of chemical and biological processes in the atmosphere.

Approximately 21% of the atmosphere's mass is composed of molecular oxygen (O_2), which is vital to almost all living things. O_2 is also important because it provides for the formation of the ozone layer (O_3), which protects all living things from high-energy ultraviolet (UV) radiation incident (striking or falling upon) on the earth's atmosphere.

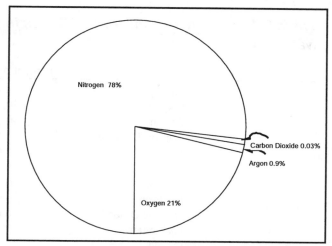

Figure 9.1 Gaseous components of air

The concentration of carbon dioxide (CO_2), in contrast to N_2 and O_2, in the atmosphere is relatively low—about 0.036% or 360 ppm by volume. Life is carbon-based, and carbon dioxide is the source of that carbon. Carbon dioxide is also a principal raw material from which green plants make the food that most living things need. Because of its thermal absorptivity, carbon dioxide works to maintain a favorable global heat balance. However, carbon dioxide is also a major contributor to the greenhouse effect (see Chapter 13).

The most visible constituent of the atmosphere is water vapor (H_2O). Like carbon dioxide, water vapor is a major greenhouse gas and absorbs thermal energy radiated from the earth's surface. Because it readily changes phase, water vapor is significant in the atmosphere. On cooling, it condenses to form large masses of air that contain precipitation.

The other gaseous constituents of the atmosphere—hydrogen, helium, xenon, and krypton (see Table 9.1)—are inert and do not appear to have any major impact on, or role in, the atmosphere.

The Atmosphere: Structure

Not only is the structure of the atmosphere characterized by the gases that comprise it, but also by physical phenomena that act on and within it. These include density and pressure, temperature, and solar and thermal radiation.

Density and Pressure

The atmosphere extends upward with continuously decreasing mass per volume (density); concentrations of the molecules that comprise the atmosphere decrease with height. The greater density near the earth's surface is due to gravitational attraction and compression of the air. As a result, more than 1/2 of the mass of the atmosphere lies below an altitude of 7 miles (11 km), and almost 99% lies below an altitude of 19 miles (30 km). At higher altitudes the air becomes thinner.

Although no clear line of demarcation between the earth's atmosphere and outer space exists, because of a continuous decrease in density in the upper regions of the atmosphere, an outermost

Table 9.1: Relative Proportions of Gases in the Lower Atmosphere *(excluding water vapor)*

Gas	Percent by Volume
Nitrogen	78.08
Oxygen	20.95
Argon	0.93
Carbon Dioxide	0.03
Neon	0.0018
Helium	0.00052
Methane	0.00015
Krypton	0.00010
Nitrous Oxide	0.00005
Hydrogen	0.00005
Ozone	0.000007
Xenon	0.000009

limit can roughly be placed at around 300 to 600 miles (480 to 960 km) from the surface of the earth.

Closely related to mass per volume or density is force per area or atmospheric pressure. This close relationship can be seen in the direct relationship between density and pressure. As atmospheric density decreases with height, atmospheric pressure decreases as well. The pressure at a particular altitude is effectively a measure of the weight of gas above that location (i.e., an object on the earth's surface literally supports a vertical column of air that overlays it).

Temperature

When measuring the temperature of the atmosphere versus altitude, distinct changes are apparent. The variations in temperature lead to distinguishing major divisions within the atmosphere. These major divisions or boundaries are not sharply defined and extend over appreciable distances, but they exist.

Near the earth's surface, the temperature of the atmosphere decreases with increasing altitude at an average rate of about 3.5°F/1000 feet (6.5°C/km) up to about 10 miles (16 km). This region is called the *troposphere* (see Figure 9.2). The atmospheric conditions of the lower troposphere are collectively called weather. Changes in the weather reflect the local variations of the atmosphere near the earth's surface (see Chapter 10).

Above the troposphere, the temperature of the atmosphere decreases non-uniformly up to an altitude of about 30 miles (50 km) (see Figure 9.2). This region of the atmosphere, from approximately 10 to 30 miles (16 to 50 km) in altitude, is called the *stratosphere*.

Beyond the stratosphere, the temperature decreases uniformly to a temperature of about -140°F (-95°C) at an altitude of 50 miles (80 km). This region of the atmosphere is called the *mesosphere*.

Above the mesosphere, the thin atmosphere is heated intensely by the sun's rays and the temperature climbs to over 1800°F (1000°C). This region extending to the outer reaches of the atmosphere is called the *thermosphere*. The temperature of the thermosphere varies directly with solar activity.

Solar and Thermal Radiation

In reality more of a physical than a structural characteristic of the earth's atmosphere, solar and thermal radiation still have a major influence on its overall character. The sun's radiated energy (solar radiation) literally showers the earth's atmosphere with huge amounts of electromagnetic energy.

Energy from the sun in the form of radiation that is incident on the earth's atmosphere is called *insolation—in*coming *sol*ar radi*ation*. Even though some incoming solar energy (radiation) is absorbed by atmospheric gases such as oxygen, ozone, carbon dioxide, and water vapor, a portion of the solar energy does reach the earth's surface.

If we assume that the sun generates incoming solar

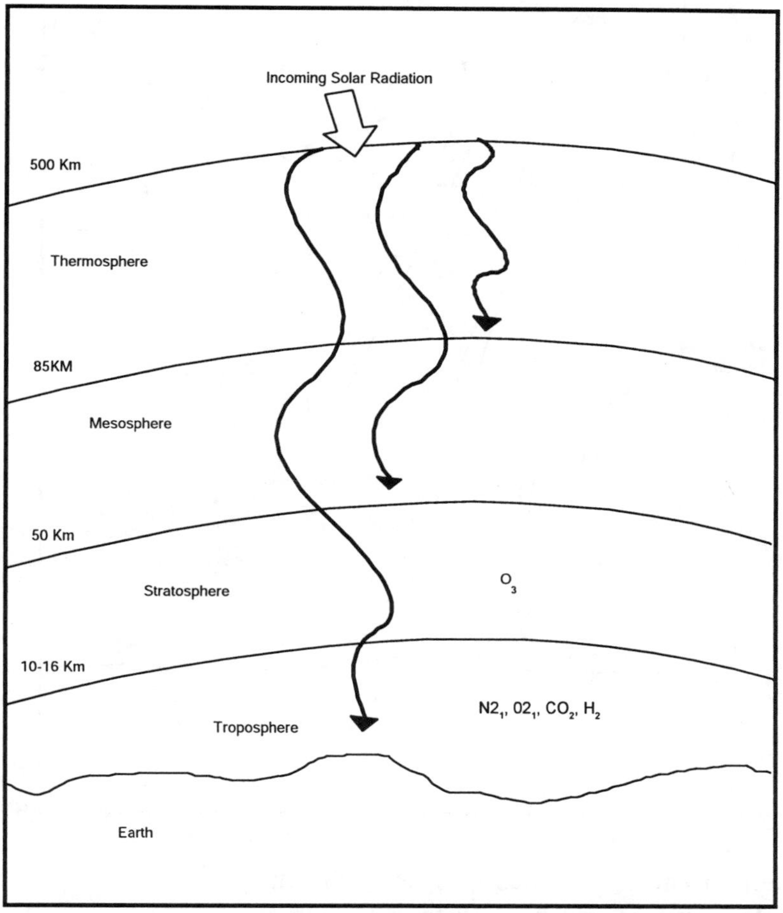

Figure 9.2 The earth's atmosphere and incoming solar radiation

radiation at 100%, we can illustrate insolation distribution (see Figure 9.3). We can see from Figure 9.3 that about 23% of solar radiation directly reaches ground level on earth. About 5% is scattered by the atmosphere to ground and 15% is absorbed by the atmosphere. 22% is able to reach the earth's surface indirectly by penetrating clouds. Approximately 2% is absorbed by clouds; 24% is reflected by clouds along with another 7% scattered by the atmosphere back into outer space (total = 33% reflected back to space).

33% of the insolation received is returned to space with no appreciable effect on the atmosphere as a result of reflection by clouds, scattering by particles in the atmosphere, and reflection from terrestrial surfaces such as water, ice, and variable ground surfaces known as *albedo* (reflectivity).

The earth and its atmosphere re-radiate solar radiation incident on the earth's surface and atmosphere at longer wavelengths within a specific emission spectrum. This thermal radiation or

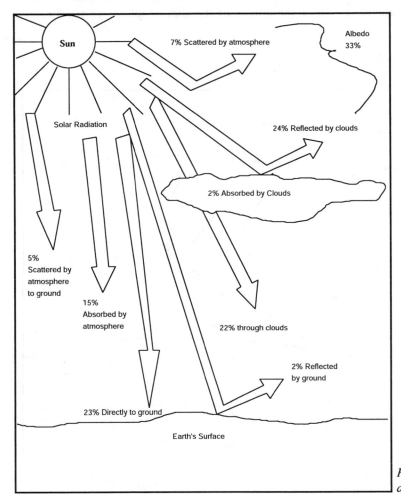

Figure 9.3 Insolation distribution

thermal energy radiates to space from the earth's surface and directly from the atmosphere within a usual infrared rage of 3 to 8 μm, with a peak of about 11 μm.

Albedo

Albedo is the ratio of light reflected (reflectivity) from a particle, planet, or satellite to that falling on it. Albedo always has a value less than or equal to 1. An object with a high albedo (near 1) is very bright, while a body with a low albedo (near 0) is dark. For example, freshly fallen snow typically has an albedo that is between 0.75 and 0.90+; that is, 75 to 95% of the solar radiation that is incident on snow is reflected. At the other extreme, the albedo

Table 9.2: The albedo of some surface types in percent reflected

Surface	Albedo
Water (low sun)	10-100
Water (high sun)	3-10
Grass	16-26
Glacier ice	20-40
Deciduous forest	15-20
Coniferous forest	5-15
Old snow	40-70
Fresh snow	75-95
Sea ice	30-40
Blacktopped tarmac	5-10
Desert	25-30
Crops	15-25

of a rough, dark surface, such as a green forest, may be as low as 0.05. The albedos of some common surfaces are listed in Table 9.2. The portion of insolation not reflected is absorbed by the earth's surface, warming it. This means the earth's albedo plays an important part in its radiation balance and influences the mean annual temperature and the climate on both local and global scales.

The Earth's Heat Balance

Approximately 50% of the solar radiation entering the atmosphere reaches the earth's surface, either directly or after being scattered by clouds, particulate matter, or atmospheric gases. The other 50% is either reflected back directly or is absorbed in the atmosphere and its energy re-radiated back into space at a later time as infrared radiation. Most of the solar energy reaching the surface is absorbed, and must be returned to space to maintain heat balance. The energy produced within the earth's interior (from the hot mantle area via convection and conduction) that reaches the earth's surface (about 1% of that received from the sun) is also lost.

Re-radiation of energy from the earth is accomplished by three energy transport mechanisms: radiation, conduction, and convection. Radiation of energy, as we've said earlier, occurs through electromagnetic radiation in the infrared region of the spectrum. The crucial importance of the radiation mechanism is that it carries energy away from earth—on a much longer wavelength than the solar energy that brings energy to the earth—and, in turn, works to maintain the earth's heat balance. The earth's heat balance is of particular interest to us in this text because it is susceptible to upset by human activities.

A comparatively small but significant amount of heat energy is transferred to the atmosphere by conduction from the earth's surface. Conduction of energy occurs through the interaction of adjacent molecules with no visible motion accompanying the transfer of heat—for example, when the whole length of a metal rod is heated when one end is held in a fire. Because air is a poor heat conductor, conduction is restricted to the layer of air in direct contact with the earth's surface. The heated air is then transferred aloft by convection, the movement of whole masses of air, which may be either relatively warm or cold. Convection is the mechanism by which abrupt temperature variations occur when large masses of air move across an area. Air temperature tends to be greater near the surface of the earth, and decreases gradually with altitude. A large amount of the earth's surface heat is transported to clouds in the atmosphere by conduction and convection before being lost ultimately by radiation, and this redistribution of heat energy plays an important role in weather and climate conditions.

The earth's average surface temperature is maintained at about 15°C because of the atmospheric greenhouse effect. The greenhouse effect occurs when the gases of the lower atmosphere trans-mit most of the visible portion of incident sunlight, in the same manner as the glass of a garden greenhouse. The warmed earth emits radiation in the infrared region, and this radiation is selec-tively absorbed by the atmospheric gases, whose absorption spectrum is similar to that of glass. This absorbed energy heats the atmosphere and helps maintain the earth's temperature. Without this greenhouse effect, the surface temperature would average around -18°C. Most of the ab-sorption of infrared energy is performed by water molecules in the atmosphere. In addition to the key role played by water molecules, carbon dioxide, although to a lesser extent, also is essential in maintaining the heat balance. Environmentalists and others concerned with environmental issues are concerned that an increase in the carbon dioxide level in the atmosphere could prevent sufficient energy loss, causing damaging increases in the earth's temperature. This phenomenon, commonly known as anthropogenic greenhouse effect (discussed in greater detail in Chapter 13), may occur from elevated levels of carbon dioxide caused by increased use of fossil fuels and the reduction in carbon dioxide absorption because of destruction of the rain forest and other forest areas.

The Atmosphere: Motion

To state that earth's atmosphere is constantly in motion is to state the obvious. Anyone observing the earth's constant weather changes is well aware of this phenomenon. But the importance of the dynamic state of our atmosphere is much less obvious.

The constant motion of the earth's atmosphere (air movement) consists of both horizontal (wind) and vertical (air currents) dimensions. The atmosphere's motion is the result of thermal energy produced from the heating of the earth's surface and the air molecules above. Because of differential heating of the earth's surface, energy flows from the equator pole-ward.

Even though air movement plays the critical role in transporting the energy of the lower atmosphere, bringing the warming influences of spring and summer and the cold chill of winter, the effects of air movements on our environment are often overlooked. All life on earth has evolved with mechanisms dependent on air movement: pollen is carried by winds for plant reproduction; animals sniff the wind for essential information; wind power was the motive force that began the earliest stages of the industrial revolution. Now we see the effects of winds in other ways, too: Wind causes weathering (erosion) of the earth's surface; wind influences ocean currents; air pollutants and contaminants such as radioactive particles transported by the wind impact our environment.

Causes of Air Motion

In all dynamic situations, forces are necessary to produce motion and changes in motion— winds and air currents. The air (made up of various gases) of the atmosphere is subject to two primary forces: (1) gravity and (2) pressure differences due to temperature variations.

Gravity (gravitational forces) holds the atmosphere close to the earth's surface. Newton's law of universal gravitation states that every body in the universe attracts another body with a force equal to:

$$F = G \ \frac{m_1 m_2}{r^2} \tag{9.1}$$

where

F	= Force
m_1 and m_2	= the masses of the two bodies
G	= universal constant of 6.67×10^{-11} N x m²/kg²
r	= distance between the two bodies

The force of gravity decreases as an inverse square of the distance between them.

Thermal conditions affect density, which in turn cause gravity to affect vertical air motion and planetary air circulation. This affects how air pollution is naturally removed from the atmosphere.

Although gravitational force is often overruled by forces in other directions, the ever-present force of gravity is vertically downward and acts on each gas molecule, accounting for the greater density of air near the earth.

Atmospheric air is a mixture of gases, so its behavior is governed by the gas laws and other physical principles. The pressure of a gas is directly proportional to its temperature. Pressure is force per unit area ($P = F/A$), so a temperature variation in air generally gives rise to a difference in pressure of force. This difference in pressure, resulting from temperature differences in the atmosphere, creates air movement—on both large and local scales. This pressure difference corresponds to an unbalanced force, and when there is a pressure difference, the air moves from a high- to a low-pressure region.

In other words, horizontal air movements (called *advective winds*) result from temperature gradients, which give rise to density gradients, and subsequently, pressure gradients. The force associated with these pressure variations (*pressure gradient force*) is directed at right angles to (perpendicular to) lines of equal pressure (called *isobars*) and is directed from high to low pressure.

Look at Figure 9.4. The pressures over a region are mapped by taking barometric readings at different locations. A line drawn through the points (locations) of equal pressure is called an isobar. All points on an isobar are of equal pressure, which means no air movement along the isobar. The wind direction is at right angles to the isobar in the direction of the lower pressure. In Figure 9.4, notice that air moves down a pressure gradient toward a lower isobar. If the isobars are close together, the pressure gradient force is large, and such areas are characterized by high wind speeds. If isobars are widely spaced, as shown in Figure 9.4, the winds are light because the pressure gradient is small.

Localized air circulation gives rise to *thermal circulation* (a result of the relationship based on a law of physics whereby the pressure and volume of a gas is directly related to its temperature ($PV \propto T$). A change in temperature causes a change in the pressure and/or volume of a gas. With a change in volume comes a change in density, since $P = m/V$, so regions of the atmosphere with different temperatures may have different air pressures and densities. As a result, localized heating sets up air motion and gives rise to thermal circulation. To gain understanding of this phenomenon, consider Figure 9.5.

Once the air has been set into motion, secondary forces (velocity-dependent forces) act. These secondary forces are (1) the earth's rotation (*Coriolis force*) and (2) contact with the rotating earth (friction). The Coriolis force, named after its discoverer, French mathematician Gaspard Coriolis (1772-1843), is the effect of rotation on the atmosphere, and on all objects on the earth's surface. In the Northern Hemisphere, it causes moving objects and currents to be deflected to the right; in the Southern Hemisphere it causes deflection to the left, because of the earth's rotation. Air, in large-scale north or south movements, appears to be deflected from its expected path. That is, air moving pole-ward in the northern hemisphere appears to be deflected toward the east; air moving southward appears to be deflected toward the west.

Figure 9.6 illustrates the Coriolis effect on a propelled particle (which is analogous to the apparent effect of an air mass flowing from Point A to Point B). From Figure 9.6, the action of the

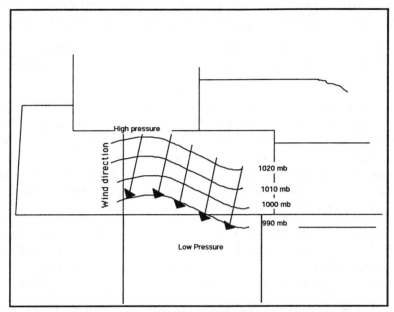

Figure 9.4 Isobars drawn through locations having equal atmospheric pressures. The air motion, or wind direction, is at right angles to the isobars and moves from a region of high pressure to a region of low pressure.

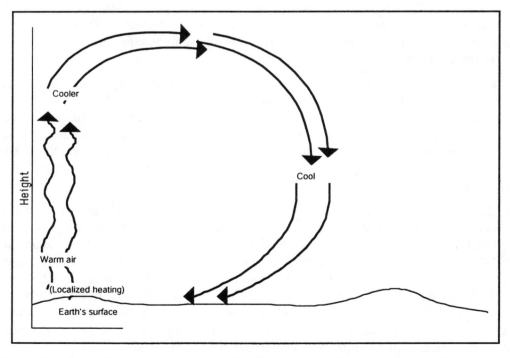

Figure 9.5 Thermal circulation of air. Localized heating, which causes the air in the region to rise, initiates the circulation. As the warm air rises and cools, cool air near the surface moves horizontally into the region vacated by the rising air. The upper, still cooler air then descends to occupy the region vacated by the cool air.

earth's rotation on the air particle, as it travels north over the earth's surface as the earth rotates beneath it from east to west, can be seen. Projected from point A to Point B, the particle will actually reach Point B′ because as it is moving in a straight line (deflected) the earth rotates, east to west, beneath it.

Friction (drag) can also cause the deflection of air movements. This friction (resistance) is both internal and external. Internal friction is generated by the friction of the air's molecules. Friction is also generated when air molecules run into each other. External friction is caused by contact with terrestrial surfaces. The magnitude of the frictional force along a surface is dependent on the air's magnitude and speed, and the opposing frictional force is in the opposite direction of the air motion.

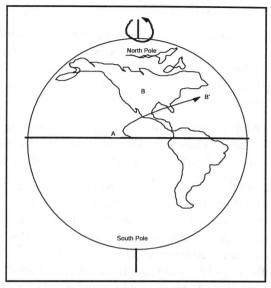

Figure 9.6 The effect of the earth's rotation on the trajectory of a propelled particle

Local and World Air Circulation

Air moves in all directions, and these movements are essential for those of us on earth: vertical air motion is essential in cloud formation and precipitation (see Chapter 10). Horizontal air movement near the earth's surface produces winds.

Wind is an important factor in human comfort, especially affecting how cold we feel. A brisk wind at moderately low temperatures can quickly make us uncomfortably cold. Wind promotes the loss of body heat, which aggravates the chilling effect, expressed through wind chill factors in winter and the heat index in summer. These two scales describe the cooling effects of wind on exposed flesh at various temperatures.

Local winds are the result of atmospheric pressure differences involved with thermal circulations due to geographic features. Land areas heat up more quickly than do water areas, giving rise to a convection cycle. As a result, during the day, when land is warmer than the water, we experience a lake or sea breeze.

At night, the cycle reverses. Land loses its heat more quickly than water, so the air over the water is warmer. The convection cycle sets to work in the opposite direction, and a land breeze blows.

In the upper troposphere (above 11 to 14 km, west to east flows) are very narrow fast-moving bands of air called jet streams. Jet streams have significant effects on surface airflows. When jet streams accelerate, divergence of air occurs at that altitude. This promotes convergence near the surface and the formation of cyclonic motion. Deceleration causes convergency aloft and subsidence near the surface, causing an intensification of high-pressure systems.

Jet streams are thought to result from the general circulation structure in the regions where great high- and low-pressure areas meet.

BASIC AIR QUALITY

The quality of the air we breathe is not normally a concern to us unless we detect something unusual about the air (its odor or taste, or that breathing is difficult or uncomfortable) or unless we have been advised by authorities or the news media that there is cause for concern. Air pollutants in the atmosphere cause great concern because of potential adverse effects on our health.

To have good air quality is a plus for any community. Good air quality attracts industry as well as people who are looking for a healthy place to live and raise a family. It is not unusual to see or hear advertisements advertising a locality's "clean or fresh air."

Note that although most people do seek an environment that has "clean or fresh air" and that is pollution free to live in, this is not always the case for everyone. A good example of this exception is in the Los Angeles Basin. Before Los Angeles became the large city that it is today, local inhabitants named the Basin area the "Valley of the Smokes" from the emissions from the campfires and settlements. This early warning about adverse climatic conditions did not stop settlement. Today, a stranger to Los Angeles need only pull into the basin area, step out of his/her air-conditioned vehicle, and take a breath of air to note the presence of the area's infamous smog.

Because of the large numbers of people who decided to make the LA Basin their homes, Los Angeles and California have enacted probably the most restrictive air pollution requirements in the U. S.

Air quality is impacted by those things we can see with the naked eye (smoke, smog, etc.), by those things that can only be seen under the microscope (pollen, microbes, dust, etc.), and by those substances we can't see (ozone, carbon dioxide, sulfur dioxide, etc.). These compounds are heavily regulated, and it seems with each passing day that the EPA or some other regulatory authority creates some new regulation for a new or old compound.

SUMMARY

We've said that air pollutants cause great concern because of potential adverse effects on human health. These adverse health effects include acute conditions such as respiratory difficulties and chronic effects such as emphysema and cancer. Although health concerns related to air pollution are usually at the top of any concerned person's list, we must keep in mind that air pollution has adverse impacts on other aspects of our environment that are important to us as well, such as on vegetation, materials, and degradation of visibility.

In any discussion of air quality, certain specific areas must be addressed. Any discussion of air quality that does not include a discussion of types of air pollutants, air pollutant sources, the mechanics of air pollution (dispersion, transformation, and depletion), emission control devices, air quality regulations, and ambient air quality evaluation methods is an empty effort. To ensure that our effort is not empty, these topics and others will be covered in detail in Chapters 11, 12, 13, and 14.

Suggested Readings

Anthes, R. A., *Meteorology*, 7th ed., Upper Saddle River, NJ: Prentice Hall, 1996.

Anthes, R. A., Cahir, J. J., Fraizer, A. B., and Panofsky, H. A., *The Atmosphere*, 3rd ed., Columbus, OH: Charles E. Merrill Publishing Company, 1984.

Ingersoll, A. P., The Atmosphere. *Scientific American*, 249(33): 162-174, 1983.

Lutgens, F. K. and Tarbuck, E. J., *The Atmosphere, An Introduction to Meteorology*. Englewood Cliffs, NJ: Prentice-Hall, 1982.

Moron, J. M., Morgan, M. D., and Wiersma, J. H., *Introduction to Environmental Science*, 2nd ed., New York: W. H. Freeman and Company, 1986.

Shipman, J. T., Adams, J. L., and Wilson, J. D., *An Introduction to Physical Science*, 5th ed., Lexington, MA: D.C. Heath and Company, 1987.

CHAPTER 10

METEOROLOGY

Just as there are people with distorted, failing, or nonexistent senses of smell, there are those at the other end of the olfactory spectrum, prodigies of the nose, the most famous of whom is probably Helen Keller. "The sense of smell," she wrote, "has told me of a coming storm hours before there was any sign of it visible. I notice first a throb of expectancy, a slight quiver, a concentration in my nostrils. As the storm draws near my nostrils dilate, the better to receive the flood of earth odors which seem to multiply and extend, until I feel the splash of rain against my cheek. As the tempest departs, receding farther and farther, the odors fade, become fainter and fainter, and die away beyond the bar of space." Other individuals have been able to smell changes in the weather, too, and, of course, animals are great meteorologists (cows, for example, lie down before a storm). Moistening, misting, and heaving the earth breathes like a great dark beast. When barometric pressure is high, the earth holds its breath and vapors lodge in the loose packing and random crannies of the soil, only to float out again when the pressure is low and the earth exhales. The keen-nosed, like Helen Keller, smell the vapors rising from the soil, and know by that signal that there will be rain or snow. This may also be, in part, how farm animals anticipate earthquakes—by smelling ions escape from the earth. —Ackerman

CHAPTER OUTLINE

- Introduction
- Meteorology: The Science of Weather
- Thermal Inversions and Air Pollution

INTRODUCTION

Do you know anyone who (like Helen Keller) claims to be able to smell changes in the weather? Maybe you've heard folks say that they "sense" a change in the weather brewing. You have heard people talk about the weather—have you heard anyone ask what the weather is like in someone else's locale? Or whether you'd heard the weather forecast for the weekend? Have you ever gotten a letter that talked about the weather in the writer's area, or have you been asked how the weather in your area was?

Because weather affects us all, physically and emotionally, we are often concerned with what changes weather is going to bring to us on a day-to-day and season-to-season basis. While you may or may not know anyone who can smell or sense weather change, people discuss the weather in conversation daily. We're interested in the weather conditions in other people's cities, and people move to one locale or another because of that area's weather conditions. But do they move because of the weather there, or because of the climate?

Let's look at climate for a moment. Have you ever asked someone how their climate is when you really wanted to know what their weather was? Probably not. We don't often confuse the two. When we talk about weather, we are generally referring to the transient changes in temperature, precipitation, and wind that affect whether we take the umbrella along or wear a heavy coat. Most people rely heavily on the local meteorologist and the daily weather forecasts—so much so that an entire branch of science is dedicated to the effort of trying to predict the weather—a difficult task, because of the extensive variables in any prediction.

Try to define climate and weather. Most people do not have a good feel for the exact meanings of and differences between "weather" and "climate." The two terms, their specific meanings and differences, and the elements that comprise them are the subject of this chapter. The fundamentals of how weather affects air pollution are essential to a basic study of pollution and how it affects our environment.

METEOROLOGY: THE SCIENCE OF WEATHER

Meteorology is the science concerned with the atmosphere and its phenomena; the meteorologist observes the atmosphere's temperature, density, winds, clouds, precipitation, and other characteristics, and endeavors to account for its observed structure and evaluation in terms of external influence and the basic laws of physics.

Weather is the state of the atmosphere, mainly with respect to its effect upon life and human activities; as distinguished from climate (long-term manifestations of weather), weather consists of the short-term (minutes to months) variations of the atmosphere. Weather is defined primarily in terms of heat, pressure, wind, and moisture—the elements of which weather is made. At high levels above the earth, where the atmosphere thins to near vacuum, there is no weather. Weather is a near-surface phenomenon. We see this clearly, daily, as we observe the ever-changing, sometimes dramatic, and often violent weather display that travels through our environment.

In the study of environmental science, and in particular the study of air quality (especially regarding air pollution in a particular area), the determining factors are directly related to the dynamics of the atmosphere—local weather. These determining factors include strength of winds, the direction they blow, temperature, available sunlight (needed to trigger photochemical reactions, which produce smog), and the length of time since the last weather event (strong winds and heavy precipitation) cleared the air.

Weather events such as strong winds and heavy precipitation that work to clean the air we breathe are obviously beneficial. However, few people would categorize weather events such as tornadoes, hurricanes, and typhoons as beneficial. Then there are those other weather events—ones that have both a positive and negative effect. One such event is *El Niño*, which we discuss in Case Study 10.1.

Case Study 10.1
El Niño

El Niño is a natural phenomenon that occurs every two to nine years on an irregular and unpredictable basis. A warming of the surface waters in the tropical eastern Pacific, El Niño causes fish to disperse to cooler waters and, in turn, causes adult birds that feed upon them to fly off in search of new food sources elsewhere.

Through a complex web of events, El Niño (which means "the child" in Spanish because it usually occurs during the Christmas season off the coasts of Peru and Ecuador) can have a devastating impact on all forms of marine life.

During a normal year, equatorial trade winds pile up warm surface waters in the western Pacific. Thunderheads unleash heat and torrents of rain. This heightens the east-west temperature difference, sustaining the cycle. The jet stream blows from north Asia to California. During an El Niño year, trade winds weaken, allowing warm waters to move east. This decreases the east-west temperature difference. The jet stream is pulled farther south than normal and picks up storms it would usually miss, carrying them to Canada or California. The warm waters eventually reach South America.

One of the first signs of El Niño's appearance is a shifting of winds along the equator in the Pacific Ocean. The normal easterly winds reverse direction and drag a large mass of warm water eastward toward the South American coastline. The large mass of warm water basically forms a barrier that prevents the upwelling of nutrient-rich cold water from the ocean bottom to the surface. As a result, the growth of microscopic algae that normally flourish in the nutrient-rich upwelling areas diminishes sharply. This decrease has further repercussions: El Niño has been linked to patterns of subsequent droughts, floods, typhoons, and other costly weather extremes around the globe. Take a look at El Niño's affect on the West Coast of the United States, where it has been blamed for hurricanes, floods, and early snowstorms. On the positive side, El Niño typically brings good news to those who live on the East Coast of the United States: a reduction in the number and severity of hurricanes.

El Niño is a phenomenon that, although not yet completely understood by scientists, causes both positive and negative results, depending upon where you live.

The Sun: The Weather Generator

The sun is the driving force behind weather. Without the distribution and re-radiation to space of solar energy, there could be no weather (as we know it) on Earth. The sun is the source of most of the earth's heat. Of the gigantic amount of solar energy generated by the sun, only a small portion reaches the earth. Most of the sun's solar energy is lost in space. A little over 40% of the sun's radiation reaching earth hits the surface and is changed to heat. The rest stays in the atmosphere or is reflected back into space.

Like a greenhouse, the earth's atmosphere admits most of the solar radiation. When solar radiation is absorbed by the earth's surface, it is re-radiated as heat waves, most of which are trapped by carbon dioxide and water vapor in the atmosphere, which work to keep the earth warm.

By now you are aware of the many functions performed by the earth's atmosphere, and you know that the atmosphere plays an important role in regulating the earth's heating supply. The atmosphere protects the earth from too much solar radiation during the day, and prevents most of the heat from escaping at night. Without the filtering and insulating properties of the atmosphere, the earth would experience severe temperature extremes similar to those on other planets.

On bright clear nights the earth cools more rapidly than on cloudy nights, because cloud cover reflects a large amount of heat back to the earth, where it is re-absorbed. The earth's air is heated primarily by contact with the warm earth. When air is warmed, it expands and becomes lighter. Air warmed by contact with the earth rises and is replaced by cold air that flows in and under it.

When this cold air is warmed, it too rises and is replaced by cold air. This cycle continues and generates a circulation of warm and cold air, which is called *convection*.

At the earth's equator, the air receives much more heat than the air at the poles. This warm air at the equator is replaced by colder air flowing in from north and south. The warm, light air rises and moves pole-ward high above the earth. As it cools, it sinks, replacing the cool surface air that has moved toward the equator.

The circulating movement of warm and cold air and the differences in heating cause local winds and breezes. Different amounts of heat are absorbed by different land and water surfaces. Soil that is dark and freshly plowed absorbs much more than grassy fields, for example. Land warms faster than does water during the day and cools faster at night. Consequently, the air above such surfaces is warmed and cooled, resulting in production of local winds.

Winds should not be confused with air currents. Wind is primarily oriented toward horizontal flow, while air currents are created by air moving upward and downward. Both wind and air currents have direct impact on air pollution, which is carried and dispersed by wind. An important factor in determining the areas most affected by an air pollution source is wind direction. Since air pollution is a global problem, wind direction on a global scale is important (see Figure 10.1)

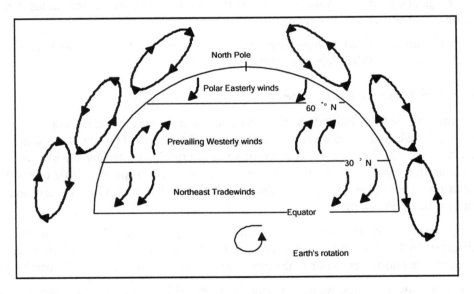

Figure 10.1 Global wind directions in the Northern Hemisphere

Another constituent associated with the earth's atmosphere is water. Water is always present in the air. It evaporates from the earth, two thirds of which is covered by water. In the air, water exists in three states: solid, liquid, and invisible vapor.

The amount of water in the air is called *humidity*. The *relative humidity* is the ratio of the actual amount of moisture in the air to the amount needed for saturation at the same temperature. Warm air can hold more water than cold. When air with a given amount of water vapor cools, its relative humidity increases; when the air is warmed, its relative humidity decreases.

Air Masses

An *air mass* is a vast body of air (so vast that it can have global implications) in which the conditions of temperature and moisture are much the same at all points in a horizontal direction. An air mass is affected by and takes on the temperature and moisture characteristics of the surface over which it forms, although its original characteristics tend to persist.

When two different air masses collide, a *front* is formed. A *cold front* marks the line of advance of a cold air mass from below, as it displaces a warm air mass. A *warm front* marks the advance of a warm air mass as it rises up over a cold one.

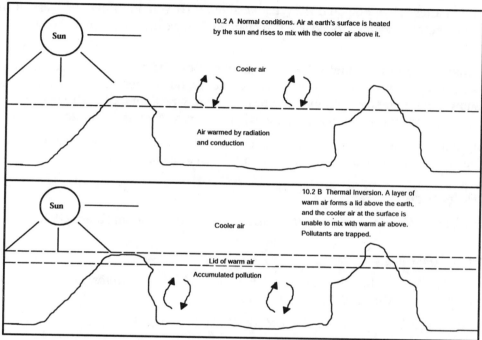

10.2 A Normal conditions. Air at earth's surface is heated by the sun and rises to mix with the cooler air above it.

Sun

Cooler air

Air warmed by radiation and conduction

10.2 B Thermal Inversion. A layer of warm air forms a lid above the earth, and the cooler air at the surface is unable to mix with warm air above. Pollutants are trapped.

Sun

Cooler air

Lid of warm air

Accumulated pollution

Figure 10.2 Normal conditions (A) and thermal inversion (B)

THERMAL INVERSIONS AND AIR POLLUTION

We have said that during the day the sun warms the air near the earth's surface. Normally, this heated air expands and rises during the day, diluting low-lying pollutants and carrying them higher into the atmosphere. Air from surrounding high-pressure areas then moves down into the low-pressure area created when the hot air rises (see Figure 10.2 A). This continual mixing of air helps keep pollutants from reaching dangerous levels in the air near the ground.

Sometimes, however, a layer of dense, cool air is trapped beneath a layer of less dense, warm air in a valley or urban basin. This is called a *thermal inversion* (see Figure 10.2 B). In effect, a warm-air lid covers the region and prevents pollutants from escaping in upward-flowing air currents. Usually these inversions trap air pollutants at ground level for a short period of time. However, sometimes they last for several days, when a high-pressure air mass stalls over an area, trapping air pollutants at ground level where they accumulate to dangerous levels.

The best known location in the United States where thermal inversions occur almost on a daily basis is in the Los Angeles basin. The Los Angeles basin is a valley with a warm climate, light winds, surrounded by mountains located near the Pacific Coast. Los Angeles itself is a large city with a large population of people and automobiles. It possesses the ideal conditions for smog—conditions worsened by frequent thermal inversions.

SUMMARY

We are intimately affected by weather in our lives. We plan our lives around it, dressing as we do because of it and working because of it—shoveling snow, raking leaves, planting gardens around it. We go places because of it, or move away from it. We make it the first topic of conversation after "How do you do?" We joke about it (If you don't like the weather in (your town's name here) just stick around and it'll change). But we can't change it. And it can change us.

This, of course, terrifies us, chiefly because in its extremes of temperature, velocity, and precipitation, we know that weather can destroy us. Tornadoes, hurricanes, too much or too little rain, heavy snows, extreme cold, and extreme heat damage wreck and kill. All we can do is take shelter from the adverse conditions, rebuild when they pass, and hope for better.

The only important way that we can affect our weather is one harmful to us—we can pump pollutants into our air so that even normal weather can affect us adversely.

Cited References

Ackerman, D., *A Natural History of the Senses*, New York: Random House, 1990.

Suggested Readings

Anthes, R. A., Cahir, J. J., Frasier, A. B., & Panofsky, H. A., *The Atmosphere*, 3rd ed., Columbus, OH: Charles E. Merrill Publishing Co., 1984.

Battan, L. J., *Weather in your Life*, New York: W. H. Freeman, 1983.

Budyko, M. I., *The Earth's Climate*, New York: Academic Press, 1982.

Gates, D. M., *Energy Exchange in the Biosphere*, New York: Harper & Row Monographs, 1962.

Ingersoll, A. P., The Atmosphere, *Scientific American* 249(3): 162-174, 1983.

Kondratyev, K. Y., *Radiation in the Atmosphere*, New York: Academic Press, 1969.

Moran, J. M. and Morgan, M.D., *Essentials of Weather*, Upper Saddle River, NJ: Prentice-Hall, 1994.

National Oceanic and Atmospheric Administration (NOAA), *U.S. Standard Atmosphere*, NOAA S/T 76-1562, 1976.

National Research Council, *Solar Variability, Weather, and Climate*. Washington, DC: National Academy Press, 1982.

National Research Council, *Understanding Climatic Change, A Program for Action*, Washington, DC: National Academy of Sciences, 1975.

CHAPTER 11

ATMOSPHERIC POLLUTANTS

LONG-TERM IMPACT OF INCREASED ATMOSPHERIC CO_2

If current patterns of carbon dioxide (CO_2) emission continue over the next century, the world's climate may heat up for several hundred years. Recent findings indicate that profound changes could occur in deep-ocean circulation, with the result that the Earth's carbon cycle would be altered. The findings suggest that if the present trends of greenhouse gas emissions prevail over the next century, atmospheric concentrations of CO_2 or equivalent gases would quadruple. Atmospheric temperatures would continue to increase, rising as much as 7 degrees C in 500-600 years. Sea levels would rise by about 2 meters from the thermal expansion of oceans alone. Most alarming, the ocean would settle into a stable pattern, its surface and deep waters no longer mixing. This could conceivably compromise marine biological activity, and by greatly reducing the ocean's ability to absorb CO_2, alter the Earth's carbon cycle. —Manabe and Stouffer

CHAPTER OUTLINE
- Introduction
- Major Air Pollutants

INTRODUCTION

In the past, and to a lesser degree in the present, belching smokestacks were a comforting sight to many people: more smoke equaled more business, which indicated that the economy was healthy. But many of us are now troubled by evidence that indicates that polluted air adversely affects our health (many toxic gases and fine particles entering the air pose health hazards—cancer, genetic defects, and respiratory disease). Nitrogen and sulfur oxides, ozone, and other air pollutants from fossil fuels are inflicting damage on our forests, crops, soils, lakes, rivers, coastal waters, and buildings. Chlorofluorocarbons (CFCs) and other pollutants entering the atmosphere are depleting the earth's protective ozone layer, allowing more harmful ultraviolet radiation to reach the earth's surface. As pointed out in the opening of this chapter, fossil fuel combustion is increasing the amount of carbon dioxide in the atmosphere, which can have severe long-term environmental impact. Historically, many felt that the air renewed itself (through interaction with vegetation and the oceans) in sufficient quantities to make up for the influx into our atmosphere of anthropogenic pollutants. Today, however, this kind of thinking is being challenged by evidence that clearly indicates that increased use of fossil fuels, expanding industrial production, and growing use of motor vehicles are having a detrimental affect on atmosphere, air, and the environment. In this chapter, we examine the types and sources of air pollutants that are related to these concerns. In Chapter 12 we examine how these pollutants are dispersed throughout the atmosphere.

MAJOR AIR POLLUTANTS

The most common and widespread anthropogenic pollutants currently emitted are sulfur dioxide (SO_2), nitrogen oxides (NO_x), carbon monoxide (CO), carbon dioxide (CO_2), volatile organic compounds (hydrocarbons), particulates, lead, and several toxic chemicals. Table 11.1 lists important air pollutants and their sources.

Table 11.1: Important Pollutants and Their Sources

Pollutant	Source
Sulfur and Nitrogen Oxides	Fossil fuel combustion
Carbon Monoxide	Primarily motor vehicles
Volatile Organic Compounds	Vehicles and industry
Ozone	Atmospheric reactions between nitrogen oxides and organic compounds

Source: EPA, *Environmental Progress & Challenges,* p. 13, 1988

National Ambient Air-Quality Standards

Before we begin our discussion of air pollutants, you should understand some of the history behind current air quality regulations. In the United States, the Environmental Protection Agency (EPA) regulates air quality under the *Clean Air Act* (CAA) and amendments which charged the federal government to develop uniform *National Ambient Air-Quality Standards* (NAAQS). These were to include a dual standard requirement of *primary standards* (covering criteria pollutants) designed to protect health, and *secondary standards* to protect public welfare. Primary standards were to be achieved by July 1975, and secondary standards in "a reasonable period of time." Pollutant levels that are protective of public welfare take priority (and are more stringent) than those for public health; achievement of the primary health standard had immediate priority. In 1971 the EPA promulgated National Ambient Air-Quality Standards (NAAQS) for six classes of air pollutants. In 1978 an air-quality standard was also promulgated for lead, and the photochemical oxidant standard was revised to an ozone (O_3) standard (the ozone permissible level was increased). The PM (particulate matter) standard was revised and was re-designated the PM10 standard in 1987. This revision reflected the need for a PM standard based on particle sizes (≤ 10 μm) that have the potential for entering the respiratory tract and affecting human health. The National Ambient Air Quality Standards are summarized in Table 11.2.

Thus, air pollutants were categorized into two groups: primary and secondary. *Primary pollutants* are emitted directly into the atmosphere where they exert an adverse influence on human health or the environmental. Of particular concern are primary pollutants emitted in large quantities: carbon dioxide, carbon monoxide, sulfur dioxide, nitrogen dioxides, hydrocarbons, and particulate matter (PM). Once in the atmosphere, primary pollutants may react with other primary pollutants or atmospheric compounds such as water vapor to form *secondary pollutants*. A secondarily pollutant that has received a lot of press and other attention is acid precipitation, which is formed when sulfur or nitrogen oxides react with water vapor in the atmosphere.

Table 11.2: National Ambient Air-Quality Standards*

Pollutant	Averaging Time	Primary Std	Secondary Std
Carbon monoxide	8 hour (9 ppm)	10 mg/m³	Same
	1 hour	40 mg/m³ (35 ppm)	
Nitrogen dioxide	Annual Average	100 µg/m³ (0.05 ppm)	Same
Sulfur dioxide	Annual Average	80 µg/m³ (0.03 ppm)	
	24 hour	365 µg/m³ (0.14 ppm)	
	3 hour	(0.5 ppm)	1300 µg/m³
PM₁₀ (= µm)	Annual arithmetic mean	50 µg/m³	Same
	24 hour	150 µm/m³	50 µg/m³
Hydrocarbons (corrected for methane)	3 hour (0.24 ppm)	160 µg/m³	Same
Ozone	1 hour (0.12 ppm)	235 µg/m³	Same
Lead	3 month avg.	1.5 µg/m³	Same

* Standards other than those based on the annual average are not to be exceeded more than once a year.
Source: EPA, *The CAA Amendments of 1990,* November 1990.

Sulfur Dioxide (SO₂)

Sulfur enters the atmosphere in the form of corrosive sulfur dioxide (SO_2) gas. Sulfur dioxide is a colorless gas possessing the sharp, pungent odor of burning rubber. On a global basis, nature and anthropogenic activities produce sulfur dioxide in roughly equivalent amounts. Its natural sources include volcanoes, decaying organic matter, and sea spray, while anthropogenic sources include combustion of sulfur-containing coal and petroleum products and smelting of nonferrous ores. According to the World Resources Institute and International Institute for Environment and Development (WRI & IIED, 1988-89), in industrial areas much more sulfur dioxide comes from human activities than from natural sources. Sulfur-containing substances are often present in fossil fuels. Since SO_2 is a product of combustion that results from materials that contain sulfur, sulfur dioxide is released into the atmosphere when such materials are burned.

The largest single source of sulfur dioxide is the burning of fossil fuels to generate electricity. As a result, sulfur dioxide is more often encountered as a serious pollutant near major industrialized areas.

In the air, sulfur dioxide converts to sulfur trioxide (SO_3) and sulfate particles (SO_4). Sulfate particles restrict visibility, and in the presence of water, form sulfuric acid (H_2SO_4), which is a highly corrosive substance that also lowers visibility.

According to McKenzie and El-Ashry (1988), global output of sulfur dioxide has increased six-fold since 1900. Most industrial nations, however, have since 1975-1985 lowered sulfur dioxide levels by 20 to 60% by shifting away from heavy industry and imposing stricter emission standards. Major sulfur dioxide reductions have come from burning coal with lower sulfur content and from using less coal to generate electricity.

Two major environmental problems have developed in highly industrialized regions of the world, where the atmospheric sulfur dioxide concentration has been relatively high: sulfurous smog and acid rain. Sulfurous smog is the haze that develops in the atmosphere when molecules of sulfuric acid accumulate, growing in size as droplets until they become sufficiently large to serve as light scatterers. The second problem, acid rain, is precipitation contaminated with dissolved acids like sulfuric acid. Acid rain has posed a threat to the environment by causing certain lakes to become void of aquatic life.

Nitrogen Oxides (NO$_x$)

Seven oxides of nitrogen are known to occur, NO, NO_2, NO_3, N_2O, N_2O_3, N_2O_4, and N_2O_5, but only two are important in the study of air pollution: nitric oxide (NO) and nitrogen dioxide (NO_2). Nitric oxide (NO) is produced by both natural and human actions. Soil bacteria are responsible for the production of most of the nitric oxide that is produced naturally and released to the atmosphere. Within the atmosphere, nitric oxide readily combines with oxygen to form nitrogen dioxide (NO_2); together, those two oxides of nitrogen are usually referred to as NO_x (nitrogen oxides). NO_x is formed naturally by lightening and by decomposing organic matter. Approximately 50% of anthropogenic NO_x is emitted by motor vehicles and about 30% comes from power plants, with the other 20% produced by industrial processes.

Scientists distinguish between two types of NO_x, thermal and fuel, depending on the mode of formation. Thermal NO_x is created when nitrogen and oxygen in the combustion of air (for example, within internal combustion engines) are heated to a high enough temperature (above 1000K) to cause nitrogen (N_2) and oxygen (O_2) in the air to combine. Fuel NO_x results from oxidation (i.e., combines with oxygen in the air) of nitrogen contained within a fuel, such as coal. Both types of NO_x generate nitric oxide first, and then, when vented and cooled, a portion of the nitric oxide is converted to nitrogen dioxide. Although both thermal NO_x and fuel NO_x can be significant contributors to total NO_x emissions, fuel NO_x is usually the dominant source, with approximately 50% coming from power plants (stationary sources) and the other half released by automobiles (mobile sources).

Nitrogen dioxide is about four times more toxic than nitric oxide, and is a much more serious air pollutant. Nitrogen dioxide, at high concentrations, is believed to contribute to heart, lung, liver, and kidney damage. In addition, because nitrogen dioxide occurs as a brownish haze (giving smog its reddish-brown color), it reduces visibility. When nitrogen dioxide combines with water vapor in the atmosphere, it forms nitric acid (HNO_2), a corrosive substance that, when precipitated out as acid rain, causes damage to plants and corrosion to metal surfaces.

WRI & IIED (1988-89) report that NO_x rose in several countries and then leveled off or declined during the 1970s. During this timeframe (see Figure 11.1), levels of nitrogen oxide have not dropped as dramatically as those of sulfur dioxide, primarily because a large part of total NO_x emissions comes from millions of motor vehicles, while most sulfur dioxide is released by a relatively small number of emission-controlled coal-burning power plants.

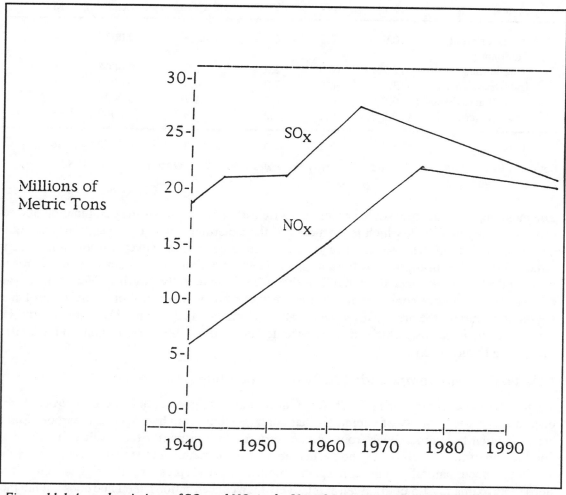

Figure 11.1 Annual emissions of SO_x and NO_x in the United States, 1940-1987 (source: EPA, National Air Pollutant Emissions Estimates, Washington, DC, 1989.)

Carbon Monoxide (CO)

Carbon monoxide is a colorless, odorless, tasteless gas that is by far the most abundant of the primary pollutants, as Table 11.3 indicates. When inhaled, carbon monoxide gas restricts the blood's ability to absorb oxygen, causing angina, impaired vision, and poor coordination. Carbon monoxide has little direct effect on ecosystems, but has an indirect environmental impact via contributing to the greenhouse effect and depletion of the earth's protective ozone layer.

Table 11.3: United States Emission Estimates, 1986 (10^{12} g/yr)

Source	SO_x	NO_x	VOC	CO	Lead	PM
Transportation	0.9	8.5	6.5	42.6	0.0035	1.4
Stationary						
source fuel	17.2	10.0	2.3	7.2	0.0005	1.8
Industrial processes	3.1	0.6	7.9	4.5	0.0019	2.5
Solid Waste disposal	0.0	0.1	0.6	1.7	0.0027	0.3
Miscellaneous	0.0	0.1	2.2	5.0	0.0000	0.8
Total	28.4	18.1	19.5	60.9	0.0086	6.8

Source: EPA, *National Air Pollutant Emission Estimates 1940-1986*, Washington, DC, 1988.

The most important natural source of atmospheric carbon monoxide is the combination of oxygen with methane (CH_4), which is a product of the anaerobic decay of vegetation (anaerobic decay takes place in the absence of oxygen). At the same time, however, carbon monoxide is removed from the atmosphere by the activities of certain soil microorganisms, so the net result is a harmless average concentration of less than 0.12-15 ppm in the Northern Hemisphere. Because stationary source combustion facilities are under much tighter environmental control than are mobile sources, the principal source of carbon monoxide that is caused by human activities is motor vehicle exhaust, which contributes up to about 70% of all CO emissions in the United States (see Figure 11.2).

Volatile Organic Compounds (VOCs—Hydrocarbons)

Volatile Organic Compounds (VOCs) (also listed under the general heading of *hydrocarbons*) encompass a wide variety of chemicals that contain exclusively hydrogen and carbon. Emissions of volatile hydrocarbons from human resources are primarily the result of incomplete combustion of fossil fuels. Fires and the decomposition of matter are the natural sources. Of the VOCs that occur naturally in the atmosphere, methane (CH_4) is present at the highest concentrations (approximately 1.5 ppm). Even at relatively high concentrations, methane does not interact chemically with other substances and causes no ill health effects. However, in the lower atmosphere, sunlight causes VOCs to combine with other gases such as NO_2, oxygen, and CO to form secondary pollutants such as formaldehyde, ketones, ozone, peroxyacetyl nitrate (PAN), and other types of photochemical oxidants. These active chemicals can irritate the eyes and damage the respiratory system, as well as damage vegetation.

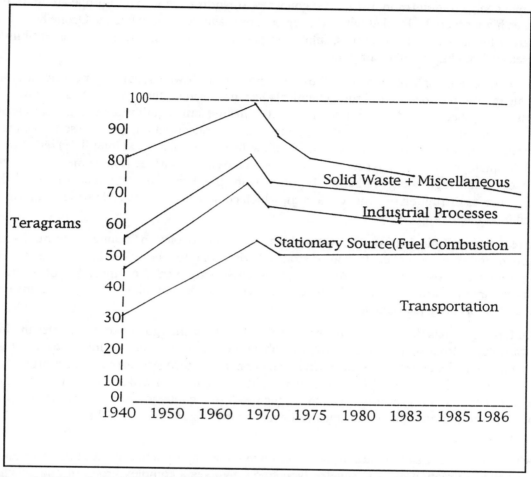

Figure 11.2 Trends in Carbon Monoxide Emissions, 1940-1986 (source: EPA, National Air Pollutant Emission Estimates 1940-1986, Washington, DC, 1988)

Ozone and Photochemical Smog

By far the most damaging photochemical air pollutant is ozone. Each ozone molecule contains three atoms of oxygen and thus is written O_3. Other photochemical oxidants (peroxyacetyl nitrate (PAN), hydrogen peroxide (H_2O_2), and aldehydes) play minor roles. All of these are secondary pollutants because they are not emitted but are formed in the atmosphere by photochemical reactions involving sunlight and emitted gases, especially NO_x and hydrocarbons.

Ozone is a bluish gas, about 1.6 times heavier than air, and relatively reactive as an oxidant. Ozone is present in a relatively large concentration in the stratosphere and is formed naturally by ultraviolet radiation. At ground level, ozone is a serious air pollutant; it has caused serious air pollution problems throughout the industrialized world, posing threats to human health and damaging foliage and building materials.

The problem of ozone pollution is very serious. According to MacKenzie and El-Ashry (1988), ozone concentrations in industrialized countries of North America and Europe are up to three times higher than the level at which damage to crops and vegetation begins. Ozone harms vegetation by damaging plant tissues, inhibiting photosynthesis, and increasing susceptibility to disease, drought, and other air pollutants.

In the upper atmosphere, good (vital) ozone is produced. However, at the same time, anthropogenic emission of ozone-depleting chemicals has increased on the ground. With this increase, concern has been raised over a potential upset of the dynamic equilibrium among stratospheric ozone reactions, with a consequent reduction in ozone concentration. This is a serious situation because stratospheric ozone absorbs much of the incoming solar ultraviolet (UV) radiation. As a UV shield, ozone helps to protect organisms on the earth's surface from some of the harmful effects of this high-energy radiation. If not interrupted, UV radiation could cause serious damage, such as disruption of genetic material, which could lead to increased rates of skin cancers and inheritable problems.

In recent times (mid-1980s), a serious problem with ozone depletion became apparent. A springtime decrease in the concentration of stratospheric ozone (ozone holes) has been observed at high latitudes, most notably over Antarctica between September and November. Scientists strongly suspect that chlorine atoms or simple chlorine compounds may play a key role in this ozone depletion problem (see Chapter 13).

On rare occasions it is possible for upper stratospheric ozone (good ozone) to enter the lower atmosphere (troposphere). Generally, this phenomenon only occurs during an event of great turbulence in the upper atmosphere. On rare incursions, atmospheric ozone reaches ground level for a short period of time. Most of the tropospheric ozone is formed and consumed by endogenous photochemical reactions, which are the result of the interaction of hydrocarbons, oxides of nitrogen, and sunlight, producing a yellowish-brown haze commonly called smog (Los Angeles-type smog).

Although the incursion of stratospheric ozone into the troposphere can cause smog formation, the actual formation of Los Angeles-type smog involves a complex group of photochemical interactions. These interactions are between anthropogenically emitted pollutants (NO and hydrocarbons) and secondarily produced chemicals (PAN, aldehydes, NO_2, and ozone). The concentrations of these chemicals exhibit a pronounced diurnal pattern, depending on their rate of emission and on the intensity of solar radiation and atmospheric stability at different times of the day (Freedman, 1989). The most important pollutant gases that contribute to LA-type smog and their diurnal pattern are illustrated in Figure 11.3.

If we look at Figure 11.3 and follow the time-line for the presence of various air pollutants in the atmosphere of Los Angeles, it is obvious that NO (emitted as NO_x) has a morning peak of concentration at 0600-0800, largely due to emissions from morning rush-hour vehicles. Hydrocarbons are emitted both from vehicles and refineries. They display a pattern similar to that of NO except that their peak concentration is slightly later. In bright sunlight the NO is photochemically oxidized to NO_2, resulting in a decrease in NO concentration and a peak of NO_2 at 0800-0900. Photochemical reactions involving NO_2 produce O atoms, which react with O_2 to form O_3. These result in a net decrease in NO_2 concentration and an increase in O_3 concentration, peaking between 1200-1500. Aldehydes (also formed photochemically) peak earlier than O_3. As

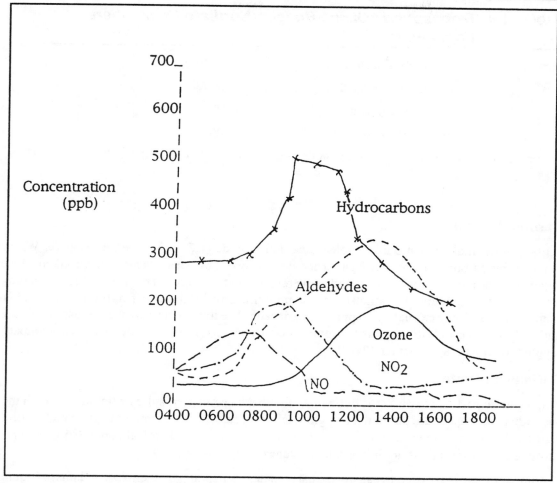

Figure 11.3: Average concentration of various air pollutants in the atmosphere of Los Angeles during days of eye irritation [adaptation from Haagen-Smit, A. J., and Wayne, L. G. "Atmospheric reactions and scavenging processes." Air Pollution, (A. C. Stern, ed.). 3rd ed., Vol. 1, pp. 235-288, New York: Academic Press, 1976.]

the day proceeds, the various gases decrease in concentration as they are diluted by fresh air masses, or are consumed by photochemical reactions. This cycle is typical of an area that experiences photochemical smog and is repeated daily (Urone, 1976).

A tropospheric ozone budget for the Northern Hemisphere is shown in Table 11.4. The considerable range of the estimates reflects uncertainty in the calculation of the ozone fluxes. On average, stratospheric incursions account for about 18% of the total ozone influx to the troposphere, while endogenous photochemical production accounts for the remaining 82%. About 31% of the tropospheric ozone is consumed by oxidative reactions at vegetative and inorganic suffocates at ground level, while the other 69% is consumed by photochemical reactions in the atmosphere (Freedman, 1989).

Table 11.4: Tropospheric Ozone Budget (Northern Hemisphere) (kg/ha-year)

Transport from stratosphere	13-20
Photochemical production	48-78
Destruction at ground	18-35
Photochemical destruction	48-55

Source: Adaptation from Hov, O. "Ozone in the troposphere: High level pollution." *Ambio* 13, 73-79, 1984.

Carbon Dioxide

Carbon-laden fuels, when burned, release carbon dioxide (CO_2) into the atmosphere. While much of this carbon dioxide is dissipated and then absorbed by ocean water, some is taken up by vegetation through photosynthesis, and some remains in the atmosphere. Today, the concentration of carbon dioxide in the atmosphere is approximately 350 parts per million (ppm) and is rising at a rate of approximately 20 ppm every decade. The increasing rate of combustion of coal and oil has been primarily responsible for this occurrence, which may eventually have an impact on global climate (see Chapter 13).

Particulate Matter

Atmospheric particulate matter is defined as any dispersed matter, solid or liquid, in which the individual aggregates are larger than single small molecules, but are smaller than about 500 μm. Particulate matter is extremely diverse and complex, since size, chemical composition, and atmospheric concentrations are important characteristics (Masters, 1991).

A number of terms are used to categorize particulates, depending on their size and phase (liquid or solid). These terms are listed and described in Table 11.5.

Dust, spray, forest fires, and the burning of certain types of fuels are among the sources of particulates in the atmosphere. Even with the implementation of stringent emission controls, which have worked to reduce particulates in the atmosphere, the U.S. Office of Technology Assessment (Postel, 1987) estimates that current levels of particulates and sulfates in ambient air may cause the premature death of 50,000 Americans every year.

Lead

Lead is emitted to the atmosphere primarily from human sources (burning leaded gasoline, for example) in the form of inorganic particulates. In high concentrations, lead can damage human health and the environment. Once lead enters an ecosystem, it remains there permanently. In humans and animals, lead can affect the neurological system and cause kidney disease. In plants, lead can inhibit respiration and photosynthesis as well as block the decomposition of microorganisms. Since the 1970s, stricter emission standards have caused a dramatic reduction in lead output.

Table 11.5: Atmospheric Particulates

Term	Description
Aerosol	General term for particles suspended in air
Mist	Aerosol consisting of liquid droplets
Dust	Aerosol consisting of solid particles that are blown into the air or are produced from larger particles by grinding them down
Smoke	Aerosol consisting of solid particles or a mixture of solid and liquid particles produced by chemical reactions such as fires
Fume	Generally the same as smoke, but often applies specifically to aerosols produced by condensation of hot vapors, especially of metals
Plume	Geometrical shape or form of the smoke coming out of a stack or chimney
Fog	Aerosol consisting of water droplets
Haze	Any aerosol other than fog that obscures the view through the atmosphere
Smog	Popular term originating in England to describe a mixture of smoke and fog; implies photochemical pollution

SUMMARY

Those who live in highly developed, heavily populated, wealthy, industrialized countries have to pay in many ways for their luxuries and high standard of living. On a personal level, two-income family groups allow people to buy what they want at the cost of suffering from the effects of stress and overwork. On a national scale, economically, we encourage more and more growth while industry alters our essential resources in harmful ways. Industry grows and advances; environmental science improves and advances, cleaning up industry's messes, constantly issuing warnings about what will happen to our earth if we don't think about what we are doing. Some of the problems that will face us in the future will be at the critical stage before our children reach adulthood. Atmospheric pollutants are the cause of several of those problems. We will all have to pay to solve them.

Cited References

Freedman, B., *Environmental Ecology*, New York: Academic Press, 1989.

Haagen-Smit, A. J. and Wayne, L. G., Atmospheric reactions and scavenging processes. *Air Pollution*, (A. C. Stern, ed.), 3rd ed., Vol. 1, New York: Academic Press, 1976.

Hov, O., Ozone in the Troposphere: High level pollution. *Ambio* 13, 73-79, 1984.

MacKenzie, J. J. and El-Ashry, T., *Ill Winds: Airborne Pollutant's Toll on Trees and Crops*, Washington, DC: World Resource Institute, 1988.

Manabe, B. and Stouffer, R., Century-Scale Effects of Increased Atmospheric CO2 on the Ocean-Atmosphere System, *Nature*, Vol. 364, No. 6434, pp. 215-218, 1993.

Postel, S., Stabilizing Chemical Cycles, Lester R. Brown (ed.), *State of the World*, New York: Norton, 1987.

Urone, P., The primary air pollutants—gaseous. Their occurrence, sources, and effects. *Air Pollution*, (A. C. Stern, ed.), Vol. 1, New York: Academic Press, 1976.

U.S. Environmental Protection Agency, *Environmental Progress & Challenges*, Washington, DC: Environmental Protection Agency, 1988a.

U.S. Environmental Protection Agency, *National Air Pollution Emission Estimates 1940-1986*, Washington, DC: Environmental Protection Agency, 1988.

U.S. Environmental Protection Agency, *National Air Pollution Emission Estimates*, Washington, DC: Environmental Protection Agency, 1989.

U.S. Environmental Protection Agency, *The CAA Amendments of 1990*, Washington, DC: Environmental Protection Agency, 1990.

Suggested Readings

Bridgman, H. A., *Global Air Pollution*, New York: John Wiley & Sons, 1994.

Colls, J., *Air Pollution*, London: Chapman & Hall, 1997.

Cooper, C. D. and Alley, *Air Pollution Control: A Design Approach*, Boston: PWS, 1986.

Dutsch, H. V., Ozone Research—Past-Present-Future. *Bulletin of the American Meteorological Society* 62:213-217, 1981.

Elsom, D. M., *Atmospheric Pollution—A Global Problem*, 2nd ed., Oxford, UK: Blackwell Publishers, 1992.

Flagan, R. C., and Seinfeld, J. H., *Fundamentals of Air Pollution Engineering*, Englewood Cliffs, NJ: Prentice Hall, 1988.

Hunton and Williams, L. B., *Clean Air Handbook*, Rockville, MD: Government Institutes, 1998.

Loiy, P. J., and Daisey, J. M., *Toxic Air Pollution*, Chelsea, Mich: Lewis Publishing, Inc., 1987.

National Research Council: *Causes and Effects of Stratospheric Ozone Reduction: An Update.* Washington, DC: National Academy Press, 1982.

Postel, S., *Air Pollution, Acid Rain, and the Future of the Forests*, Washington, DC: Worldwatch Institute, 1984.

Revelle, R., Carbon Dioxide and World Climate, *Scientific American* 247:35-43, 1982.

Stern, A. C., and Wohlers, H. C., *Fundamentals of Air Pollution*, New York: Academic Press, 1984.

Turco, R. P., *Earth Under Siege: Air Pollution and Global Change*, New York: Oxford University Press, 1996.

Waldbott, G. L., *Health Effects of Environmental Pollution*, St. Louis: C.V. Mosby, 1978.

Wark, K., and Warner, C.F., *Air Pollution, Its Origin and Control*, New York: Harper and Row, 1981.

Warneck, P., *Chemistry of the Natural Atmosphere*, San Diego, CA: Academic Press, 1988.

Williamson, S. J., *Fundamentals of Air Pollution*, Reading, MA: Addison-Wesley, 1973.

CHAPTER 12

ATMOSPHERIC AIR DISPERSION

GHOSTS OF THE MIDWEST

The winds come from the west, sluggish and foggy when the Pacific is at peace, the air up high moving steadily, relentlessly eastward, the lower draft butting against sea cliffs, drifting slowly over the mussel coves.

In day the mass of air is an off-white soup of sulfides and hydrocarbons, infused, over the metropolises, with filaments of russet brown.

At sunrise, in cities like Los Angeles, the hue is a resplendent purple that eventually makes its way east over the desertlands of Nevada and Arizona, settling as a haze in the Grand Canyon...before funneling through mountain passes or lofting into the turbulence above.

There the winds eddy and dip or wash back like an undertow.

But much of the air is high enough by now to have an expansive view of the heartland: the lakes, the quilted farms, the crushed stone of old mining territory, the scattered herbicide factories. It points east and north—like the weather vane.

No one is quite certain what all is in the wind from California, or for that matter in the currents from Oregon and Washington, from New Mexico or Colorado.

While the overall trend of air is eastward, it can move every which way, and in the heartland the wind comes not only from the west but more importantly up from the Gulf states. Mixing together, these western and southern air masses pick up the formulations of the Midwest and then proceed eastward.

The central part of America, then, while once thought to be pristine, is rather like a huge pot that accepts traces of what the west and south give, mixes in its own unmonitored concoctions, and overflows with its own nebulous broth.

Though the winds vary according to their height, and according to heat and terrain, they generally sweep across the entire American landscape. —Michael E. Brown

CHAPTER OUTLINE
- Introduction
- The Atmosphere and Meteorology
- Dispersion Models

INTRODUCTION

The air pollutants discussed in Chapter 11 are released from both stationary and mobile sources. Scientists have gathered much information on the sources, quantity, and toxicity levels of these pollutants. The measurement of air pollution is an important scientific skill, and practitioners of this skill usually have a good foundation in the pertinent related sciences and modeling aspects applicable to their studies and analyses of air pollutants in the ambient atmosphere. However, to get at the very heart of air pollution, the air pollution practitioner must also be well-versed in how to determine the origin of the pollutants, on the dispersal process in the atmosphere, on the impact on new sources, and the benefits of controls.

Air pollution practitioners must constantly deal with one fact: air pollutants rarely sit still at their release locations. Instead, various atmospheric wind flow conditions and turbulence, local topographic features, and other physical conditions work to disperse these pollutants. So, along with having a thorough knowledge and understanding of the pollutants in question, the air pollution practitioner has a definite need for detailed knowledge of the atmospheric processes that govern their subsequent dispersal and fate. Up to this point we have discussed wind and its formation (Chapter 10), which are important in air pollutant dispersion. In this chapter we discuss atmospheric dispersion of air pollutants in greater detail, and the factors associated with this phenomenon, including dispersion modeling.

THE ATMOSPHERE AND METEOROLOGY

In Chapter 10 we discussed the basics of meteorology, which set the stage for providing a comprehensive presentation of air pollution (Chapters 10 through 14). In this section we expand and build on the information provided in Chapter 10 to emphasize the natural association between our atmosphere, air pollution, and meteorology.

Weather

As we said in Chapter 10, the air contained in the earth's atmosphere is not still. Constantly in motion, air masses warmed by solar radiation rise at the equator and spread toward the colder poles, where they sink and flow downward, eventually returning to the equator. Near the earth's surface, as a result of the earth's rotation, major wind patterns develop. During the day, the land warms more quickly than the sea; at night, the land cools more quickly. Local wind patterns are driven by this differential warming and cooling between the land and adjacent water bodies. Normally, onshore breezes bring cooler, denser air from over the land masses out over the waters during the night. Precipitation is also affected by wind patterns. Warm, moisture-laden air rising from the oceans is carried inland, where the air masses eventually cool, causing the moisture to fall as rain, hail, sleet, or snow.

Even though pollutant emissions may remain relatively constant, air quality varies tremendously from day to day. The determining factors have to do with weather.

Weather conditions have a significant impact on air quality and air pollution, both favorable and unfavorable, especially on local conditions. For example, on hot, sun-filled days, when the weather is calm with stagnating high pressure cells, air quality suffers because these conditions allow the buildup of pollutants at the ground level. When local weather conditions include cool, windy, stormy weather with turbulent low pressure cells and cold fronts, these conditions allow the upward mixing and dispersal of air pollutants.

Weather has a direct impact on pollution levels in both mechanical and chemical ways. Mechanically, precipitation works to cleanse the air of pollutants (transferring the pollutants to rivers, streams, lakes, or the soil). Winds transport pollutants from one place to another. Winds and storms often dilute pollutants with cleaner air, making pollution levels less annoying in the area of their release. Air and its accompanying pollution (in a low pressure cell) are also carried aloft as the air is heated by the sun. When wind accompanies this rising air mass, the pollutants are diluted with fresh air. In a high pressure cell the opposite occurs, with air and the pollutants it carries sinking toward the ground. When there is no wind, these pollutants are trapped and concentrated near the ground, where serious air pollution episodes may occur.

Weather can also chemically affect pollution levels. Winds and turbulence mix pollutants together in a sort of giant chemical broth in the atmosphere. Energy from the sun, moisture in the clouds, and the proximity of highly reactive chemicals may cause chemical reactions, which lead to the formation of secondary pollutants. Many of these secondary pollutants may be more dangerous than the original pollutants.

In the preceding section, we discussed wind and its affect on dispersion of air pollutants; in the following section, we discuss other important factors that play a role in air contaminant dispersion.

Initially, dispersion of air pollutants from point or area sources depends on meteorological phenomena such as wind. Other factors (turbulence, adiabatic lapse rate, mixing height, topography, temperature inversions, and plume rise and transport) have affects on air contaminant dispersion as well.

Turbulence

In the atmosphere, the degree of *turbulence* (which results from wind speed and convective conditions related to the change of temperature with height above the earth's surface) is directly related to *stability* (a function of vertical distribution of atmospheric temperature). The stability of the atmosphere refers to the susceptibility of rising air parcels to vertical motion; consideration of atmospheric stability or instability is essential in establishing the dispersion rate of pollutants. When specifically discussing the stability of the atmosphere, we are referring to the lower boundary of the earth where air pollutants are emitted. The degree of turbulence in the atmosphere is usually classified by *stability class*. Ambient and adiabatic lapse rates are a measure of atmospheric stability.

Stability is divided into three classes: stable, unstable, and neutral. (1) A *stable atmosphere* is marked by air cooler at the ground than aloft, by low wind speeds, and consequently, by a low degree of turbulence. A plume of pollutants released into a stable lower layer of the atmosphere can remain relatively intact for long distances. Thus we can say that stable air discourages the dispersion and dilution of pollutants. (2) An *unstable atmosphere* is marked by a high degree of turbulence. A plume of pollutants released into an unstable atmosphere may exhibit a characteristic looping appearance produced by turbulent eddies. (3) A *neutrally stable atmosphere* is an intermediate class between stable and unstable conditions. A plume of pollutants released into a neutral stability condition is often characterized by a coning appearance because the edges of the plume spread out in a V-shape.

The importance of the "state of the atmosphere" and stability's effects cannot be overstated. Since the ease with which pollutants can disperse vertically into the atmosphere is mainly determined by the rate of change of air temperature with height (altitude), air stability is a primary factor in determining where pollutants will travel and how long they will remain aloft. Stable air discourages the dispersion and dilution of pollutants. Conversely, in unstable air conditions, rapid vertical mixing takes place, encouraging pollutant dispersal and increasing air quality.

Adiabatic Lapse Rate

With an increase in altitude in the troposphere, the temperature of the ambient air usually decreases. The rate of temperature change with height is called the *lapse rate*. On average, temperature decreases -0.65°C/100m or -6.5°C/km. This is the *normal lapse rate*.

In a dry environment, when a parcel of warm dry air is lifted in the atmosphere, it undergoes adiabatic expansion and cooling. This adiabatic cooling results in a lapse rate of -1°C/100m or 1-10°C/km, the *dry adiabatic lapse rate*.

When the ambient lapse rate exceeds the adiabatic lapse rate, the ambient rate is said to be *superadiabatic* and the atmosphere is highly unstable. When the two lapse rates are exactly equal, the atmosphere is said to be *neutral*. When the ambient lapse rate is less than the dry adiabatic lapse rate, the ambient lapse rate is termed *subadiabatic* and the atmosphere is stable (Peavy et al., 1985).

The cooling process within a rising parcel of air is assumed to be adiabatic (occurring without the addition or loss of heat). A rising parcel of air (under adiabatic conditions) behaves like a rising balloon, with the air in that distinct parcel expanding as it encounters air of lesser density until its own density is equal to that of the atmosphere that surrounds it. This process is assumed to occur with no heat exchange between the rising parcel and the ambient air.

Mixing

Within the atmosphere, for effective pollutant dispersal to occur, turbulent mixing is important. Turbulent mixing, the result of the movement of air in the vertical dimension, is enhanced by vertical temperature differences. The steeper the temperature gradient and the larger the vertical air column in which the mixing takes place, the more vigorous the convective and turbulent mixing of the atmosphere.

Topography

Near point and area sources (i.e., on a local scale with a geographical area encompassing less than 100 miles), *topography* may affect air motion. In the United States, most large urban centers are located along sea and lake coastal areas. Contained within these large urban centers is much heavy industry. Local airflow patterns in these urban centers have a significant impact on pollution dispersion processes. Topographic features also affect local weather patterns, especially in large urban centers located near lakes, seas, and open land. Breezes from these features affect vertical mixing and pollutant dispersal. Seasonal differences in heating and cooling land and water surfaces may also precipitate the formation of inversions near the sea or lakeshore.

River valley areas are also geographical locations that routinely suffer from industry-related pollution. Many early settlements began in river valleys because of the readily available water supply and the ease of transportation afforded to settlers by river systems within such valleys.

Along with settlers came industry—the type of industry that invariably produces air pollutants. These air pollutants, because of the terrain and physical configuration of the valley, are not easily removed from the valley.

Winds that move through a typical river valley are called *slope winds*. Slope winds, like water, flow downhill into the valley floor. At valley floor level, slope winds transform to *valley winds,* which flow down-valley with the flow of the river. Down-valley winds are lighter than slope winds. The valley floor becomes flooded with a large volume of air that intensifies the surface inversion that is normally produced by radiative cooling. As the inversion deepens over the course of the night, it often reaches its maximum depth just before sunrise with the height of the inversion layer dependent on the depth of the valley and the intensity of the radiative cooling process.

Hills and mountains can also affect local airflow. These natural topographical features tend to decrease wind speed (because of their surface roughness) and form physical barriers preventing the air movement.

Temperature Inversions

Temperature inversions (extreme cases of atmospheric stability) create a virtual lid on the upward movement of atmospheric pollution (see Chapter 10). Two types of inversions are important from an air quality standpoint: radiation and subsidence inversions.

Radiation inversions prompt the formation of fog, and simultaneously trap gases and particulates, creating a concentration of pollutants. They are characteristically a nocturnal phenomenon caused by the cooling of the earth's surface. On a cloudy night, the earth's radiant heat tends to be absorbed by water vapor in the atmosphere. Some of this is re-radiated back to the surface. However, on clear winter nights, the surface more readily radiates energy to the atmosphere and beyond, allowing the ground to cool more rapidly. The air in contact with the cooler ground also cools, and the air just above the ground becomes cooler than the air above it, creating an inversion close to the ground, lasting for only a matter of hours. These radiation inversions usually begin to form at the worst time of the day for human concerns in large urban areas—when early evening traffic begins to build up, trapping automobile exhaust at ground level and causing elevated concentrations of pollution for commuters. During evening hours, photochemical reactions cannot take place, so the biggest problem can be the accumulation of carbon monoxide. At sunrise, the sun warms the ground and the inversion begins to break up. Pollutants that have been trapped in the stable air mass are suddenly brought back to earth in a process known as *fumigation*. Fumigation can cause a short-lived, high concentration of pollution at ground level (Masters, 1991).

The second type of inversion is the *subsidence inversion* usually associated with a high pressure system. Known as *anticyclones*, they may significantly affect the dispersion of pollutants over large regions. A subsidence inversion is caused by the characteristic sinking motion of air in a high pressure cell. Air in the middle of a high pressure zone descends slowly. As the air descends, it is compressed and heated. It forms a blanket of warm air over the cooler air below, thus creating an inversion (located anywhere from several hundred meters above the surface to several thousand meters) that prevents further vertical movement of air.

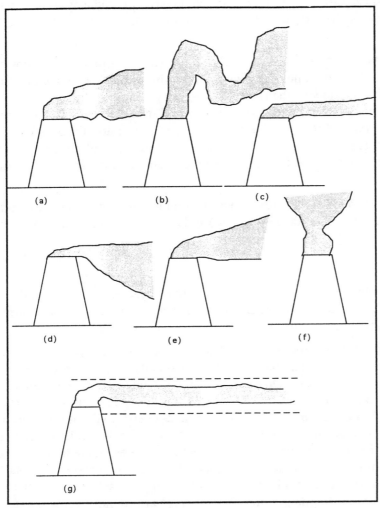

Figure 12.1 Seven types of classic plume behavior

Plume Rise

One way to quickly determine the stability of the lower atmosphere is to view the shape of a smoke trail or plume from a tall stack located on flat terrain (see Figure 12.1). Visible plumes usually consist of pollutants emitted from a smokestack into the atmosphere. The formation and fate of the plume itself depend on a number of related factors: (1) the nature of the pollutants, (2) meteorological factors, (3) source obstructions, and (4) local topography, especially downwind. Overall, maximum ground-level concentrations will occur in a range from the vicinity of the smokestack to some distance downwind.

Figure 12.1 shows the classic types of plume behavior. When the atmosphere is slightly stable, a typical plume "cones" as indicated in Figure 12.1a. When the atmosphere is highly unstable, a "looping" plume like the one shown in 12.1b forms. In the looping plume, the stream of emitted pollutants undergoes rapid mixing, and the wind causes large eddies that may carry the entire plume down to the ground, causing high concentrations close to the stack before dispersion is complete. In an extremely stable atmosphere, a "fanning" plume spreads horizontally (Figure 12.1c), with little mixing. When an inversion layer occurs a short distance above the plume source, the plume is said to be "fumigating" (Figure 12.1 d). When inversion conditions exist below the plume source, the plume is said to be "lofting" (Figure 12.1 e). When conditions are neutral, the plume issuing from a smokestack tends to rise directly into the atmosphere (Figure 12.1 f). When an inversion layer prevails both above and below the plume source, the plume issuing from a smokestack tends to be "trapped" (Figure 12.1 g).

But pollutant dispersion is rarely coming from a single point source (smokestack plume). In large urban areas, many plumes are generated and collectively combine into a large plume (city plume) whose dispersion represents a huge environmental challenge: the high pollutant concentrations from the city plume frequently affect human health and welfare.

Air quality problems associated with dispersion of city plumes are compounded by the presence of an already contaminated environment. Even though conventional processes that normally work to disperse emissions from point sources occur within the city plume, because of microclimates within the city, and the volume of pollutants they must handle, they are usually less effective. Other compounding conditions present in areas where city plumes are generated (topographical barriers, surface inversions, and stagnating anticyclones) work to intensify the city plume and result in high pollutant concentrations.

Transport

People living east of the Mississippi River might be surprised to find out that they are breathing air contaminated by pollutants from various sources many miles from their location. Most people view pollution as "out of sight, out of mind." As far as they are concerned, if they don't see it, it doesn't exist. For example, assume that a person on a farm in rural Arkansas heaps together a huge pile of assorted rubbish to be burned. The person preparing this huge bonfire probably gives little thought to the long-range transport of any contaminants that might be generated from that bonfire. This person simply has trash he or she no longer wants or needs, and an easy disposal solution seems to be to burn it.

This particular massive pile of rubbish contains various elements: discarded rubber tires, old compressed gas bottles, assorted plastic containers, paper, oils and greases, wood, and old paint cans. The person burning it doesn't consider this to be hazardous material—just household trash. When the pile of rubbish is ignited and allowed to burn, the fire-starter stands clear and watches as a huge plume of smoke forms and is carried away by a westerly flow of wind. The fire-starter looks far off to his/her right, downwind, and notices that the smoke disappears (or is out of view) just over the property line. Obviously, for our fire-starter, the dilution processes and the enormity of the atmosphere dissipated the smoke plume. The fire-starter doesn't give it a second thought. However, elevated levels of pollutants from many such fires may occur hundreds to thousands of miles downwind of the combination point sources producing such plumes. The result is that people living many miles from such pollution generators end up breathing contaminated air, transported over distance to their location.

DISPERSION MODELS

Air quality models are used to predict or describe the fate of airborne gases, particulate matter, and ground-level concentrations downwind of point sources. To determine the significance of air quality impact to a particular area, the first consideration is normal background concentrations, those pollutant concentrations from natural sources and/or distant, unidentified man-made sources. Every geographical area has a "signature" or background level of contamination considered an annual mean background concentration level of certain pollutants—an area, for example, might normally have a particulate matter reading of 30-40 $\mu g/m^3$. If particulate matter readings are significantly higher than the background level, this suggests an additional source. To establish background contaminations for a particular source under consideration, air quality data related to that site and its vicinity must be collected and analyzed.

The EPA recognized that in calculating the atmospheric dispersion of air pollutants, some means by which consistency could be maintained in air quality analysis had to be established. Thus, the EPA promulgated two guidebooks to assist in modeling for air quality analyses: *Guidelines on Air Quality Models (1978)* and *Industrial Source Complex (ISC) Dispersion Models User's Guide (1986)*.

In performing dispersion calculation, particularly for health effect studies, the EPA and other recognized experts in the field recommend following a four-step procedure:

1)   Estimate the rate, duration, and location of the release into the environment.

2)   Select the best available model to perform the calculations.

3)   Perform the calculations and generate downstream concentrations, including lines of constant concentration (isopleths) resulting from the source emission(s).

4)   Determine what effect, if any, the resulting discharge has on the environment, including humans, animals, vegetation, and materials of construction. These calculations often include estimates of the so-called vulnerability zones—that is, regions that may be adversely affected because of the emissions (Holmes et al., 1993).

Before beginning any dispersion determination activity, you must first determine the acceptable ground-level concentration of the waste pollutant(s). Local meteorological conduits and local topography must be considered, and having an accurate knowledge of the constituents of the waste gas and its chemical and physical properties is paramount (EPA, 1989).

Air quality models provide a relatively inexpensive means of determining compliance and predicting the degree of emission reduction necessary to attain ambient air-quality standards. Under the 1977 Clean Air Act Amendments, the use of models is required for the evaluation of permit applications associated with permissible increments under the so-called Prevention of Significant Deterioration (PSD) requirements, which require localities "to protect and enhance" air that is not contaminated (Godish, 1997).

Several dispersion models have been developed. Really equations, these models are mathematical descriptions of the meteorological transport and dispersion of air contaminants in a particular area and permit estimates of contaminant concentrations, either in plume from a ground-level or an elevated source (Carson and Moses, 1969). User-friendly modeling programs are now available that produce quick, accurate results from the operator's pertinent data.

This chapter's intent is not to develop each dispersion model in detail, but rather to recommend the one with the greatest applicability today. Probably the best atmospheric dispersion workbook for modeling published to date is by Turner (1970) for the EPA, and most of the air dispersion models used today are based on the *Pasquill-Gifford Model*.

SUMMARY

Breathing is automatic—so automatic that we cannot voluntarily stop it for a time long enough to hurt us. Because breathing is a largely unconscious action, most people don't pay attention to the air they breathe unless they sense something wrong with how it smells or feels. But as testing devices for pollutants, our noses are not very sensitive. Wishful thinking may allow us to believe that our air is okay and will remain so and that our atmosphere is self-cleaning and capable of absorbing the abuse we heap on it. We ignore the truth—that often the air we breathe contains elements we would be surprised and alarmed to know that we had inhaled. Those pollutants, carried by a variety of routes and methods, enter and affect our environments, our lives, and our bodies.

Cited References

Brown, M. H., *The Toxic Cloud*, New York: Harper & Row, 1987.

Carson, J. E. and Moses, H., "The Validity of Several Plume Rise Formulas," *J. Air Pol. Cont. Assoc.*, 19 (11): 862 (1969).

Godish, T., *Air Quality*, 3rd ed., Boca Raton, FL: Lewis Publishers, (1997).

Holmes, G., Singh, B. R., and Theodore, L., *Handbook of Environmental Management & Technology*, New York: John Wiley & Sons, 1993.

Masters, G. M., *Introduction to Environmental Engineering and Science*, Englewood Cliffs, NJ: Prentice-Hall, 1991.

Peavy, H. S., Rowe, D. R., and Tchobanglous, G., *Environmental Engineering*, New York: McGraw-Hill, Inc., 1985.

U.S. Environmental Protection Agency, *Workbook of Atmospheric Dispersion Estimates*, by D. B. Turner, Washington, DC: Environmental Protection Agency, 1970.

U.S. Environmental Protection Agency, *Guidelines on Air Quality Models*, Washington, DC: Environmental Protection Agency, 1978.

U.S. Environmental Protection Agency, *Industrial Source Complex (ISC) Dispersion Models User's Guide*, Washington, DC: Environmental Protection Agency, 1986.

Suggested Readings

Crawford, M., *Air Pollution Control Theory*, New York: McGraw-Hill, 1976.

Perkins, H. C., *Air Pollution*, New York: McGraw-Hill, 1974.

Scorer, R. S., *Meteorology of Air Pollution Implications for the Environment and Its Future*, New York: Ellis Horwood, 1990.

Seinfeld, J. H., *Air Pollution, Physical and Chemical Fundamentals*, New York: McGraw-Hill, 1975.

Stern, A. C. and Wohlers, H. C., *Fundamentals of Air Pollution*, New York: Academic Press, 1984.

Turner, D. B., *Workbook of Atmospheric Dispersion Estimates*, Washington, DC: Environmental Protection Agency, 1970.

U.S.C. 7401, *Clean Air Act*, (including the Clean Air Act Amendments of 1990, P. L. 101-549).

U.S. Environmental Protection Agency, *Guidelines on Air Quality Models (Revised)*, Washington, DC: Environmental Protection Agency, 1986.

CHAPTER 13

ATMOSPHERIC CHANGE: GLOBAL CLIMATE CHANGE

OZONE: THE JEKYLL AND HYDE OF CHEMICALS

In Robert Lewis Stevenson's classic horror novel, Dr. Jekyll and Mr. Hyde, *Jekyll and Hyde are different aspects of the same person. Dr. Jekyll's kind, compassionate character is countered by Mr. Hyde's evil, dispassionate nature. The chemical ozone has the same potential for good and evil within a single entity.*

Ozone (O_3) is a molecule containing three atoms of oxygen. In the earth's stratosphere, about 50,000 to 120,000 feet high, ozone molecules band together to form a protective layer that shields the earth from some of the sun's potentially destructive ultraviolet (UV) radiation. Stratospheric ozone (ozone in its kindly Dr. Jekyll incarnation), formed in the atmosphere by radiation from the sun, provides us with an enormously beneficial function. Life as we know it on earth could have evolved only with the protective ozone shield in place.

The Centers for Disease Control in Atlanta look at ozone more critically, however. They point out that ozone (in its evil Mr. Hyde form) is an extraordinarily dangerous pollutant. Only two-hundredths of a gram of ozone is a lethal dose. A single 14-ounce aerosol can filled with ozone could kill 14,000 people. Ozone is nearly as effective at destroying lung tissue as mustard gas. Not only is ozone a poisonous gas for us on earth, it is a main contributor to air pollution, especially smog.

CHAPTER OUTLINE

- Introduction
- Global Warming
- Acid Precipitation
- Photochemical Smog

INTRODUCTION

How serious a matter is environmental air pollution? Simply answered, you can literally bet your life on its seriousness. And that is exactly what we are doing—betting our lives. Environmental pollution transcends national boundaries and threatens the global ecosystem. And while we should not panic, we should be very concerned about our environmental problems.

We have altered our environment in dramatic fashion, especially over the past 200 hundred years. In this chapter, we focus on human activities that profoundly affect the environment. These activities are not secret or mysterious; in fact they are obvious, and most of us take part daily in some of these activities somehow. As Graedel and Crutzen (1989) put it, our activities are changing our atmosphere. To summarize these activities: (1) Our industrial activities emit a variety of atmospheric pollutants; (2) Our practice of burning large quantities of fossil fuel introduces pollutants into the atmosphere; (3) Our transportation practices emit pollutants into the atmosphere; (4) Our mismanagement and alteration of land surfaces (e.g. deforestation) lead to atmospheric problems; (5) Our practice of clearing and burning massive tracts of vegetation produces atmospheric contaminants; and (6) Our agricultural practices produce chemicals such as methane that impact the atmosphere. These man-made alterations to the earth's atmosphere have produced profound effects, including increased acid precipitation, localized smog events, greenhouse gases, ozone depletion, and increased corrosion of materials induced by atmospheric pollutants.

We need to understand the man-made mechanisms at work destroying our environment—what we are collectively doing to our environment—and we must be aware that our environment is finite. It is not inexhaustible or indestructible. We must also clearly identify and understand the causal as well as remedial factors involved. Recognizing one particular salient point is absolutely essential: Life on earth and the nature of the earth's atmosphere are connected, literally chained together. The atmosphere drives the earth's climate and ultimately determines its suitability for life. We must work to preserve the quality of our atmosphere.

Only through a cool-headed, scientific, intellectual, informed mindset will we be able to solve our environmental dilemma. To save our environment (and ourselves), we must develop an accomplishable vision of an environmentally healthy world. And it is something we can accomplish.

In this chapter, we discuss those issues relevant to environmental pollution of our atmosphere and air quality on earth. We will discuss global warming, acid precipitation, photochemical smog, and stratospheric ozone depletion.

GLOBAL WARMING

> *Humanity is conducting an unintended, uncontrolled, globally pervasive experiment whose ultimate consequences could be second only to nuclear war. The Earth's atmosphere is being changed at an unprecedented rate by pollutants resulting from human activities, inefficient and wasteful fossil fuel use and the effects of rapid population growth in many regions. These changes are already having harmful consequences over many parts of the globe. —Toronto Conference Statement, June 1988*

The preceding quotation clearly states the issue. But what is global warming? It is a long-term rise in the average temperature of the earth. Here's a second question, one many people use to question the validity of the concept of global warming as an environmental hazard: Is global warming actually occurring? The answer to this accompanying question is of enormous importance to all life on earth, and is the subject of intense debate throughout the globe.

For the sake of discussion, let's assume that it is. With this assumption in place, we must ask other questions, ones that deal with the why, how, and what: (1) Why is global warming occurring? (2) How can we be sure it is occurring? (3) What will be the ultimate effects? (4) What can and are we going to do about it? These questions are difficult to answer. The real danger is that we may not be able to definitively answer them before it is too late—when we've reached the point where the process has progressed beyond our power to effect prevention or mitigation.

This situation raises a huge red flag, but there is still time before it begins to wave in the climate change that would be inevitable without action.

Exactly what is the nature of the problem of global warming? We may not provide all the answers, but we are about to launch into a discussion of the entire phenomena and its potential impact on the earth.

The Greenhouse Effect

To help understand the earth's Greenhouse Effect, here's an explanation most people (especially gardeners) are familiar with. In a garden greenhouse, the glass walls and ceilings are largely transparent to short-wave radiation from the sun, which is absorbed by the surfaces and objects inside the greenhouse. Once absorbed, the radiation is transformed into long-wave (infrared) radiation (heat), which is radiated back from the interior of the greenhouse. But the glass does not allow the long-wave radiation to escape, instead absorbing it. With the heat trapped inside, the interior of the greenhouse becomes much warmer than the air outside.

With the earth and its atmosphere, much the same greenhouse effect takes place (see Figure 13.1). The short-wave and visible radiation that reaches earth is absorbed by the surface as heat. The long heat waves are then radiated back out toward space, but the atmosphere instead absorbs many of them. This is a natural and balanced process, and indeed is essential to our life on earth. The problem comes when changes in the atmosphere radically change the amount of absorption, and therefore the amount of heat retained. Scientists speculate that this may have been happening in recent decades as various air pollutants have caused the atmosphere to absorb more heat.

That this phenomenon takes place at the local level with air pollution, causing heat islands in and around urban centers, is not questioned. The main contributors to this effect are the greenhouse gases: water vapor, carbon dioxide, carbon monoxide, methane, volatile organic compounds (VOCs), nitrogen oxides, chlorofluoro-carbons (CFCs), and surface ozone. These gases delay the escape of infrared radiation from the earth into space, causing a general climatic warming. Note that scientists stress that this a natural process—indeed, the earth would be 33°C cooler than it is presently if the "normal" greenhouse effect did not exist (Hansen et al., 1986).

The problem with the earth's greenhouse effect is that human activities are now rapidly intensifying this natural phenomenon, which may lead to global warming. There is much debate, confusion, and

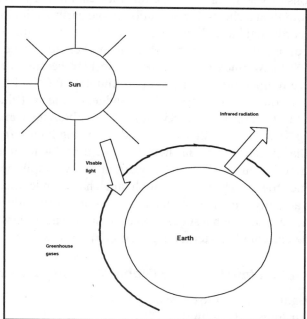

Figure 13.1 Greenhouse effect

Source: *Global Warming: Emission Reductions Possible as Scientific Uncertainties Are Resolved.* Government Accounting Office (1990).

speculation about the potential consequences. Scientists are not entirely sure of, nor do they agree about, whether the recently perceived worldwide warming trend is because of greenhouse gases, is due to some other cause, or is simply a wider variation in the normal heating and cooling trends they have been studying. But if it continues unchecked, the process may lead to significant global warming, with profound effects. Human impact on the Greenhouse Effect is real; it has been measured and detected. The rate at which the Greenhouse effect is intensifying is now more than five times what it was during the last century (Hansen et al., 1989).

The Greenhouse Effect and Global Warming

Those who support the theory of global warming base their assumptions on our alteration of the earth's normal greenhouse effect, which provides necessary warmth for life. They blame human activities (burning of fossil fuels, deforestation, and use of certain aerosols and refrigerants) for the increased quantities of greenhouses gases. These gases have increased the amounts of heat trapped in the earth's atmosphere, gradually increasing the temperature of the whole globe.

Many scientists note that (based on recent or short-term observation) the last decade has been the warmest since temperature recordings began in the late 19th century, and that the more general rise in temperature in the last century has coincided with the Industrial Revolution, with its accompanying increase in the use of fossil fuels. Other evidence supports the global warming theory. For example, in the Arctic and Antarctica, places that are synonymous with ice and snow, we see evidence of receding ice and snow cover.

Taking a long-term view, scientists look at temperature variations over thousands or even millions of years. Having done this they cannot definitively show that global warming is anything more than a short-term variation in the earth's climate. They base this assumption on historical records that have shown that the earth's temperature does vary widely, growing colder with ice ages and then warming again. On another side of the argument, some people point out that the 1980s saw nine of the twelve warmest temperatures ever recorded, and the earth's average surface temperature has risen approximately 0.6°C (1°F) in the last century (EPA, 1995). At the same time, still others offer as evidence the fact that the same decade also saw three of the coldest years: 1984, 1985, and 1986. Although we are somewhat uncertain about global warming, if we assume that we are indeed seeing long-term global warming then we must determine what is causing it. But here we face a problem. Scientists cannot be sure of the Greenhouse Effect's causes. The global warming may simply be part of a much longer trend of warming since the last ice age. Although much has been learned in the past two centuries of science, little is actually known about the causes of the worldwide global cooling and warming that have sent the earth through a succession of major ice ages, as well as smaller ones. We simply don't have the extreme long-term data to support our theories.

Factors Involved with Global Warming/Cooling

Right now, scientists are able to point to six factors that could be involved in long-term global warming and cooling:

1) Long-term global warming and cooling could result if changes in the earth's position relative to the sun were to occur (i.e., the earth's orbit around the sun), with higher temperatures occurring when the two are closer together and lower when they are further apart.

2)   Long-term global warming and cooling could result if major catastrophes were to occur (meteor impacts or massive volcanic eruptions) that throw pollutants into the atmosphere, blocking out solar radiation.

3)   Long-term global warming and cooling could result if changes in albedo (reflectivity of the earth's surface) were to occur. If the earth's surface were more reflective, for example, the amount of solar radiation radiated back toward space instead of absorbed would increase, lowering temperatures.

4)   Long-term global warming and cooling could result if the amount of radiation emitted by the sun were to change.

5)   Long-term global warming and cooling could result if the shape and relationship of the land and oceans were to change.

6)   Long-term global warming and cooling could result if the composition of the atmosphere were to change.

"If the composition of the atmosphere changes"—this possibility, of course, relates directly to our present concern: Have human activities had a cumulative impact large enough to affect the total temperature and climate of earth? Although we are somewhat concerned and alert to the problem, we are not certain. So what are we doing about global warming? We answer this question in the next section.

How Is Climate Change Measured?

Worldwide, scientists are trying to establish ways to test or measure whether or not greenhouse-induced global warming is occurring. Scientists are currently looking for signs that collectively are called a greenhouse "signature" or "footprint." If it is occurring, eventually it will be obvious to everyone—but what we really want is advance warning. Thus scientists are currently attempting to collect and decipher a mass of scientific evidence to find those signs that will give us clear advance warning. According to Franck and Brownstone (1992) these signs are currently believed to include changes in:

- *global temperature patterns*, with continents being warmer than oceans, lands near the Arctic warming more than the tropics, and the lower atmosphere warming while the higher stratosphere becomes cooler.
- *atmospheric water vapor*, with increasing amounts of water evaporating into the air as a result of the warming, more in the tropics than in the higher latitudes. Since water vapor is a "greenhouse gas," this would intensify the warming process.
- *sea surface temperature*, with a fairly uniform rise in the temperature of oceans at their surface and an increase in the temperature differences among oceans around the globe.
- *seasonality*, with changes in the relative intensity of the seasons, with the warming effects especially noticeable during the winter and in higher latitudes (p. 143).

In a measured, scientific way, these signs give a general overview of some of the changes that would be expected to occur with global warming. Note, however, that from a viewpoint of life on earth, changes resulting from long-term global warming would be drastic. Probably the most dramatic—and the effect with the most far-reaching results—would be sea level rise. This potential problem is discussed in the following section.

Global Warming and Sea Level Rise

In the past few decades, human activities (burning fossil fuels, leveling forests, and producing synthetic chemicals such as CFCs) have released into the atmosphere huge quantities of carbon dioxide and other greenhouse gases. These gases are warming the earth at an unprecedented rate. If current trends continue, they are expected to raise the earth's average surface temperature by at least 1.5°C to 4.5°C in the next century, with warming at the poles perhaps two to three times as high as warming at the middle latitudes (Wigley et al., 1986).

If we assume global warming is inevitable and/or is already underway, what must we do? Obviously we cannot jump off the planet and head for greener pastures. It therefore makes good sense to understand the dynamics of change that are evolving around us and to take whatever prudent actions we can to mitigate the situation. This is the attitude and approach we must take with the effect global warming is having on rise in sea level. This rise is already underway, and with it will come increased storm damage, pollution, and subsidence of coastal lands.

Consider the following information taken from the EPA's 1995 report, *The Probability of Sea Level Rise.*

1) *Global warming is most likely to raise sea level 15cm by the year 2050 and 34 cm by the year 2100. There is also a 10 percent chance that climate change will contribute 30 cm by 2050 and 65 cm by 2100. These estimates do not include sea level rise caused by factors other than greenhouse warming.*

2) *There is a 1 percent chance that global warming will raise sea level 1 meter in the next 100 years and 4 meters in the next 200 years. By the year 2200, there is also a 10 percent chance of a 2-meter contribution. Such a large rise in sea level could occur either if Antarctic ocean temperature warms 5°C and Antarctic ice streams respond more rapidly than most glaciologists expect, or if Greenland temperatures warm by more than 10 °C. Neither of these scenarios is likely.*

3) *By the year 2100, climate change is likely to increase the rate of sea level rise by 4.1 mm/yr. There is also a 1-in-10 chance that the contribution will be greater than 10 mm/yr, as well as a 1-in-10 chance that it will be less than 1 mm/yr.*

4) *Stabilizing global emissions in the year 2050 would be likely to reduce the rate of sea level rise by 28 percent by the year 2100, compared with what it would be otherwise. These calculations assume that we are uncertain about the future trajectory of greenhouse gas emissions.*

5) *Stabilizing emissions by the year 2025 could cut the rate of sea level rise in half. If a high global rate of emissions growth occurs in the next century, sea level is likely to rise 6.2 mm/yr by 2100; freezing emissions in 2025 would prevent the rate from exceeding 3.2 mm/yr. If less emissions growth were expected, freezing emissions in 2025 would cut the eventual rate of sea level rise by one-third.*

6) *Along most coasts, factors other than anthropogenic climate change will cause the sea to rise more than the rise resulting from climate change alone. These factors include compaction and subsidence of land, groundwater depletion, and natural climate variations. If these factors do not change, global sea level is likely to rise 45 cm by the year 2100, with a 1 percent chance of a 112 cm rise. Along the coast of New York, which typifies the United States, sea level is likely to rise 26 cm by 2050 and 55 cm by 2100. There is also a 1 percent chance of a 55 cm rise by 2050 and a 120 cm rise by 2100.*

Along with the EPA's findings reported above, additional lines of evidence corroborate that global mean sea level has been rising during at least the last 100 years. According to Broecker (1987), this evidence is apparent in tide gauge records, erosion of 70% of the world's sandy coasts and 90% of America's sandy beaches, and the melting and retreat of mountain glaciers. Edgerton (1991) points out that the correspondence between the two curves of rising global temperatures and rising sea levels during the last century appears to be more than coincidental.

There are major uncertainties in estimates of future sea level rise. The problem is further complicated by our lack of understanding of the mechanisms contributing to relatively recent rises in sea level. In addition, different outlooks for climatic warming dramatically affect estimates. In all this uncertainty, one thing is sure: estimates of sea level rise will undergo continual revision and refinement as time passes and more data is collected.

Major Physical Effects of Sea Level Rise

With increased global temperatures, global sea level rise will occur at a rate unprecedented in human history (Edgerton, 1991). Changes in temperature and sea level will be accompanied by changes in salinity levels. For example, a coastal freshwater aquifer is influenced by two factors: pumping and mean sea level. In pumping, if withdrawals exceed recharge, the water table is drawn down and saltwater penetrates inland. With mean sea level, the problem occurs if sea level rises and the coastline moves inland, reducing aquifer area. Additional problems brought about by changes in temperature and sea level are seen in tidal flooding, oceanic currents, biological processes of marine creatures, runoff and landmass erosion patterns, and saltwater intrusion.

To understand the significance, refer to Figure 13.2. You can easily see one of the most important direct physical effects of sea level rise on a coastal beach system. At current rates of sea level rise of 1 to 2 mm/year, significant coastal erosion is already produced. Two major factors contribute to beach erosion. First, deeper coastal waters enhance wave generation, thus increasing their potential for overtopping barrier islands. Second, shorelines and beaches will attempt to establish new equilibrium positions according to what is known as the *Bruun rule*; these adjustments will include a recession of shoreline and a decrease in shore slope (Bruun, 1962, 1986). Figure 13.2 shows how the Bruun rule works.

Major Direct Human Effects of Sea Level Rise

Along with the physical effects of sea level rise, in one way or another, directly or indirectly, accompanying effects have a direct human side, especially regarding human settlements. Human settlements include the infrastructure that accompanies them: highways, airports, waterways, water supply and wastewater treatment facilities, landfills, hazardous waste storage areas, bridges, and associated maintenance systems, and they cause intrusion of saltwater into groundwater supplies (Edgerton, 1991). To point out that this infrastructure will be placed under tremendous strain by a rising sea level coupled with other climatic change is to understate the possible consequences. Indeed, the impact on infrastructure is only part of the direct human impact. For example, there is widespread agreement among scientists that any significant change in world climate resulting from warming or cooling will (1) disrupt world food production for many years, (2) lead to a sharp increase in food prices, and (3) cause considerable economic damage.

Just how much of a rise in sea level are we talking about? According to the EPA (1995), "if the experts on whom we relied fairly represent the breadth of scientific opinion, the odds are fifty-fifty that greenhouse gases will raise sea level at least 15 cm by the year 2050, 25 cm by 2100, and 80 cm by 2200" (p. 123).

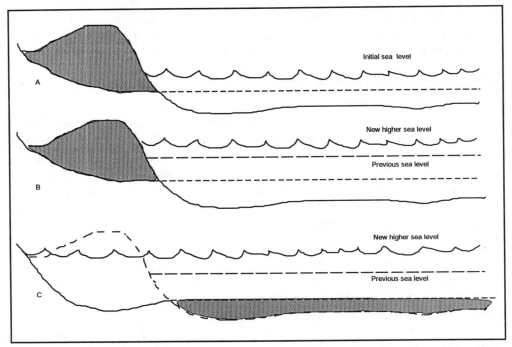

Figure 13.2 The Bruun Rule. (A) initial condition; (B) immediate inundation when sea level rises; (C) subsequent erosion due to sea level rise. A rise in sea level immediately results in shoreline retreat from indundation. A 1-meter rise in sea level also implies that the offshore bottom must also rise 1 meter. Waves will erode the necessary sand from the upper part of the beach [as shown in part (C)]. Adapted from Barth and Titus (1984).

ACID PRECIPITATION

In the evening, when you stand on your porch and look out on your terraced lawn and flourishing garden of perennials during a light rainfall, you probably feel a sense of calm and relaxation. It's probably the sound of raindrops against the roof of the house and porch, against the foliage and lawn, the sidewalk, the street, and the light wind through the boughs of the evergreens that soothes you. Whatever it is that makes you feel this way, rainfall is a major ingredient.

But someone knowledgeable and/or trained in environmental science might take another view of such a seemingly welcome and peaceful event. He or she might wonder whether the rainfall is as clean and pure as it should be. Is this actually rainfall, or is it rain carrying acids as strong as lemon juice or vinegar with it, capable of harming both living and nonliving things like trees, lakes, and manmade structures?

Maybe such a question was unknown before the Industrial Revolution, but today the purity of rainfall is a major concern for many people, especially regarding acidity. Most rainfall is slightly acidic because of decomposing organic matter, the movement of the sea, and volcanic eruptions, but the principal factor is atmospheric carbon dioxide, which causes carbonic acid to form. Acid rain (pH <5.6) (in the pollution sense) is produced by the conversion of the primary pollutants sulfur dioxide and nitrogen oxides to sulfuric acid and nitric acid, respectively (see Figure 13.3 for an explanation of pH). These processes are complex, depending on the physical dispersion processes and the rates of the chemical conversions. The basic cycle is shown in Figure 13.4.

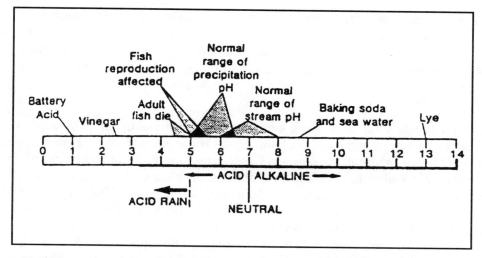

Figure 13.3 Measuring activity: pH scale (source: Water Fact Sheet, *U.S. Geological Society, U.S. Dept. of Interior, 1987)*

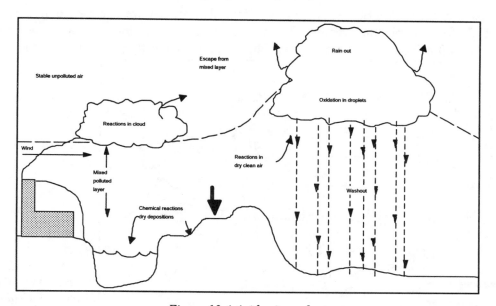

Figure 13.4 Acid rain cycle

Contrary to popular belief, acid rain is not a new phenomenon, nor does it result solely from industrial pollution. Natural processes such as volcanic eruptions and forest fires produce and release acid particles into the air. The burning of forest areas to clear land in Brazil, Africa, and elsewhere also contributes to acid rain. However, the rise in manufacturing that began with the Industrial Revolution literally dwarfs all other contributions to the problem.

The main culprits are emissions of sulfur dioxide from the burning of fossil fuels such as oil, coal, and nitrogen oxide, formed mostly from internal combustion engine emissions, which is readily transformed into nitrogen dioxide. These mix in the atmosphere to form sulfuric acid and nitric acid.

In dealing with atmospheric acid deposition, the earth's ecosystems are not completely defense-less; they can deal with a certain amount of acid through natural alkaline substances in soil or rocks that buffer and neutralize acid. The American Midwest and southern England are areas with highly alkaline soil (limestone and sandstone) that provides some natural neutralization. Those areas with thin soil and those laid on granite bedrock, however, have little ability to neutralize acid rain.

Scientists continue to study how living beings are damaged and/or killed by acid rain. This complex subject has many variables. We know from various episodes of acid rain that pollution can travel over very long distances. Lakes in Canada and New York are feeling the effects coal-burning in the Ohio Valley. For this and other reasons, the lakes of the world are where most of the scientific studies have taken place. In lakes, the smaller organisms often die off first, leaving the larger animals to starve. Sometimes the larger animals (fish) are killed directly: as lake water becomes more acidic, it dissolves heavy metals, leading to concentrations at toxic and often lethal levels. If you have ever observed thousands of fish belly-up at a local lake you know it is not a pleasant sight or smell. But loss of life in lakes also disrupts the system of life on the land and in the air around them.

In some parts of the United States, the acidity of rainfall has fallen well below 5.6. In the north-eastern U.S., for example, the average pH of rainfall is 4.6, and it is not unusual to have rainfall with a pH of 4.0, which is 100 times more acidic than distilled water (see Figure 13.5).

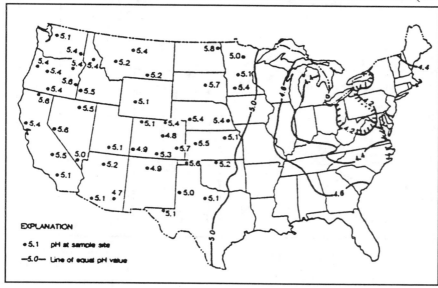

Figure 13.5 Profile of acid rain deposition in the U.S. (source: Water Fact Sheet, U.S. Geological Survey, Dept. of Interior, 1987)

Despite intensive research into most aspects of acid rain, scientists still have many areas of uncertainty and disagreement. That is why the progressive, forward-thinking countries emphasize the importance of further research into acid rain.

PHOTOCHEMICAL SMOG

When various hydrocarbons, oxides of nitrogen, and sunlight come together, they can initiate a complex set of reactions that produce a number of secondary pollutants known as photochemical oxidants or photochemical smog. Photochemical smog (discussed earlier in Chapter 11) of the type most people are familiar with was first noticed in Los Angeles in the early 1940s. Determining its true cause has taken many years. According to Black-Covilli (1992), at first it was thought to arise from dust and smoke emitted from factories and incinerators. Accordingly, Los Angles County officials issued a ban on all outdoor burning of trash and initiated steps toward control of industrial smoke emission. Before long, though, county authorities determined that their initial efforts were not working; the smog continued unabated. Then they went after another suspected culprit, sulfur dioxide (SO_2) given off by oil refineries and by combustion of sulfur-bearing coal. They placed controls on sulfur dioxide emissions, but still gained no benefit.

The biochemist Dr. Arie Haagen-Smit, as a result of chance, during research aimed at finding the compounds responsible for the pleasant tastes and odors of fruit, found the cause of the smog problem, which showed beyond a doubt that the internal combustion engine was the principal source.

How do internal combustion engines produce smog? A few of the finer details are yet unclear, but the following, in simplified form, appears to be what happens. It begins with the high temperatures in the internal combustion engine, which cause atmospheric oxygen and nitrogen to react, producing nitric oxide ($N_2 + O_2 \rightarrow 2NO$). At the same time, varying quantities of fuel in the engine fail to burn completely. This results in a mixture of aldehydes, ketones, olefins, and aromatic hydrocarbons that is expelled from the exhaust. The exhaust enters the atmosphere, where ultraviolet radiation from the sun causes a complex series of reactions to take place. These reactions involve atmospheric oxygen, nitric oxide, and organic compounds. The result is that nitrogen dioxide (NO_2) and ozone (O_3) are formed, both of which are highly toxic and irritating. In addition, this reaction also causes the formation of other constituents of photochemical smog, including formaldehyde, peroxybenzohl nitrate, peroxyacetyl nitrate (PAN), and acrolein.

Photochemical smog is known to cause many annoying respiratory effects: coughing, shortness of breath, airway constriction, headache, chest tightness, and eye, nose, and throat irritation (Masters, 1991).

STRATOSPHERIC OZONE DEPLETION

Ozone (discussed earlier in Chapter 11) is formed in the stratosphere by radiation from the sun, and helps to shield life on earth from some of the sun's potentially destructive ultraviolet (UV) radiation.

In the early 1970s, scientists suspected that the ozone layer was being depleted. By the 1980s it became clear that the ozone shield was indeed thinning in some places, and that at times it even has a seasonal hole in it, notably over Antarctica. The exact causes and actual extent of the depletion are not yet fully known, but most scientists believe that various chemicals in the air are responsible.

Most scientists identify the family of chlorine-based compounds, most notably chlorofluorocarbons (CFCs) and chlorinated solvents (carbon tetrachloride and methyl chloroform), as the primary culprits involved in ozone depletion. In 1974, Molina and Rowland hypothesized the CFCs containing chlorine were responsible for ozone depletion. They pointed out that chlorine molecules are highly active and readily and continually break apart the three-atom ozone into the two-atom form of oxygen generally found close to earth, in the lower atmosphere.

The Interdepartmental Committee for Atmospheric Sciences (1975) estimates that a 5% reduction in ozone could result in nearly a 10% increase in cancer. This already frightening scenario was made even more frightening in 1987 when evidence showed that CFCs destroy ozone in the stratosphere above Antarctica every spring. The ozone hole had become larger, with more than half of the total ozone column wiped out and essentially all ozone gone from some regions of the stratosphere (Davis and Cornwell, 1991).

In 1988, Zurer reported that on a worldwide basis, the ozone layer shrank approximately 2.5% in the preceding decade. This obvious thinning of the ozone layer, with its increased chances of skin cancer and cataracts, is also implicated in suppression of the human immune system and damage to other animals and plants, especially aquatic life and soybean crops. The urgency of the problem spurred the 1987 signing of the *Montreal Protocol* by 24 countries, which required signatory countries to reduce their consumption of CFCs 20% by 1993, and 50% by 1998, marking a significant achievement in solving a global environmental problem.

SUMMARY

Little doubt exists that we are putting our environment, and thus ourselves, at serious risk. While sometimes we are not certain of the exact causes, these changes we observe in our world are measurable. As active members of an increasingly global economy, we increase our risk by pretending it will go away. Sooner or later we will have to face these problems—and technology will be how we solve them.

Cited References

Black-Covilli, L. L., Basic Air Quality, in *Fundamentals of Environmental Science and Technology*, Porter-C. Knowles, ed., Rockville, Maryland: Government Institutes, Inc., 1992.

Broecker, W., "Unpleasant Surprises in the Greenhouse?" *Nature* 328:123-126, 1987.

Bruun, P., Sea Level Rise as a Cause of Shore Erosion, *Proceedings of the American Society of Engineers and Journal Waterways Harbors Division* 88: 117-130, 1962.

Bruun, P., Worldwide Impacts of Sea Level Rise on Shorelines, *Effects of Changes in Stratospheric Ozone and Global Climate*, vol. 4, New York: UNEP/EPA, pp. 99-128, 1986

Davis, M. L. and Cornwell, D. A., *Introduction to Environmental Engineering*, New York: McGraw-Hill, Inc., 1991.

Edgerton, L., *The Rising Tide: Global Warming and World Sea Levels*, Washington, DC: Island Press, 1991.

Franck, I. and Brownstone, D., *The Green Encyclopedia*, New York: Prentice-Hall, 1992.

Graedel, T. E., and Crutzen, P. J., The Changing Atmosphere, *Scientific American*, pp. 58-68, September, 1989.

Hansen, J. E. et al., "Climate Sensitivity to Increasing Greenhouses Gases," *Greenhouse Effect and Sea Level Rise: A Challenge for This Generation*, ed., M. C. Barth & J. G. Titus, New York: Van Nostrand Reinhold, 1986.

Hansen, J. E., et al., "Greenhouse Effect of Chloroflurocarbons and Other Trace Gases," *Journal of Geophysical Research* 94 November pp. 16,417-16,421, 1989.

Interdependent Committee for Atmospheric Sciences, *The Possible Impact of Fluorocarbons and Hydrocarbons on Ozone* Washington, DC: U.S. Government Printing Office, p. 3., May 1975.

Masters, G. M., *Introduction to Environmental Engineering and Science*, Englewood Cliffs, NJ: Prentice-Hall, 1991.

Molina, M. J. and Rowland, F. S., Stratospheric Sink for Chlorofluoromethanes: Chlorine Atom Catalyzed Destruction of Ozone, *Nature*, vol. 248, pp. 810-812, 1974.

U.S. Environmental Protection Agency *The Probability of Sea Level Rise*, Washington, DC: Environmental Protection Agency, 1995.

Wigley, T. M., Jones, P. D., and Kelly P. M., Empirical Climate Studies: Warm World Scenarios and the Detection of Climatic Change Induced by Radioactively Active Gases, *The Greenhouse Effect, Climatic Change, and Ecosystems*, ed. B. Bolin et al., New York: Wiley, 1986.

Zurer, P. S., Studies on Ozone Destruction Expand Beyond Antarctic, *C & E News*, pp. 18-25, May 1988.

Suggested Readings

Adams, D. D. and Page, W. P. (eds.), *Acidic Deposition—Environmental Economic and Policy Issues*, New York: Plenum Publishers, 1985

Armentrout, P., *The Ozone Layer*, New York: Rourke Publishing Group, 1997.

Bridgman, H. A., *Global Air Pollution—Problems in the 1990s*, New York: John Wiley & Sons, 1994.

Brown, P., *Global Warming: Can Civilization Survive?*, New York: Blandford Press, 1997.

Crawford, M., *Air Pollution Control Theory*, New York: McGraw-Hill, 1976.

Dimitriades, B., and Whisman, M., Carbon Monoxide in Lower Atmosphere Reactions, *Environmental Science and Technology*, 5:213, 1971.

Lyman, F., *The Greenhouse Trap*, Boston, MA: Beacon Press, 1990.

Magill, P. L., Holden, F. R., and Ackley, *Air Pollution Handbook*, New York: McGraw-Hill, 1956.

Manabe, S., Wetherald, R. T., and Stauffer, R. J., Summer dryness due to an increase in atmospheric CO_2 concentration, *Climate Changer*, 3, 347-386, 1981.

Manabe, S., and Wetherald, R. T., "Reduction in summer soil wetness induced by an increase in atmospheric carbon dioxide," *Science*, 232, 626-628, 1986.

Manabe, S., and Wetherald, R. T., "Large-scale changes of soil wetness induced by an increase in atmospheric carbon dioxide," *J. Atmos. Sci.*, 44, 1211-1235, 1987.

National Research Council, *Rethinking the Ozone Problem in Urban and Regional Air Pollution*, Washington: DC, National Academy Press, 1991.

Perkins, H. C., *Air Pollution*, New York: McGraw-Hill, 1974.

Samuel, J. H., *Global Warming and the Built Environment*, London, UK: Chapman & Hall, 1996.

Schneider, S. H., "Detecting Climate Change Signals: Are There Any 'Fingerprints'?" *Science*, 263:341-347, 1994.

Seinfeld, J. H., *Air Pollution, Physical and Chemical Fundamentals*, New York: McGraw-Hill, 1975.

Wark, K., and Warner, C. F., *Air Pollution, Its Origin and Control*, 2nd ed., New York: Harper & Row, 1981.

CHAPTER 14

AIR POLLUTION CONTROL TECHNOLOGY

The application of control technology to air pollution problems assumes that a source can be reduced to a predetermined level to meet a regulation of some unknown minimum value. Control technology cannot be applied to an uncontrollable source, such as a volcano, nor can it be expected to control a source completely to reduce emissions to zero. The cost of controlling any given air pollution source is usually an exponential function of the percentage of control and therefore becomes an important consideration in the level of control equipment .—Boubel et al.

CHAPTER OUTLINE

- Introduction
- Air Pollution Control: Choices
- Air Pollution Control Equipment and Systems
- Removal of Dry Particulate Matter
- Removal of Gaseous Pollutants: Stationary Sources
- Removal of Gaseous Pollutants: Mobile Sources

INTRODUCTION

Chapters 9 through 13 set the foundation for the discussion presented in this chapter. Now that you have a clear picture of the problems that air pollution control is trying to solve, the time has come to examine the measures used to control it.

Two important factors related to the topic are presented in the opening sentence of this chapter's introductory quote: control technology and regulation. Neither is more important than the other—in fact, in many ways they drive each other.

Air pollution control begins with regulation. Regulations (for example, to clean up, reduce, or eliminate a pollutant emission source) are generated because of certain community concerns. Buonicore et al. (1992) point out that regulations usually evolve around three considerations:

1)   Legal limitations imposed for the protection of public health and welfare.

2)   Social limitations imposed by the community in which the pollution source is or will be located.

3)   Economic limitations imposed by marketplace constraints (p.1).

The goal an engineer assigned to mitigate an air pollution problem sets out meet is to ensure that the design control methodology used will bring the source into full compliance with applicable regulations. To accomplish this feat, environmental engineers must first understand the problem(s), then rely heavily on technology to correct the situation. Various air pollution control technologies are available to environmental engineers or scientists working to mitigate air pollution source problems. By analyzing the problem carefully and applying the most effective method for the situation, the engineer or scientist can ensure that a particular pollution source is brought under control and the responsible parties are in full compliance with regulations.

In this chapter we discuss the various air pollution control technologies available to the environmental scientist/engineer in mitigating air pollution source problems. In doing so, the environmental practitioner will be able to ensure that a particular pollutant source is brought under control and that the responsible parties are in full compliance with applicable regulations.

AIR POLLUTION CONTROL: CHOICES

Assuming that the design engineer has a complete knowledge of the contaminant and the source, all available physical and chemical data on the effluent from the source, and the regulations of the control agencies involved, he or she must then decide which control methodology to employ. Since only a few control methods exist, the choice is limited. Control of atmospheric emissions from a process will generally consist of one of four methods, depending on the process, types, fuels, availability of control equipment, etc. The four general control methods are (1) elimination of the process entirely or in part, (2) modification of the operation to a fuel which will give the desired level of emission, (3) installation of control equipment between the pollutant source and the receptor, and (4) relocation of the operation.

Because of tremendous costs involved with eliminating or relocating a complete process (which makes either of these options the choice of last resort), let's take a look at the first and last control methods first. Eliminating a process is no easy undertaking, especially it is the process for which the facility exists. Relocation is not always an answer, either. Consider the real-life situation presented in Case Study 14.1.

Case Study 14.1
Cedar Creek Composting Facility

Cedar Creek Composting (CCC) Facility was built in 1970. A 44-acre site designed to receive and process to compost wastewater biosolids from six local wastewater treatment plants, CCC composted biosolids at the rate of 17.5 dry tons per day. CCC employed the aerated static pile (ASP) method to produce pathogen-free, humus-like material that could be beneficially used as an organic soil amendment. The final compost product was successfully marketed under a registered trademark name.

Today, Cedar Creek Composting Facility is no longer in operation. The site was shut down in early 1997. From an economic point of view, CCC was highly successful. When a fresh pile of compost had completed the entire composting process, dumptruck after dumptruck would line the street outside the main gate, waiting in hope of buying loads of the popular product. Economics was not the problem. In fact, CCC could not produce enough compost fast enough to satisfy the demand.

If economics was not the problem, what was? There are two parts to this answer. The first problem was social limitations imposed by the community where the compost site was located.

In 1970, the 44 acres CCC occupied were located in an out-of-town, rural area. CCC's only neighbor was a regional one-plane airport on its eastern border. CCC was completely surrounded by woods on the other three sides. The nearest town was two miles away. But by the mid-1970s, things started to change. Population growth and its accompanying urban sprawl quickly turned forested lands into housing complexes and shopping centers. CCC's western border soon became the site of a two-lane road that was upgraded to four and then to six lanes. CCC's northern fence separated it from a large shopping mall. On the southern end of the facility, acres of houses, playgrounds, swimming pools, tennis courts, and a golf course were built. CCC became an island surrounded by urban growth. Further complicating the situation was the airport; it expanded to the point that by 1985, three major airlines used the facility.

CCC's ASP composting process was not a problem before the neighbors moved in and started to complain. The first complaints came from the airport. The airport complained that dust from the static piles of compost was interfering with air traffic control.

The new, expanded highway brought several thousand new commuters right up alongside CCC's western fenceline. Commuters started complaining whenever the compost process was in operation; they complained primarily about the odor—a thick, earthy smell that permeated everything. A handful of commuters actually liked the smell; they thought it comforting, bringing them thoughts of gardens, farms and nature. But a handful of admirers does not offset a stadium full of complainers.

After an enormous housing project was completed and people took up residence there, complaints were raised on a daily basis. The new homeowners complained about the earthy odor and the dust that blew from the compost piles onto their properties whenever the wind blew from the direction of the site.

Then there were the shoppers. They drove into the mall, got out of their cars, breathing normally, got a hit of that heavy, earthy air and gagged. "What is that horrendous smell?" they asked. When they found out what it was and where it came from, the phones at City Hall rang off the hook.

City Hall received several thousand complaints over the first few months before they took any action. The city environmental engineer was told to approach CCC's management and see if some resolution of the problems could be effected. CCC management listened to the engineer's concerns but stated that there wasn't a whole lot that the site could do to rectify the problem.

As you might imagine, this was not the answer the city leaders were hoping to get. Feeling the increasing pressure from local inhabitants, commuters, shoppers, and airport management people, the city brought the state representatives into the situation.

The two state representatives for the area immediately began a campaign to close down the CCC facility. But CCC was not powerless in this struggle. CCC had been there first, and the developers and the people in those new houses had not been forced to buy land right next to the facility. Besides, CCC had the EPA on its side. CCC was taking a waste product no one wanted, one that traditionally ended up in the local landfill (taking up valuable space), and was turning it into a beneficial reuse product. CCC was helping conserve and protect the local environment. Can there be a more noble goal?

The city leaders didn't really care about noble goals, but they did care about the concerns of their constituents—the voters. They continued their assault through the press, electronic media, legislatively, and by any other means they could bring to bear.

CCC management understood the problem and felt the pressure. They had to do something, and they did. Their environmental engineering division was assigned the task of coming up with a plan to mitigate not only CCC's odor problem, but also its dust problem. After several months of research and a pilot study, CCC's environmental engineering staff came up with a solution. The solution included enclosing the entire facility within a self-contained structure. The structure would be equipped with a state-of-the-art ventilation system and two-stage odor scrubbers. The engineers estimated that the odor problem could be reduced by 90% and the dust problem reduced by 98.99%. CCC management thought they had a viable solution to the problem and they were willing to spend the 5.2 million dollars to retrofit the plant.

After CCC presented their mitigation plan to the city counsel, the counsel members made no comment, but said that they needed time to study the plan.

Three weeks later, CCC received a letter from the mayor stating that CCC's efforts to come up with a plan to mitigate the odor and dust problems were commendable and to be applauded, but were unacceptable.

From the mayor's letter, CCC could see that the focus of attack had now changed from a social to a legal issue. The mayor pointed out that he and the city leaders had a legal responsibility to ensure the good health and well-being of local inhabitants, and that certain legal limitations would be imposed and placed on the CCC facility to protect their health and welfare.

Compounding the problem was the airport. Airport officials also rejected CCC's plan to retrofit the compost facility. Their complaint stated that the dust generated at the compost facility was a hazard to flight operations, and even though the problem would be reduced substantially by engineering controls, the chance of control failure was always possible, and then an aircraft could be in danger. From the airport's point of view, this was unacceptable.

Several years went by, with local officials and CCC management contesting each other over the compost facility. In the end, CCC management decided they had to shut down and move to another location. So they closed the facility.

After shutdown, CCC management staff immediately started looking for another site to build a new wastewater biosolids to compost facility. They are still looking. To date, their search has located several pieces of property relatively close to the city (but far enough away to preclude any dust and odor problems) but they've had problems finalizing any deal. Buying the land is not the problem; getting the required permits from various county agencies to operate the facility is. CCC officials were turned down in each and every case.

CCC officials are still looking for a location for their compost facility, but they are not optimistic about their chances of finding one.

The alternative of the second control method, modification of the operation to a fuel which will give the desired level of emission, often looks favorable to those who have weighed the high costs associated with air pollution control systems. Modifying the process to eliminate as much of the pollution problem as possible at the source is generally the first approach to be examined.

Often, the easiest way to modify a process for air pollution control is to change the fuel. If a power plant (for example) emits large quantities of sulfur dioxide and fly ash, conversion to cleaner-burning natural gas is cheaper than installing the necessary control equipment to reduce the pollutant emissions to permitted values.

But changing from one fuel to another causes its own problems related to costs, availability, and competition. Today's fuel prices are high, and no one counts on the trend reversing. Finding a low-sulfur fuel isn't easy, especially since many industries own their own dedicated supplies (which are not available for use in other industries). And with regulation compliance threatening everyone, everyone wants his share of any available low cost, low sulfur fuel. With limited supplies available, the law of supply and demand takes over and prices go up.

Some industries employ other process modification techniques. These may include evaluation of alternative manufacturing and production techniques, substitution of raw materials, and improved process control methods (Buonicore et al., 1992).

When elimination of the process, entirely or in part, relocation of the operation, or modification of the operation to a fuel that will give the desired level of emission is not possible, the only alternative control method left is the third method, installation of control equipment between the pollutant source and the receptor. The purpose of installing pollution control equipment or a control system is to remove the pollution from the polluted carrier gas. To accomplish this, the polluted carrier gas must pass through a control device or system, which collects or destroys the pollutant and releases the cleaned carrier gas to the atmosphere (Boubel et al., 1994).

Since making the choice of what type of air pollution control methodology to employ often comes down to selecting pollution control equipment and systems as the only alternative that is feasible and practicable, the rest of this chapter will focus on these air pollution control equipment (devices) and systems.

AIR POLLUTION CONTROL EQUIPMENT AND SYSTEMS

Several considerations must be factored into to any selection decision for air pollution control equipment or systems. Careful consideration must be given to costs. Air pollution equipment/ systems are not inexpensive. Obviously, the equipment/system must be designed to comply with applicable regulatory emission limitations. The operational and maintenance history/record (costs of energy, labor, and repair parts should also be factored in) of each equipment/system must be evaluated. Emission control equipment must be operated on a continual basis, without interruptions, because any interruption could be subject to severe, costly regulatory penalty.

Probably the major factor to consider in the equipment/system selection process is what type of pollutant or pollutant stream is under consideration. If the pollutant is conveyed in a carrier gas, for example, factors such as carrier gas pressure, temperature, viscosity, toxicity, density, humidity, corrosiveness, and inflammability must all be considered before any selection is made.

Many of the other important factors that must be considered when selecting air pollution control equipment are listed below.

1)   Best available technology (BAT)
2)   Reliability
3)   Lifetime and salvage value
4)   Power requirements

5)   Collection efficiency
6)   Capital cost, including operation and maintenance costs
7)   Track record of equipment/system and manufacturer
8)   Space requirements and weight
9)   Power requirements
10)   Availability of space parts and manufacture's representatives

In addition, process considerations dealing with gas flow rate and velocity, pollutant concentration, allowable pressure drop, and the variability of gas and pollutant flow rates (including temperature) must also be considered.

The type of pollutant (gaseous or particulate) is also an important factor to take into consideration. Certain pertinent questions must be asked and answered. If the pollutant, for example, is gaseous, how corrosive, inflammable, reactive, and toxic is it?

After these factors have been evaluated, the focus shifts to the selection of the best air pollution control equipment/system that is affordable, practical, and permitted by regulatory requirements—depending, of course, on the type of pollutant to be removed.

In the following sections, we discuss two types of pollutants (dry particulates and gaseous pollutants) and the various air pollution control equipment/processes available for their removal.

REMOVAL OF DRY PARTICULATE MATTER

To select the proper air pollution control equipment/system for particulate removal, a basic understanding of particulate matter is required.

Particulate Matter

Constituting a major class of air pollutants, particulates have a variety of shapes and sizes, and as either liquid droplet or dry dust, they have a wide range of physical and chemical characteristics. Dry particulates are emitted from a variety of different sources, including both combustion and noncombustion sources in industry, mining, construction, incinerators and internal combustion engines. Dry particulates are also emitted from natural sources: volcanoes, forest fires, pollen, and windstorms.

All particles and particulate matter exhibit certain important characteristics, which, along with process conditions, must be considered in any engineering strategy to separate and remove them from a stream of carrier gas. Particulate size-range and distribution, particle shape, corrosiveness, agglomeration tendencies, abrasiveness, toxicity, reactivity, inflammability, and hygroscopic tendencies must all be examined in light of equipment limitations.

In an air pollution control system, particulates are separated from the gas stream by application of one or more forces in gravity settlers, centrifugal settlers, fabric filters, electrostatic precipitators, or wet scrubbers. The particles are then collected and removed from the system.

When a flowing fluid (engineering and science applications consider both liquid and gaseous states fluid) approaches a stationary object such as a metal plate, a fabric thread, or a large water droplet, the fluid flow will diverge around that object. Particles in the fluid (because of inertia) will not follow stream flow exactly, but will tend to continue in their original directions. If the particles have enough inertia and are located close enough to the stationary object, they will collide with the object, and can be collected by it. This is an important phenomenon and is depicted in Figure 14.1.

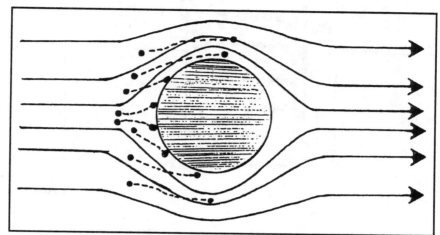

Figure 14.1 Particle collection of a stationary object

Particles are collected by *impaction*, *interception*, and *diffusion*. Impaction occurs when the center of mass of a particle that is diverging from the fluid strikes a stationary object. Interception occurs when the particle's center of mass closely misses the object, but, because of its finite size, the particle strikes the object. Diffusion occurs when small particulates happen to diffuse toward the object while passing near it. Particles that strike the object by any of these means are collected if short-range forces (chemical, electrostatic, and so forth) are strong enough to hold them to the surface (Cooper & Alley, 1990).

Air Pollution Control Equipment for Particulates

Different classes of particulate control equipment include gravity settlers, cyclones, electrostatic precipitators, wet (Venturi) scrubbers, and baghouse (fabric) filters. In this section we briefly introduce each of the major types of particulate control equipment and point out their advantages and disadvantages.

Gravity Settlers

Gravity settlers have long been used by industry for removing solid and liquid waste materials from gaseous streams. Simply constructed (see Figures 14.2 and 14.3), a gravity settler is actually nothing more than an enlarged chamber in which the horizontal gas velocity is slowed, allowing par-

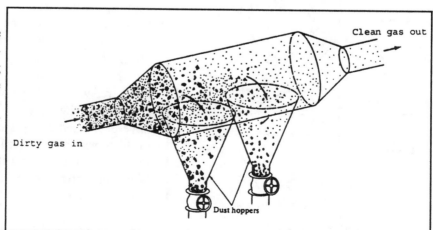

Figure 14.2 Gravitational settling chamber (source: EPA, Control Techniques for Gases and Particulates, *1971)*

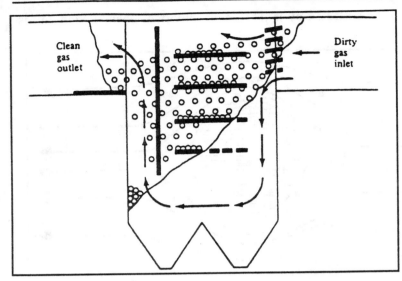

Figure 14.3 Baffled gravitation settling chamber (source: EPA, Control Techniques for Gases and Particulates, 1971)

ticles to settle out by gravity. Gravity settlers have the advantage of having low initial cost and are relatively inexpensive to operate—there's not a lot to go wrong. Although simple in design, gravity settlers require a large space for installation and have relatively low efficiency, especially for removal of small particles (< 50 μm).

Cyclone Collectors

The *cyclone* (centrifugal) *collector* removes particles by causing the entire gas stream to flow in a spiral pattern inside a tube, and is the collector of choice for removing particles greater than 10 μm in diameter. By centrifugal force, the larger particles more outward and collide with the wall of the tube. The particles slide down the wall and fall to the bottom of the cone, where they are removed. The cleaned gas flows out the top of the cyclone (see Figure 14.4). Cyclones have low construction costs, and require relatively small space for installation. However, note that the overall particulate collection efficiency of cyclones is low, especially on particles below 10 μm in size, and they do not handle sticky materials well. The most serious problems encountered with cyclones are with airflow equalization and their tendency to plug. Cyclones have been used successfully at feed and grain mills, cement plants, fertilizer plants, petroleum refineries, and other applications involving large quantities of gas containing relatively large particles.

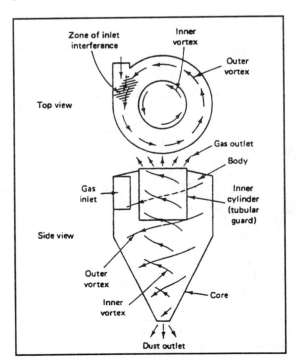

Figure 14.4 Conventional reverse-flow cyclone (source: USHEW, Control Techniques for Particulate Air Pollutants, 1969)

Electrostatic Precipitators

The *electrostatic precipitator* is usually used to remove small particles from moving gas streams at high collection efficiencies. Widely used in power plants for removing fly ash from the gases prior to discharge, an electrostatic precipitator applies electrical force to separate particles from the gas stream. A high voltage drop is established between electrodes, and particles passing through the resulting electrical field acquire a charge. The charged particles are attracted to and collected on an oppositely charged plate, and the cleaned gas flows through the device. Periodically, the plates are cleaned by rapping to shake off the layer of dust that accumulates, and the dust is collected in hoppers at the bottom of the device (see Figure 14.5). Although electrostatic precipitators have the advantages of low operating costs, capability for operation in high temperature applications (to 1300°F), low pressure drop, and potentially extremely high particulate (coarse and fine) collection efficiencies, they have the disadvantages of high capital costs and space requirements.

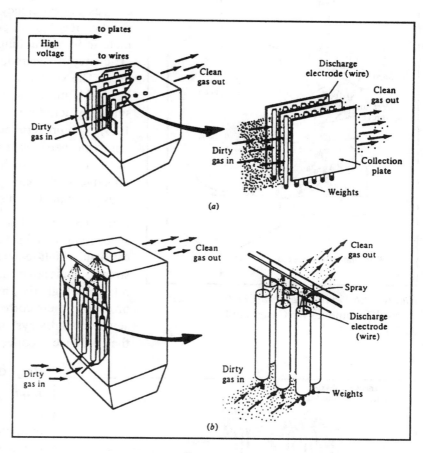

Figure 14.5 Electrostatic precipitators: (a) plate type, and (b) tube type (source: EPA, Control Techniques for Gases and Particulates, *1971)*

Wet (Venturi) Scrubbers

Wet scrubbers (or collectors) have found widespread use in cleaning contaminated gas streams (e.g., foundry dust emissions, acid mists, and furnace fumes) because of their ability to effectively remove particulate and gaseous pollutants. Wet scrubbers vary in complexity from simple spray chambers to remove coarse particles to high efficiency systems (venturi-types) to remove fine particles. Whichever system is used, operation employs the same basic principles of inertial impingement or impaction and interception of dust particles by droplets of water. The larger, heavier water droplets are easily separated from the gas by gravity. The solid particles can then

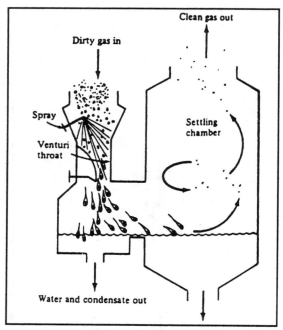

Figure 14.6 Venturi wet scrubber (source: EPA, Control Techniques for Gases and Particulates, *1971)*

be independently separated from the water, or the water can be otherwise treated before reuse or discharge. Increasing either the gas velocity or the liquid droplet velocity in a scrubber increases the efficiency because of the greater number of collisions per unit time. For the ultimate in wet scrubbing, where high collection efficiency is desired, the *venturi* scrubber is used. The venturi operates at extremely high gas and liquid velocities with a very high pressure drop across the venturi throat (see Figure 14.6). Venturi scrubbers are most efficient for removing particulate matter in the size range of 0.5 to 5μm, which makes them especially effective for the removal of sub-micron particulates associated with smoke and fumes.

Although wet scrubbers require relatively small space requirements, have low capital cost, and can handle high-temperature, high-humidity gas streams, their power and maintenance costs are relatively high, they may create water disposal problems, their corrosion problems are more severe than dry systems, and the final product they produce is collected wet.

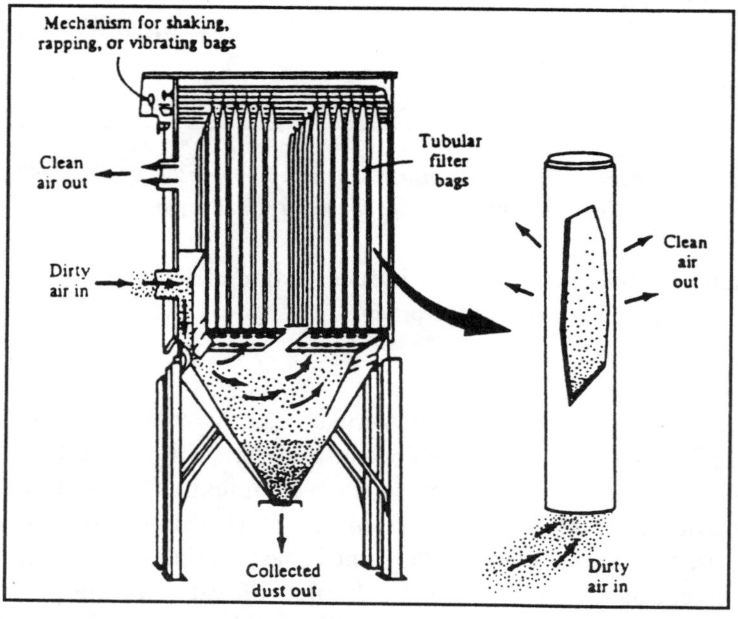

Baghouse (Fabric) Filters

Baghouse filters (or fabric filters) are the most commonly used air pollution control filtration system. In much the same manner as the common vacuum cleaner, fabric filter material, capable of removing most particles as small as 0.5 μm and substantial quantities of particles as small as 0.1 μm, is formed into cylindrical or envelope bags and suspended in the baghouse (see Figure 14.7). The particulate-laden gas stream is forced through the fabric filter, and as the air passes through the fabric, particulates accumulate on the cloth, providing a cleaned air stream. As particulates build up on the inside surfaces of the bags,

Figure 14.7 Typical simple fabric filter baghouse design (source: EPA, Control Techniques for Gases and Particulates, *1971)*

the pressure drop increases. Before the pressure drop becomes too severe, the bags must be relieved of some of the particulate layer. The particulates are periodically removed from the cloth by shaking or by reversing the airflow.

Fabric filters are relatively simple to operate, provide high overall collection efficiencies (up to 99+%) and are very effective in controlling sub-micrometer particles, but they do have limitations. These include relatively high capital costs, high maintenance requirements (bag replacement, etc), high space requirements, and flammability hazards for some dusts.

REMOVAL OF GASEOUS POLLUTANTS: STATIONARY SOURCES

In the removal of gaseous air pollutants, the principal gases of concern are the sulfur oxides (SO_x), carbon oxide (CO_x), nitrogen oxides (NO_x), organic and inorganic acid gases, and hydrocarbons (HC). Four major treatment processes are currently available for control of these and other gaseous emissions: absorption, adsorption, condensation, and combustion (incineration).

The decision of which single or combined air pollution control technique to use for stationary sources is not always easy. Gaseous pollutants can be controlled by a wide variety of devices, and choosing the most cost-effective, efficient units requires careful attention to the particular operation for which the control devices are intended. Specifically, the choice of control technology depends on the pollutant(s) to be removed, the removal efficiency required, pollutant and gas stream characteristics, and specific characteristics of the site (Peavy et al., 1985).

In making the difficult and often complex decisions on which air pollution control technology to employ, it is helpful to follow guidelines based on experience and set forth by Buonicore (1992) in the prestigious engineering text: *Air Pollution Engineering Manual*. Table 14.1 summarizes these guidelines.

Table 14.1: Comparison of Air Pollution Control Technologies

Treatment technology	Concentration and efficiency	Comments
Incineration	(< 100 ppmv) 90-95% efficient (> 100 ppmv) 95-99% efficient	Incomplete combustion may require additional controls.
Carbon Adsorption	(> 200 ppmv) 90+% efficiency (> 1000 ppmv) 95+% efficiency	Recovered organics may need additional treatment. Can increase cost.
Absorbtion	(< 200 ppmv) 90-95% efficiency (> 200 ppmv) 95+% efficiency	Can blowdown stream be accommodated at site?
Condensation	(> 2000 ppmv) 80+% efficiency	Must have low temperature or high pressure for efficiency

NOTE: Typically, only incineration and absorption technologies can achieve greater than 99% gaseous pollutant removal consistently (p. 15).

Absorption

Absorption (or *scrubbing*) is a major chemical engineering unit operation that involves bringing contaminated effluent gas into contact with a liquid absorbent so that one or more constituents of the effluent gas are selectively dissolved into a relatively nonvolatile liquid.

Absorption units are designed to transfer the pollutant from a gas phase to a liquid phase. The absorption unit accomplishes this by providing intimate contact between the gas and the liquid, providing optimum diffusion of the gas into the solution. The actual removal of a pollutant from the gas stream takes place in three steps: (1) diffusion of the pollutant gas to the surface of the liquid; (2) transfer across the gas/liquid interface; and (3) diffusion of the dissolved gas away from the interface into the liquid (Davis and Cornwell, 1991).

Several types of absorbers are available, including spray chambers (and towers or columns), plate or tray towers, packed towers, and venturi scrubbers. Pollutant gases commonly controlled by absorption include sulfur dioxide, hydrogen sulfide, hydrogen chloride, chlorine, ammonia, and oxides of nitrogen.

The two most common absorbent units in use today are the plate and packed tower systems. *Plate towers* contain perforated horizontal plates or trays designed to provide large liquid-gas interfacial areas. The polluted air stream is usually introduced at one side of the bottom of the tower or column and rises up through the perforations in each plate; the rising gas prevents the liquid from draining through the openings rather than through a downpipe. During continuous operation, contact is maintained between air and liquid, allowing gaseous contaminants to be removed, with clean air emerging from the top of the tower.

The *packed tower* scrubbing system (see Figure 14.8) is predominately used to control gaseous pollutants in industrial applications, where it typically demonstrates a removal efficiency of 90 to 95%. Usually configured in vertical fashion (Figure 14.8), the packed tower is literally "packed" with devices (see Figure 14.9) of large surface-to-volume ratio and a large void ratio that offers minimum resistance to gas flow. In addition, packing should provide even distribution of both fluid-phases, be sturdy enough for self-support in the tower, and be low cost, available, and easily handled (Hesketh, 1991).

The flow through a packed tower is typically counter-current, with gas entering at the bottom of the tower and liquid entering at the top. Liquid flows over the surface of the packing in a thin film, affording continuous contact with the gases.

Though highly efficient for removal of gaseous contaminants, packed towers may create liquid disposal problems, become easily clogged when gases with high particulate loads are introduced, and have relatively high maintenance costs.

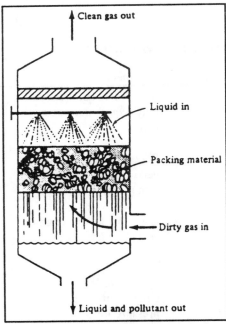

Figure 14.8 Typical countercurrent-flow packed tower (source: EPA, Control Techniques for Gases and Particulates, *1971)*

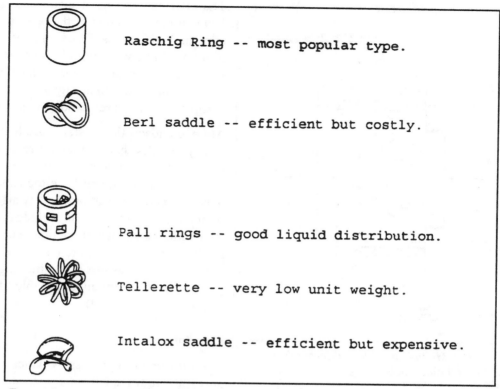

Figure 14.9 Various packing used in packed tower scrubbers (adapted from Air Pollution Manual: Control Equipment, Part II, *American Industrial Hygiene Association, 1968)*

Adsorption

Adsorption is a mass transfer process that involves passing a stream of effluent gas through the surface of prepared porous solids (adsorbents). The surfaces of the porous solid substance attract and hold the gas (the adsorbate) by either physical or chemical adsorption. In *physical adsorption* (a readily reversible process), a gas molecule adheres to the surface of the solid because of an imbalance of electron distribution. In *chemical adsorption* (not readily reversible), once the gas molecule adheres to the surface, it reacts chemically with it.

Several materials possess adsorptive properties. These materials include activated carbon, alumina, bone char, magnesia, silica gel, molecular sieves, strontium sulfate, and others. The most important adsorbent for air pollution control is activated charcoal. The surface area of activated charcoal will preferentially adsorb hydrocarbon vapors and odorous organic compounds from an airstream.

In the adsorption system, in contrast to the absorption system where the collected contaminant is continuously removed by flowing liquid, the collected contaminant remains in the adsorption bed. The most common adsorption system is the *fixed bed adsorber,* which can be contained in either a vertical or horizontal cylindrical shell. The adsorbent (usually activated carbon) is arranged in beds or trays in layers about 0.5 in. thick. Multiple beds may be arranged as shown in Figure 14.10. In multiple bed systems, one or more beds are adsorbing vapors, while the other bed is being regenerated.

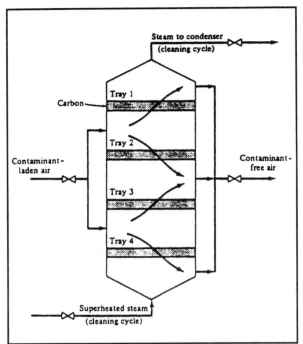

Figure 14.10 Multiple fixed-bed absorber (source: EPA, Air Pollution Engineering Manual, *2nd ed., 1973)*

The efficiency of most adsorbers is near 100% at the beginning of the operation and remains high until a *breakpoint* or *breakthrough* occurs. When the adsorbent becomes saturated with adsorbate, contaminant begins to leak out of the bed, signaling that the adsorber should be renewed or regenerated.

Although adsorption systems are high efficiency devices that may allow recovery of product, have excellent control and response to process changes, and have the capability of being operated unattended, they also have some disadvantages. These include the need for exotic, expensive extraction schemes if product recovery is required, relatively high capital cost, and gas stream pre-filtering needs (to remove any particulate capable of plugging the adsorbent bed).

Condensation

Condensation is a process by which volatile gases are removed from the contaminant stream and changed into a liquid. In air pollution control, a *condenser* can be used in two ways: either for pretreatment to reduce the load problems with other air pollution control equipment, or for effectively controlling contaminants in the form of gases and vapors.

Condensers condense vapors to liquid phase by either increasing the system pressure without a change in temperature, or by decreasing the system temperature to its saturation temperature without a pressure change. Condensation is affected by the composition of the contaminant gas stream. When (for some reason) gases are present in the stream and condense under different conditions, condensation is hindered.

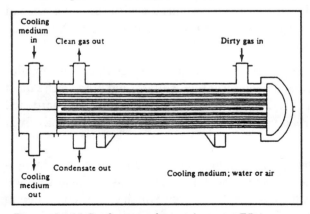

Figure 14.11 Surface condenser (source: EPA, Control Techniques for Gases and Particulates, *1971)*

There are two basic types of condensation equipment—surface and contact condensers. A *surface condenser* is normally a shell-and-tube heat exchanger (see Figure 14.11). The surface condenser uses a cooling medium of air or water where the vapor to be condensed is separated from the cooling medium by a metal wall. Coolant flows through the tubes, while the vapor is passed over and condenses on the outside of the tubes, and drains off to storage (EPA, 1971).

In a *contact condenser* (which resembles a simple spray scrubber), the vapor is cooled by spraying liquid directly on the vapor stream (see Figure 14.12). The cooled vapor condenses, and the water and condensate mixture are removed, treated, and disposed.

In general, contact condensers are less expensive, more flexible, and simpler than surface condensers, but surface condensers require much less water and produce many times less wastewater that must be treated than do contact condensers. Condensers are used in a wide range of industrial applications, including petroleum refining, petrochemical manufacturing, basic chemical manufacturing, dry cleaning, and degreasing.

Combustion

Even though *combustion* (or incineration) is a major source of air pollution, it is also, if properly operated, a beneficial air pollution control system in which the objective is to convert certain air contaminants (usually CO and hydrocarbons) to innocuous substances such as carbon dioxide and water (EPA, 1973).

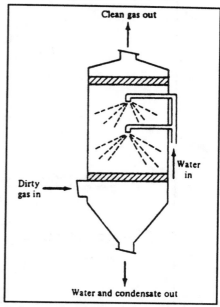

Figure 14.12 Contact condenser (source: EPA, Control Techniques for Gases and Particulates, *1971)*

Combustion is a chemical process defined as rapid, high-temperature gas-phase oxidation. The combustion equipment used to control air pollution emissions is designed to push these oxidation reactions as close to complete combustion as possible, leaving a minimum of unburned residue. The operation of any combustion operation is governed by four variables: oxygen, temperature, turbulence, and time. For complete combustion to occur, oxygen must be available, put into contact with sufficient temperature (turbulence), and held at this temperature for a sufficient time. These four variables are not independent—changing one affects the entire process.

Depending upon the contaminant being oxidized, equipment used to control waste gases by combustion can be divided into three categories: direct-flame combustion (or flaring), thermal combustion (afterburners), or catalytic combustion.

Direct-Flame Combustion (Flaring)

Direct-flame combustion devices (flares) are the most commonly used air pollution control devices by which waste gases are burned directly (with or without the addition of a supplementary fuel). Common flares include steam-assisted, air-assisted, and pressure head types. Flares are normally elevated from 100 to 400 feet to protect the surroundings from heat and flames. Often designed for steam injection at the flare top (see Figure 14.13), flares commonly use steam in this appli-

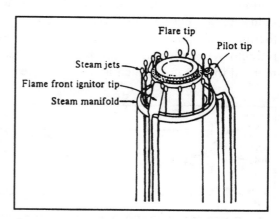

Figure 14.13 Close-up of steam injection type flare head (source: Control Techniques for Gases and Particulates, *1971)*

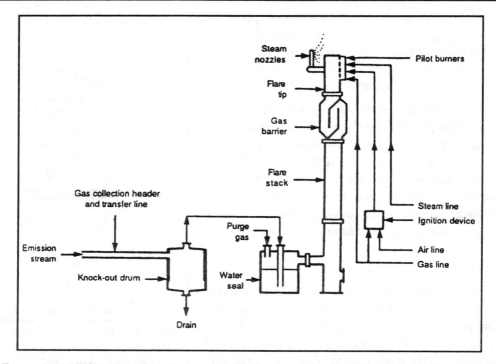

Figure 14.14 Schematic of a steam-assisted flare system (source: EPA, Handbook—Control Technologies for Hazardous Air Pollution, *1986)*

cation because it provides sufficient turbulence to ensure complete combustion, which prevents production of visible smoke or soot. Flares are also noisy, which can cause problems for adjacent neighborhoods, and some flares produce oxides of nitrogen, thus creating a new air pollutant. Figure 14.14 shows a steam-assisted flare system commonly used in industry.

Thermal Combustion (Afterburners)

The *thermal incinerator* or *afterburner* is usually the unit of choice in cases where the concentration of combustible gaseous pollutants is too low to make flaring practical. Widely used in industry, the thermal combustion system typically operates at high temperatures. Within the thermal incinerator, the contaminant airstream passes around or through a burner and into a refractory-line residence chamber, where oxidation occurs (see Figure 14.15). Flue gas from a thermal incinerator, which is relatively clean, is at high temperature and contains recoverable heat energy. Figure 14.16 shows a schematic of a typical thermal incinerator system.

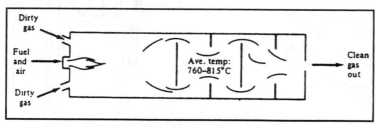

Figure 14.15 A thermal incinerator (source: EPA, Control Techniques for Gases and Particualtes, *1971)*

Catalytic Combustion

Catalytic combustion operates by passing a preheated contaminant-laden gas stream through a catalyst bed (usually a thinly-coated platinum mesh

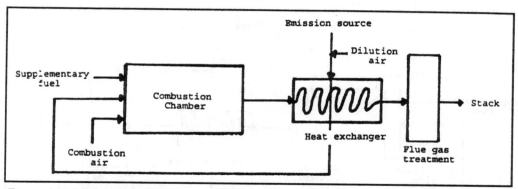

Figure 14.16 Schematic of a thermal incinerator system (adapted from Corbitt, Handbook of Environmental Engineering, *1990)*

mat, honeycomb, or other configuration designed to increase surface area), which promotes the oxidization reaction at lower temperatures (see Figure 14.17). The metal catalyst is used to initiate and promote combustion at much lower temperatures than those required for thermal combustion (metals in the platinum family are recognized for their ability to promote combustion at low temperature). Catalytic incineration may require 20 to 50 times less residence time than thermal incineration. Catalytic incinerators normally operate at 700°-900°F. At this reduced temperature range, a saving in fuel usage and cost is realized; however, this may be offset by the cost of the catalytic incinerator itself.

Other advantages of catalytic incinerators over thermal incinerators include lower fuel requirements, lower operating temperatures, little or no insulation requirements, reduced fire hazards, and reduced flashback problems (Buonicore and Davis, 1992.)

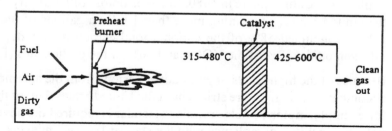

Figure 14.17 A catalytic incinerator (source: EPA, Control Techniques for Gases and Particualtes, *1971)*

A schematic diagram of a catalytic incinerator is presented in Figure 14.18. A heat exchanger is an option for systems with heat transfer between two gas streams (recuperative heat exchange). The need for dilution air, combustion air, and/or flue gas treatment is based on site-specific conditions. Catalysts are subject to both physical and chemical deterioration, and their usefulness is suppressed by sulfur-containing compounds. For best performance, catalyst surfaces must be clean and active.

Catalytic incineration is used in a variety of industries to treat effluent gases, including emissions from paint and enamel bake ovens, asphalt oxidation, coke ovens, formaldehyde manufacture, and varnish cooking.

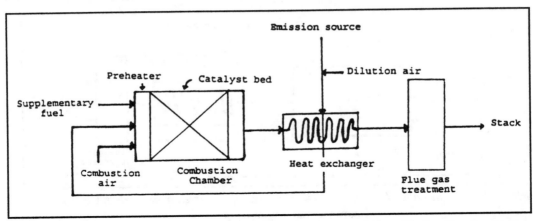

Figure 14.18 Schematic of catalytic incinerator system (adapted from Corbitt, Handbook of Environmental Engineering, *1990)*

REMOVAL OF GASEOUS POLLUTANTS: MOBILE SOURCES

Mobile sources of gaseous pollutants include locomotives, ships, airplanes, and automobiles. However, automobiles are by far the most important, both in terms of total emissions and in location of emissions relative to people. According to the *Twelfth Annual Report of the Council on Environmental Quality* (1982), transportation accounted for 55% of all major air pollutants emitted to the atmosphere in 1980. In 1986, almost 140 million passenger cars were registered in the United States, consuming more than 1 billion gallons of fuel. Total emissions from these vehicles included 58% of the nation's total carbon monoxide, 38% of the lead, 34% of the nitrogen oxide, 27% of the VOCs, and 16% of the particulates (EIA, 1988; EPA, 1988).

Because of the high levels of pollutant emissions from the automobile, emission standards have become increasingly more stringent in the United States. Under the 1970 Clean Air Act (CAA), for example, motor vehicle emissions standards required the development of new control technology to achieve compliance with the standard emission levels.

Mobile source pollution problems can be solved by either of two means: replacing the internal combustion engine (e.g., with electrical power or mass transit) or use of direct pollutant control systems. At the present time, replacement is not feasible and technology in this regard is still in its infancy. Even if the technology were currently available for replacement, replacement would be inordinately difficult to undertake. Direct pollutant control systems (those that control emissions from the crankcase, carburetor, fuel tank, and exhaust) are what we rely on. We discuss these systems in the following section.

Control of Crankcase Emissions

Crankcase emissions can be controlled by technology called *positive crankcase ventilation (PCV)*. In this control technology, *hydrocarbon blowby gases* (gases which go past the piston rings into the crankcase) are recirculated to the combustion chamber for reburning (see Figure 14.19). The National Air Pollution Control Administration (1970) estimated that incorporation of the PCV system reduced crankcase hydrocarbon emissions to negligible levels.

Control of Evaporative Emissions

Changes in ambient temperatures (diurnal losses), hot soak, and running losses result in *evaporative emissions*. *Diurnal losses* are caused by expansions of the air-fuel mixture in a partially filled fuel tank, which expels gasoline vapor into the atmosphere. *Hot soak* emissions occur after the engine is shut off as heat from the engine causes increased evaporation of fuel. *Running* (or operating) *losses* occur during driving as the fuel is heated by the road surface, and when fuel is forced from the fuel tank while the vehicle is being operated and the fuel tank becomes hot. In 1971, the direct control measure instituted to control hydrocarbon emissions was the installation of a canister filled with activated charcoal that adsorbs hydrocarbon emissions. Adsorbed vapors are purged from charcoal into the engine during high-power operating conditions. In California, vapor control systems at service stations reduce potential refueling vapor losses (Perkins, 1974).

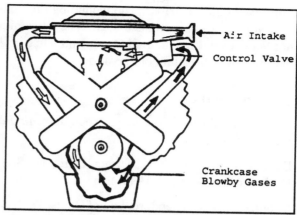

Figure 14.19 Positive crankcase ventilation (PCV) of blowby gases (adapted from Godish, Air Quality, *1997, p. 290)*

Catalytic Converters

Beginning in 1975, all new U.S. automobiles were required to be equipped with *catalytic converters* to meet the more restrictive tailpipe emission standards. Three types of catalytic converters are used for this purpose: oxidizing, reducing, and three-way.

Oxidizing catalytic converters work to accelerate the completion of the oxidation of CO and hydrocarbons so that CO is converted to carbon dioxide and water vapor. Platinum and palladium are used as catalysts. These converters can be poisoned by lead; thus, only unleaded gasoline should be used with this type.

Reducing Catalytic Converters use rhodium and ruthenium to accelerate the reduction of NO_x to N_2.

Three-Way Catalytic Converters are favored by U.S. automobile manufacturer's because they enable them to meet compliance requirements of the Clean Air Act. Three-Way Catalytic Converters oxidize hydrocarbons and carbon monoxide to carbon dioxide, while reducing NO_x to N_2. They are effective in controlling emissions and have the advantage of allowing the engine to operate at normal conditions where engine performance and efficiency are greatest (Demmler, 1977).

SUMMARY

While regulations drive industry to work to clean up its act, this process is not static. People and industry are constantly pushing the envelope in one way or another. When one pollutant comes to our attention, is regulated, then controlled, because of new processes, population growth, new test findings, and/or new technologies, we discover new problems, or can now clean up something we couldn't before. The technology to control the pollutants we create is growing. The need for environmentalists to design new techniques to meet new standards and problems will not fade away. Like the Red Queen and Alice in *Alice in Wonderland*, we have to run as fast as we can just to maintain our position.

Cited References

American Industrial Hygiene Association, *Air Pollution Manual: Control Equipment, Part II*, Detroit: AIHA, 1968.

Boubel, R. W., Fox, D. L., Turner, D. B., and Stern, A. C., *Fundamentals of Air Pollution*, New York: Academic Press, 1994.

Buonicore, A. J., Theodore, L. and Davis, W. T., *Air Pollution Engineering Manual*, Buonicore & Davis (eds.), New York: Van Nostrand Reinhold, 1992.

Buonicore, A. J., and Davis, W. T., (eds.), *Air Pollution Engineering Manual*, New York: Van Nostrand Reinhold, 1992.

Cooper, C. D., and Alley, F. C., *Air Pollution Control: A Design Approach*, Prospect Heights, Illinois: Waveland Press, Inc., 1990.

Corbitt, R. A., *Standard Handbook of Environmental Engineering*, New York: McGraw-Hill, 1990.

Davis, M. L., and Cornwell, D. A., *Introduction to Environmental Engineering*, New York: McGraw-Hill, 1991.

Demmler, A. W., "Automotive Catalysis," *Auto Eng.*, 85(3): 29,32 (1977).

Godish, T., *Air Quality*, 3rd ed., Boca Raton, FL: Lewis Publishers, 1997.

Hesketh, H. E., *Air Pollution Control: Traditional and Hazardous Pollutants*, Lancaster, PA: Technomic Publishing Company, 1991.

National Air Pollution Control Administration: *Control Techniques for Hydrocarbon and Organic Solvent Emissions for Stationary Sources*, document B, publ. AP-68, Washington, D.C., 1970.

Peavy, H. S., Rowe, D. R., and Tchobanglous, G., *Environmental Engineering*, New York: McGraw-Hill, 1985.

Perkins, H. C., *Air Pollution*, New York: McGraw-Hill, 1974.

Twelfth Annual Report of the Council of Environmental Quality, Washington, D.C., 1982.

USHEW, *Control Techniques for Particulate Air Pollutants*, Washington, D.C.: National Air Pollution Control Administration, 1969.

U.S. Energy Information Agency, *Annual Energy Review 1987*, Washington, D.C.: Energy Information Agency, Department of Energy, 1988.

U.S. Environmental Protection Agency, *Annual Report of the Environmental Protection Agency to the Congress of the United States in Compliance with Section 202(b)(4),*

Public Law 90-148, Washington, D.C., 1971.

U.S. Environmental Protection Agency, *Air Pollution Engineering Manual*, 2nd ed., U.S. Environmental Protection Agency, AP-40, Research Triangle Park, N.C., 1973.

U.S. Environmental Protection Agency, *Handbook—Control Technologies for Hazardous Air Pollutants*, U.S. Environmental Protection Agency, Center for Environmental Research Information, EPA 625/6-86/014, Cincinnati, Ohio, 1986.

U.S. Environmental Protection Agency, *National Air Pollutant Emission Estimates 1940-1986*, Environmental Protection Agency, Washington, D.C., 1988.

Suggested Readings

Brunner, C. R., *Handbook of Incineration Systems*, New York: McGraw-Hill, 1991.

Bubenick, D. V., "Control of fugitive emissions," in S. Calvert and H. Englund (eds.), *Handbook of Air Pollution Technology*, New York: Wiley, 1984.

Clavert, S., "How to choose a particulate scrubber," *Chem. Eng.*, 84 (18): 54-68.

Control Technology for Hazardous Air Pollutants, Rockville, MD: Government Institutes, 1992.

Counce, R. M., and Perona, J. J., "Scrubbing of gaseous nitrogen oxides in packed towers," *Journal of the American Institute of Chemical Engineers*, 29 (1), January 1983, pp. 26-32.

Crawford, M., *Air Pollution Control Theory*, New York: McGraw-Hill, 1976.

Heumann, W. J., (ed.), *Industrial Air Pollution Control Systems*, New York: McGraw-Hill, 1997.

Higgins, T. E., (ed.), *Pollution Prevention Handbook*, Boca Raton, FL: CRC/Lewis Publishers, 1995.

Hunton & Williams, *Clean Air Handbook*, Rockville, MD: Government Institutes, 1998.

Masters, G. M., *Introduction to Environmental Engineering and Science*, Englewood Cliffs, NJ: Prentice Hall, 1991.

Powell, J. D., and Brennan, R. P., *The Automobile, Technology and Society*, Englewood Cliffs, NJ: Prentice Hall, 1988.

Public Law 101-549, 101st Congress—*Clean Air Act Amendments of 1990*, November 1990.

Stern, A. C., (ed.), *Air Pollution*, Vols. 1, 2 and 3, New York: Academic Press, 1968.

Straus, W., *Industrial Gas Cleaning*, Elmsford, New York: Pergamon Press, Inc., 1966.

Theodore, L., and Buonicore, A. J., *Industrial Air Pollution Control Equipment for Particulates*, West Palm Beach, FL: CRC Press, 1986.

Theodore, L., and Buonicore, A. J., *Selection, Design, Operation and Maintenance: Air Pollution Control Equipment*, Roanoke, VA: ETS, 1982.

Vatauk, W. M., *Estimating Costs of Air Pollution Control*, Chelsea, MI: Lewis Publishers, 1990.

Vesilind, P. A., and Pierce, J. J., *Environmental Engineering*, Ann Arbor, Michigan: Ann Arbor Science, 1982.

Williamson, S. J., *Fundamentals of Air Pollution*, Reading, MA: Addison-Wesley, 1973.

PART III
WATER QUALITY

U. S. WATERSHEDS HAVE WATER QUALITY PROBLEMS

The USEPA released its first comprehensive assessment of the 2,100 watersheds in the U.S. (October 1997) and bears the bad news: 21% have poor water quality, 36% have moderate water quality, and only 16% have good water quality. The report, issued as part of its right-to-know initiatives, also indicated that one in fourteen watersheds are vulnerable to further degradation, primarily from runoff, not point-source discharges. —Environmental Technology, *Nov/Dec 1997, p. 10*

CHAPTER 15

CHARACTERISTICS OF WATER

EARTH'S BLOOD

The watery environment in which single-cell organisms live provides them food and removes their wastes, a function that the human circulatory system provides for the 60-100 trillion cells in the human body. The circulatory system brings each cell its daily supply of nutritive amino acids and glucose, and carries away waste carbon dioxide and ammonia, to be filtered out of our systems and flushed away through micturition and excretory functions. The heart, the center of our circulatory system, keeps blood moving on its predetermined circular path—so essential that if the pump fails, we quickly fail as well: we die.

As single-celled organisms, humans sometimes assume they do not need a watery environment in which to live—but they aren't paying close attention to the world around them. Actually, those of us who live on the earth are as dependent upon the earth's circulatory system as we are on our own circulatory system. As our human hearts pump blood, circulating it through a series of vessels, and as our lives are dependent upon that flow of blood, so life on earth is, and our own lives are dependent on, the earth's water cycle, and on water, in every aspect of our lives.

This cycle is so automatic that we generally ignore it until we are slapped in the face by it. Just as we don't control or pay attention to the beating of our hearts unless the beat skips or falters, unless we are confronted by flood or drought, unless our plans are disrupted by rain, we ignore the water cycle, preferring to believe that the water we drink comes out of the faucet and not from the earth, placed there by a process we only dimly comprehend. But water is as essential to us and to the earth as blood is in our bodies, and the constant cycle water travels makes our lives possible.

Earth's blood, water, is pumped, not by a heart, but by the hydrological cycle—the water cycle. A titanic force of nature, the water cycle is beyond our control—a fact that we ignore until weather patterns shift and suddenly inundated rivers flow where they will and not within human-engineered banks, flood-walls, dikes, and levees. In the water cycle, water evaporated from the oceans falls as rain, hail, sleet, or snow and it strikes the earth again; the cycle continues.

In cities, in summer, rain strikes hot cement and asphalt and swiftly evaporates, or runs into storm drains, swiftly rejoining the cycle. In fields, rain brings essential moisture to crops, and sinking deeper into the earth, ends as groundwa-

ter. If water strikes a forested area, the forest canopy breaks the force of the falling drops. The forest floor, carpeted in twigs, leaves, moss, and dead and decaying vegetation, keeps the soil from splashing away in erosion as the water returns to the depths of the earth, or runs over the land to join a stream.

Whenever water strikes the earth, it flows along one of four pathways, which carry water through the cycle as our veins, arteries, and capillaries carry our blood to our cells.

Water may evaporate directly back into the air, it may flow overland into a stream as runoff, it may soak into the ground and be taken up by plants for evapotranspiration, or it may seep down to groundwater. Whatever pathway water takes, one fact is certain: water is dynamic, vital, constantly on the move. And like human blood, which sustain our lives, earth's blood, to sustain us as well, must continue to flow.

CHAPTER OUTLINE

- Introduction
- Water Resources
- Water Use
- Characteristics of Water

INTRODUCTION

While we pursue the second environmental medium in Part III of this text, notice that we are only shifting between related concepts—air, water, and soil are interrelated; we must not lose sight of these essential connections.

Water is an essential compound in the maintenance of all forms of life on Earth. This fact has resulted in the development of direct relationships between the abundance of water and the quality of the water—remember, having mega-gallons of water in storage, readily available, is of little use if the water is unsuitable for human consumption.

In the pre-Columbian era, on the North American continent, the natural water cycle was able to deliver clean water to the landscape. Today, the natural water cycle has been changed and tainted in a number of ways. We have dredged, dammed, channeled, tampered with, and sometimes eliminated the ecological niches where water cleans itself. We have simplified the pathways that water takes through the American landscape. As a result, water that is no longer able to clean itself naturally must be regulated and cleaned using various advanced technologies.

In response to a national crisis in water quality, water has been regulated since 1972. The purpose of these water regulations is to restore and maintain the chemical, physical, and biological integrity of the waterways. By 1995 the discharge of pollutants into streams, rivers, and lakes was to cease. The goal was to make these freshwater bodies fishable and swimmable.

The means to accomplish this goal was to be provided by technology. Every U.S. city was required to build a wastewater treatment plant with secondary treatment capability. Every industry had to incorporate the best available technology (BAT) to decrease the discharge of pollutants into the waterways.

In the years that followed 1972, the stranglehold that pollutant wastes had on the nation's surface water bodies was eased. A generation after the Clean Water Act was promulgated, about 30% of the stream miles and lake acres in the U.S. are still polluted (Outwater, 1997). A great deal must still be done.

Despite our best legislative efforts, our waterways are still impaired. The question has become: Will our waterways ever return to their natural state as they were in pre-Columbian times? No, they will not.

A mindset pervades regulatory and environmental activist circles, hoping that by changing the way we manage our vast public lands—by restoring those elements of the natural world that made the water pristine in pre-Columbian waterways—we can have clean water again. Much can be said in favor of this mindset. We have to start somewhere; we have to learn to respect water for what it really is: the land's blood.

But more is required. With the explosive growth of populations and the accompanying increased need for more natural resources and habitable land, we will have to work at restoring and maintaining water quality. The authors of this text take the view that this can only be accomplished by the correct and judicial use of technology to clean our water. Water must be available for human use and in abundant supply. Available water must have specific characteristics; water quality is defined in terms of those characteristics (Tchobanoglous & Schroeder, 1987).

Water quality is assessed in terms of its physical, chemical, and biological characteristics. In this chapter we discuss these characteristics. In Chapter 16 we discuss the characteristics of freshwater bodies: streams, lakes, and rivers. In Chapter 17 we discuss water pollution, and in Chapter 18 we discuss the water pollution control technology available to ensure that water performs its most vital function—that of quenching our thirst.

WATER RESOURCES

To find out where potable (drinkable) water is readily found for human consumption, look at a map of the world, one clearly indicating the world's population centers (cities). Take a look at the United States, for example. The first American settlers built their settlements along rivers. Rivers provided the water that settlers needed to sustain life, the principal source of power for early industry, and an easy means of transportation.

Most of the earliest settlements in the United States occurred on the East Coast. In most cases (the early Jamestown, Virginia settlement is an exception—the potable water quality there was poor), settlers along this Eastern seaboard area were lucky. They had settled along river systems of excellent quality. These rivers were ideally suited for paper and textile manufacturing, which were among the earliest industries started.

As more settlers arrived in North America, they began to branch out inland from the earliest settlements and, in many cases, they found that finding potable water was not so easy. The further west they traveled, the higher the salinity of rivers and streams, especially with long rivers and streams that flowed through and over areas of relatively soluble rock formations.

In western regions of the United States (e.g., in deserts), the U.S. map shows only sparse settlement, because those arid areas lack water to support a larger population. These regions are occupied by fewer people and different (more tolerant of water shortage) species than other biomes. A real-estate saying you may have heard is "Location! Location! Location!" Is location everything? Though we don't think of it consciously, when you examine the beginnings of human settlements, "Location, Location, Location!" really meant, "Water! Water! Water!" To be suitable as a living place, potable water was essential. On land, the availability of a regular supply of potable water is the most important factor affecting the presence or absence of many life forms.

The Major Sources of Drinking Water

Approximately 326 million cubic miles of water cover the earth, but only about 3% of this total is fresh. Most of this fresh water supply is locked up in polar ice caps, glaciers, lakes, flows through soil, and river and stream systems. Out of this 3%, only 0.027% is available for human consumption.

The data contained in Table 15.1 show where the world's water is distributed.

Table 15.1: World Water Distribution

Location	% of Total
Land areas	
Freshwater lakes	0.009
Saline lakes and inland seas	0.008
Rivers (average instantaneous volume)	0.0001
Soil moisture	0.005
Groundwater (above depth of 4000 m)	0.61
Ice caps and glaciers	2.14
	2.8
Atmosphere (water vapor)	0.001
Oceans	97.3
Total all locations (rounded)	100

Source: Adapted from Peavy et al., (1985), p. 12.

Table 15 illustrates the fact that the major sources of drinking water are from surface water and groundwater.

Surface Water

Surface water [water on the earth's surface as opposed to sub-subsurface water (groundwater)] is mostly a product of precipitation: rain, snow, sleet, or hail. Surface water is exposed or open to the atmosphere and results from the movement of water on and just under the earth's surface (overland flow). This overland flow is the same thing as surface runoff, which is the amount of rainfall that passes over the earth's surface. Specific sources of surface water include:

- Rivers
- Streams
- Lakes
- Impoundments (man-made lakes made by damming a river or stream)
- Very shallow wells that receive input via precipitation
- Springs that are affected by precipitation (i.e., their flow or quantity is directly affected by precipitation)
- Rain catchments (drainage basin)
- Tundra ponds or muskegs (peat bogs)

As a source of potable water, surface water does have some advantages:

1) It is usually easy to locate—you do not need a geologist or hydrologist to find it.

2) It is not normally tainted with chemicals precipitated from the earth's strata.

Surface water also has its disadvantages. The biggest disadvantage of using surface water as a source of potable water is that it is easily contaminated (polluted) by microorganisms that can cause waterborne diseases and by chemicals that enter from surrounding runoff and upstream discharges. Problems can also occur with water rights.

If you are familiar with the battles that took place (some are still being waged) in the western United States between cattle ranchers and homesteaders over rangeland, you are familiar with the cause: rights to surface water. Today, in most places in the U.S., significant removal of water from a river, stream, spring or lake requires a legal permit.

Most surface water is the result of surface runoff. The amount and flow rate of surface runoff is highly variable. This variability comes into play for two main reasons: (1) human interference (influences) and (2) natural conditions. In some cases, surface water runs quickly off land. Generally this is undesirable (from a water resources standpoint) because it does not provide enough time for water to infiltrate into the ground and recharge groundwater aquifers. Other problems associated with quick surface water runoff are erosion and flooding. Probably the only good thing that can be said about surface water that quickly runs off land is that it does not have enough time (usually) to become contaminated with high mineral content. Surface water running slowly off land may be expected to have all the opposite effects.

Surface water travels over the land to what amounts to a predetermined destination. What factors influence how surface water moves?

Surface water's journey over the face of the earth typically begins at its *drainage basin*, sometimes referred to as its *drainage area, catchment,* and/or *watershed*. For a groundwater source, this is known as the *recharge area*—the area from which precipitation flows into an underground water source.

A surface water drainage basin is usually an area measured in square miles, acres, or sections, and if a city takes water from a surface water source, the size of (and what lies within) the drainage basin is essential information for assessment of water quality.

Water doesn't run uphill. Surface water runoff (like the flow of electricity) follows the path of least resistance. Generally speaking, water within a drainage basin will naturally (by the geological formation of the area) be shunted toward one primary watercourse (a river, stream, creek, brook) unless some manmade distribution system diverts the flow.

Various factors directly influence the surface water's flow over land. The principal factors are:

- rainfall duration
- rainfall intensity
- soil moisture
- soil composition
- vegetation cover
- ground slope
- human interference

Rainfall Duration. The length of the rainstorm affects the amount of runoff. Even a light, gentle rain will eventually saturate the soil if it lasts long enough. Once the saturated soil can absorb no more water, rainfall builds up on the surface and begins to flow as runoff.

Rainfall Intensity. The harder and faster it rains, the more quickly soil becomes saturated. With hard rains, the surface inches of soil quickly become inundated, and with short, hard storms, most of the rainfall may end up as surface runoff, because the moisture is carried away before significant amounts of water are absorbed into the earth.

Soil Moisture. If the soil is already laden with water from previous rains, the saturation point will be reached sooner than if the soil were dry. Frozen soil also inhibits water absorption: up to 100% of snow melt or rainfall on frozen soil will end as runoff, because frozen ground is impervious.

Soil Composition. Runoff amount is directly affected by soil composition. Hard rock surfaces will shed all rainfall, obviously, but so will soils with heavy clay composition. Clay soils possess small void spaces that swell when wet. When the void spaces close, they form a barrier that does not allow additional absorption or infiltration. On the opposite end of the spectrum, course sand allows easy water flow-through, even in a torrential downpour.

Vegetation Cover. Runoff is limited by ground cover. Roots of vegetation and pine needles, pine cones, leaves, and branches create a porous layer (a sheet of decaying natural organic substances) above the soil. This porous organic sheet readily allows water into the soil. Vegetation and organic waste also act as a cover to protect the soil from hard, driving rains. Hard rains can compact bare soils, close off void spaces, and increase runoff. Vegetation and groundcover work to maintain the soil's infiltration and water-holding capacity. Note that vegetation and groundcover also reduce evaporation of soil moisture as well.

Ground Slope. Flatland water flow is usually so slow that large amounts of rainfall can infiltrate the ground. Gravity works against infiltration on steeply sloping ground, where up to 80% of rainfall may become surface runoff.

Human Influences. Various human activities have a definite impact on surface water runoff. Most tend to increase the rate of water flow. For example, canals and ditches are usually constructed to provide steady flow, and agricultural activities generally remove groundcover that would work to retard the runoff rate. On the opposite extreme, man-made dams are generally built to retard the flow of runoff.

Human habitations—paved streets, parking lots, tarmac, and buildings—create surface runoff potential because their surfaces are impervious to infiltration. Since all these surfaces hasten the flow of water, they also increase the possibility of flooding, often with devastating results. Because of urban increases in runoff, a whole new industry has developed: storm water management.

Paving over natural surface acreage has another serious side effect. Without enough area available for water to infiltrate the ground and percolate into the soil to eventually reach and replenish/recharge groundwater sources, those sources may eventually fail, with devastating impact on local water supply.

Groundwater

Approximately three feet of water falls each year on every square foot of land. About six inches of this goes back to the sea. Evaporation takes up about two feet. What remains, approximately six inches, seeps into the ground, entering and filling every interstice, each hollow and cavity, like an absorbent. Although comprised of only 1/6 of the total (1,680,000 miles of it), if it could be ladled up and spread out over the earth's surface, it would blanket all land to a depth of one thousand feet.

This gigantic water source (literally an ocean beneath our feet) forms a reservoir that feeds all the natural fountains and springs of earth. Eventually, it works its way to the surface. Some comes out clean and cool, a liquid blue-green phantom; and some, occupying the deepest recesses, pressurizes and shoots back to the surface in white, foamy, wet chaos, as geysers.

Fortunately, most of the rest lies within easy reach, just beneath the surface.

This is groundwater. —Spellman, 1998

Water falling to the ground as precipitation normally follows one of four courses. Some runs off directly to rivers and streams, some infiltrates to ground reservoirs, some evaporates, and the rest is taken up by vegetation and transpired. Groundwater is "invisible" and may be thought of as a temporary natural reservoir (ASTM, 1969). Almost all groundwater is in constant motion toward rivers or other surface water bodies.

Groundwater is defined as water below the earth's crust, but above a depth of 2500 feet below the surface. Thus, if water is located between the earth's crust and the 2500-foot level, it is considered potable fresh water. In the United States, it is estimated "that at least 50% of total available fresh water storage is in underground aquifers" (Kemmer, 1979).

Groundwater is usually obtained from wells, or from springs that are not influenced by surface water or local hydrologic events.

Groundwater, in relationship to surface water, has several advantages:

1) Unlike surface water, groundwater is not easily contaminated.

2) Groundwater sources are usually lower in bacteriological contamination than surface water.

3) Supply usually remains stable throughout the year.

4) In the United States, for example, groundwater is available in most locations.

When comparing groundwater with surface water sources there are also some disadvantages in using groundwater:

1) If there is contamination, it is often hidden from view.

2) Groundwater is usually loaded with minerals (increased level of hardness) because it is in contact with them for a longer period of time.

3) When groundwater supplies are contaminated, removing the contaminant is very difficult.

4) Because it must be pumped from the ground, operating costs are usually higher.

5) If the groundwater source is located near coastal areas, it may be subject to salt water intrusion.

WATER USE

In the United States, rainfall averages approximately $4{,}250 \times 10^9$ gallons a day. About two-thirds of this returns to the atmosphere through evaporation from the surface of rivers, steams, and lakes or from transpiration from plant foliage. This leaves approximately $1{,}250 \times 10^9$ gallons a day to flow across or through the earth to the sea (Kemmer, 1979).

Water is being used at the present time in the United States at a rate of 1.6×10^{12} liters per day, which amounts to almost a tenfold increase in liters used since the turn of the century.

Of the billions of gallons of water that are available for use in the United States, where is this water used? The National Academy of Sciences (1962) estimates that approximately (1) 310 billion gallons per day (bgd) are withdrawn; (2) 142 bgd are used for irrigation; (3) 142 bgd are used for industry (principally utility cooling water—100 bgd); (4) 26 bgd are used in municipal applications; (5) 90 bgd are consumed by irrigation (principally loss to ground and evaporation); and (6) 220 bgd are returned to streams. From the preceding list it is evident that much of this increase in use is accounted for by high agricultural and industrial use, each of which accounts for more than 40% of total consumption. Municipal use comprises the remaining 10-12% (Manahan, 1997).

We are primarily concerned with water use for municipal applications. Municipal water demand is usually classified according to the nature of the user. These classifications are:

Domestic. Domestic water is supplied to houses, schools, hospitals, hotels, restaurants, etc. for culinary, sanitary, and other purposes. Use varies with the economic level of the consumer, the range being 20-100 gallons per capita per day. Note that these figures include water used for watering gardens and lawns and washing cars.

Commercial and industrial. Commercial and industrial water is supplied to stores, offices, and factories. The importance of commercial and industrial demand is based, of course, on whether large industries use water supplied from the municipal system. Large industrial facilities can make heavy demands on a municipal system. Large industrial facilities demand a quantity directly related to the number of persons employed, to the actual floor space or area of each establishment, and to the number of units manufactured or produced.

Public use. Water for public use is the water furnished to public buildings and used for public services. This includes water for schools, public buildings, fire protection, and for flushing streets.

Loss and waste. Water that is lost or wasted (i.e., unaccounted for) is attributable to leaks in the distribution system, inaccurate meter readings, and for unauthorized connections. Loss and waste of water can be expensive. To reduce these costs, a regular program that includes maintenance of the system and replacement and/or recalibration of meters is required (McGhee, 1991).

Manahan (1997) points out that water is not destroyed, but it can be lost for practical use. The three ways in which this may occur are the following:

Evaporative losses: occur during spray irrigation, and when water is used for evaporative cooling.

Infiltration of water into the ground, often in places and ways that preclude its later use as groundwater.

Degradation from pollutants, such as salts picked up by water used for irrigation (p. 133).

CHARACTERISTICS OF WATER

When we attempt to characterize water, we normally choose the obvious features: appearance, taste, and smell. These physical attributes are important, but so are chemical and biological characteristics.

The chemical characteristics of water are important because even though water may appear, smell, and even taste okay, it doesn't necessarily mean that some chemical contaminant is not present. Today, with all the pesticides that are used in agriculture and various other industrial activities, chemical contamination is a very real possibility.

The biological characteristics of water are extremely important to anyone who might drink it. As we stated earlier, even before Typhoid Mary was cooking up her daily food preparations and passing on typhoid to her unsuspecting victims, people had suspected a relationship between microorganisms and disease. Today, we know with certainty that waterborne diseases are a real threat to human health.

We also know that contaminated water (whether it be physically, chemically, or biologically contaminated) does not start out that way. Precipitation in the form of rain, snow, hail, or sleet (aside from what we contaminate ourselves) contains very few (if any) impurities. It may pick up trace amounts of mineral matter, gases, and other substances as it forms and falls through the earth's atmosphere. The precipitation, however, has virtually no bacterial content—no water-borne disease.

When precipitation reaches the earth's surface, many opportunities for the introduction of mineral and organic substances, microorganisms, and other forms of pollution are presented. When water runs over and through the ground surface, it may pick up particles of soil. This attaches a physical property to water that can be readily seen—its cloudiness or turbidity. Water, as it courses its way along the earth's surface, also picks up organic matter and bacteria. As it seeps down into the soil and through the underlying material to the water table, most of the suspended particles are filtered out. This natural filtration may be a two-edged sword: it can be partially effective in removing bacteria and other particulate matter, but it may change the chemical properties of water as it comes in contact with mineral deposits.

Substances that alter the quality of water or its characteristics as it moves over or below the surface of the earth are of major concern to environmental practitioners. Thus, in the following sections, we discuss the physical, chemical, and biological characteristics of water.

Physical Characteristics of Water

What makes water wet? Why is water wet? David Clary (1997), a chemist at University College London, points out that water does not start to behave like a liquid until at least six molecules form a cluster. He found that groups of five water molecules or fewer have planar structures, forming films one molecule thick. However, when a sixth molecule is added, the cluster switches to a three-dimensional cage-like structure, and suddenly it has the properties of water—it becomes wet.

Beyond the physical property of wetness, other physical characteristics of interest to us include water's solids, turbidity, color, taste and odor, and temperature—all of which are apparent to our senses of smell, taste, sight, and touch.

Solids

Other than gases, all contaminants of water contribute to its total solids (filterable + nonfilterable solids) content. *Solids* are classified by size and state, chemical characteristics, and size distribution. Solids can be dispersed in water in both suspended and dissolved forms. Solids are also size classified as suspended, settleable, colloidal, or dissolved, depending on their behavioral attributes. *Colloidal material* in water is sometimes beneficial and sometimes harmful. Beneficial colloids are those that provide a dispersant effect by acting as protective colloids. Some colloids (silica-based types) can be troublesome, forming very hard scale when deposited on heat transfer surfaces (Kemmer, 1988).

Chemically, solids are also characterized as being *volatile* (solids that volatize at 550°C) or *nonvolatile*.

In determining the distribution of solids, we compute the percentage of solids by size range. Typically, solids include inorganic solids (silt and clay from riverbanks) and organic matter such as plant fibers and microorganisms from natural or man-made sources. In flowing water, many of these contaminants result from the erosive action of water flowing over surfaces. Suspended material is not normally found in groundwater because of the filtering effect imparted by the soil.

Suspended material present in water is objectionable because it provides adsorption sites for biological and chemical changes. These adsorption sites provide attached microorganisms a protective barrier against the chemical action of chlorine used in the disinfection process. Suspended solids in water may also be objectionable because they may be degraded biologically into unwanted by-products. Obviously, the removal of these solids is of primary concern in the production of clean, safe drinking water.

In the water treatment process, the most effective means of removing solids from water is by filtration. However, not all solids (colloids and other dissolved solids) can be removed by filtration.

Turbidity

One of the first conditions we notice about water is its clarity, which we measure by *turbidity*, an assessment of the extent to which light is either absorbed or scattered by suspended material in water. Absorption and scattering are influenced by both the size and surface characteristics of the suspended material.

In surface water, most turbidity results from the erosion of very small colloidal material (rock fragments, silt, clay, and metal oxides from the soil). Microorganisms and vegetable material may also contribute to turbidity. In running water, turbidity interferes with light penetration and photosynthetic reactions that are critical to aquatic plants. In water treatment, turbidity is a useful indicator of water quality.

Color

The *color* of water is a physical characteristic often used to judge water quality. Pure water is colorless. Water takes on color when foreign substances—organic matter from soils, vegetation, minerals, and aquatic organisms—are present. For the most part, color in water is a mixture of colloidal organic compounds that represent breakdown products of high molecular weight substances produced by living cells. Consider, for example, water with a yellowish or tea color. The source of this yellow color is decayed vegetation leached from the watershed by runoff. These organic materials are broadly classified as humic substances (Kemmer, 1988). Color can also be contributed to water by municipal and industrial wastes.

Color in water is classified as either *true color* or *apparent color*. Water whose color is partly due to dissolved solids that remain after removal of suspended matter is known as true color. Color contributed by suspended matter is said to have apparent color. In water treatment, true color is the most difficult to remove.

Colored water is generally unacceptable to the general public. People tend to prefer clear, uncolored water. Another problem with colored water is the affect it has on manufacturing, textiles, food preparation/processing, papermaking, and laundering. The color of water has a profound effect on its marketability for both domestic and industrial use.

In water treatment, color is not usually considered unsafe or unsanitary, but is a treatment problem in regard to exerting chlorine demand, which reduces the effectiveness of chlorine as a disinfectant. Some of the processes used in treating colored water include filtration, softening, oxidation, chlorination, and adsorption.

Taste and Odor

Taste and *odor* are used jointly in the vernacular of freshwater science. In drinking water, odor and taste are usually not a consideration until the consumer complains. The problem is, of course, that most consumers find any taste and odor in water aesthetically displeasing. Taste and odor do not normally present a health hazard, but they can cause the customer to seek out water that might taste and smell better, but may also be unsafe to drink. The fact is, consumers expect that water should be tasteless and odorless—and if it isn't, they consider it substandard. If a consumer can taste or smell the water, he or she automatically associates that with contamination.

Water contaminants are attributable to contact with natural substances (rocks, vegetation, soil, etc.) or human use. Taste and odor are caused by foreign matter—organic compounds, inorganic salts, or dissolved gases. Again, these substances may come from domestic, agricultural, or natural sources. Some substances found naturally in groundwater, while not necessarily harmful, may impart a disagreeable taste or undesirable property to the water. Magnesium sulfate, sodium sulfate, and sodium chloride are a few of these (Corbitt, 1990).

When water has a taste but no accompanying odor, the cause is usually inorganic contamination. Water that tastes bitter is usually alkaline, while salty water is commonly the result of the salts mentioned earlier. However, when water has both taste and odor, the likely cause is organic materials. The list of possible organic contaminants is too long to record here, but petroleum-based products lead the list of offenders. Taste- and odor-producing liquids and gases in water are produced by biological decomposition of organics. A prime example of one of these is hydrogen sulfide, known best for its characteristic "rotten-egg" taste and smell. In addition, a number of other distinct smells are commonly encountered (see Table 15.2).

Objectionable tastes and odors also caused by biological activity include those contributed by various algae species (e.g., *Diatomaceae:* Asterionella, Synerdra; *Protozoa*: Synura, Dionbyron; *Cyanophyceae*: Anabaea, Aphanizomenon; and *Chlorophyceae*: Volvox, Staurastrum), diatoms, and actinomycetes that produce organic by-products, such as essential oils, which can be observed by microscopic examination. The release of these materials into water, particularly when large populations of organisms die, produces objectionable tastes and odors (Kemmer, 1988).

Table 15.2: Categories of Offensive Odors Often Encountered in Water

Compound	Descriptive Quality
Amines	Fishy
Ammonia	Ammoniacal
Diamines	Decayed flesh
Hydrogen sulfide	Rotten egg
Mercaptans	Skunk secretion
Organic sulfides	Rotten cabbage
Skatole	Fecal

Source: Adaptation from: *The Chemical Senses*, Moncrief, R. W., (1967).

In water treatment, one of the common methods used to remove taste and odor is oxidation (usually with potassium permanganate and chlorine) of the problem material. Another standard treatment method is to feed powdered activated carbon to the flow prior to filtration. The activated carbon has numerous small openings that adsorb the components that cause odor and taste.

Temperature

Most consumers prefer drinking waters that are consistently cool and do not have temperature fluctuations of more than a few degrees. Groundwater from mountainous areas generally meets these criteria. Water with a temperature between 10° to 15°C (50° and 60°F) is generally the most palatable (Corbitt, 1990).

Heat is added to surface and groundwater in many ways. Some of these are natural, some artificial. A problem associated with heat or excessive temperature in surface waters is that it affects the solubility of oxygen in water, the rate of bacterial activity, and the rate at which gases are transferred to and from the water.

In the actual examination of water (for its suitability for consumption), temperature is not normally a parameter used in evaluation. However, temperature is one of the most important parameters in natural surface-water systems, which are subject to great temperature variations. It affects a number of important water quality parameters. Temperature has an effect on the rate at which chemicals dissolve and react. When water is cold, more chemicals are required for efficient coagulation and flocculation to take place. When water temperature is high, the result may be a higher chlorine demand because of increased reactivity, and because of an increased level of algae and other organic matter in raw water. Temperature also has a pronounced effect on the solubilities of gases in water.

Ambient temperature (temperature of the surrounding atmosphere) has the most profound and universal effect on the temperature of shallow natural water systems. When water is used by industry to dissipate process waste heat, the discharge points in surface waters may experience dramatic localized temperature changes. Other sources of increased temperatures in running water systems result because of clear-cutting practices in forests (where protective canopies are removed) and also from irrigation flows returned to a body of running water.

Many people hold a misconception related to water temperature. Although the temperature of groundwater seems relatively "cool" in summer and warm in winter, its temperature remains nearly constant throughout the year. Human perception of temperature is relative to air temperature—the slight temperature fluctuation is not usually detectable. Contrary to popular belief, colder water is not obtained by drilling deeper wells. Beyond the 100 ft. depth mark, the temperature of groundwater actually increases steadily at the rate of about 0.6°C (1°F) for each 100 ft. or so of depth. This rate may increase dramatically in volcanic regions.

Chemical Characteristics of Water

Other parameters used to define water quality are the water's chemical characteristics. Water's chemical composition is changed by the nature of the rocks that form the earth's crust. As surface water seeps down to the water table, it dissolves and carries portions of the minerals contained in/by soils and rocks. Because of this, groundwater usually is heavier in dissolved mineral content than surface water.

Each chemical constituent that water may contain (see Table 15.3 for principal constituents and attributes) affects water use in some manner, by either restricting or enhancing specific uses.

Water, commonly called the *universal solvent*, is a solvent because of its chemical characteristics. Water analysts test a water supply to determine the supply's chemical characteristics, to determine if other harmful substances are present, to determine if substances are present that will enhance corrosion (of metals in water heaters, for example), and to determine if the chemicals responsible for staining fixtures and clothing are present in the water. The exact analyses to be conducted on a municipal water supply are mandated by Public Health Service Drinking Water Standards promulgated by the U.S. Department of Health, Education and Welfare.

Table 15.3: Chemical Constituents and Attributes of Water

Constituents	Attributes
Calcium	Silica
Magnesium	TDS
Sodium	Hardness
Potassium	Color
Iron	pH
Manganese	Turbidity
Bicarbonate	Temperature
Carbonate	
Sulfate	
Chloride	
Fluorine	
Nitrate	

Along with the elements and attributes listed in Table 15.3 and toxic substances, water-quality managers are concerned with the presence of total dissolved solids (TDS), alkalinity, hardness, fluorides, metals, organics, and nutrients that might be present in a water supply. These chemical parameters are discussed in the following sections.

Total Dissolved Solids (TDS)

Total dissolved solids (TDS) come from the minerals dissolved in water from rocks and soil as water passes over and through it. The measurement of TDS constitutes a part of the total solids in water; it is the residue remaining in a water sample after filtration or evaporation, and is expressed in mg/l.

TDS is an important water quality parameter and is commonly used to measure salinity. Roughly, fresh water has a TDS of less than 1500 mg/l (drinking water has a recommended maximum TDS level of 500 mg/l), brackish water has a TDS up to 5000 mg/l, and saline water has a TDS above 5000 mg/l. Seawater contains 30,000-34,000 mg/l TDS (Tchobanoglous & Schroeder, 1987).

Dissolved solids may be organic or inorganic. Water may come into contact with these substances within the soil, on surfaces, and/or in the atmosphere. The organic dissolved constituents of water come from degradation (decay) of products of vegetation, from organic chemicals, and from organic gases.

Dissolved solids can be removed from water by distillation, electrodialysis, reverse osmosis, or ion exchange. Removing these dissolved minerals, gases, and organic constituents is desirable because they may cause physiological effects and produce aesthetically displeasing color, taste, or odor.

You might think removing all these dissolved substances from water is desirable, but this is not a prudent move. Pure, distilled water tastes flat. Also, water has an equilibrium state with respect to dissolved constituents. If water is out of equilibrium or undersaturated, it will aggressively dissolve materials it comes into contact with. Because of this particular problem, substances that are readily dissolvable are sometimes added to pure water to reduce its tendency to dissolve plumbing fixtures.

Alkalinity

Alkalinity, imparted to water by bicarbonate, carbonate, or hydroxide components, is a measure of water's ability to absorb hydrogen ions without significant pH change. Stated another way, alkalinity is a measure of the buffering capacity of water. Alkalinity measures the ability of water to neutralize acids and is the sum of all titratable bases down to about pH 4.5. The bicarbonate, carbonate, and hydroxide constitutes of alkalinity originate from carbon dioxide (from the atmosphere and as a by-product of microbial decomposition of organic material) and from their mineral origin (primarily from chemical compounds dissolved from rocks and soil).

Highly alkaline waters have no known significant impact on human health but are unpalatable. The principal problem with alkaline water is the reactions that occur between alkalinity and certain substances in the water. The resultant precipitate can foul water system appurtenances. Alkalinity levels also affect the efficiency of certain water treatment processes, especially the coagulation process.

Hardness

Water *hardness* is familiar to those individuals who have washed their hands with a bar of soap and found that they needed more soap to "get up a lather." For this reason, originally hardness referred to the soap-consuming power of water. Hardness is the presence in water of multivalent cations, most notably calcium and magnesium ions. Hardness is classified as *carbonate hardness* and *noncarbonate hardness*. The carbonate that is equivalent to the alkalinity is termed carbonate hardness. Hardness is either temporary or permanent. Carbonate hardness (temporary hardness) can be removed by boiling. Noncarbonate hardness cannot be removed by boiling and is classified as permanent.

Hardness values are expressed as an equivalent amount or equivalent weight of calcium carbonate (*equivalent weight* of a substance is its atomic or molecular weight divided by n). Water with a hardness of less than 50 ppm is *soft*. Above 200 ppm, domestic supplies are usually blended to reduce the hardness value. The U.S. Geological Survey uses the following classification:

Range of Hardness [mg/liter (ppm) as $CaCO_3$]	Descriptive Classification
1 to 50	Soft
51 to 150	Moderately hard
151 to 300	Hard
Above 300	Very Hard

Hardness has an economic impact, in soap consumption as well as with problems in tanks and pipes. Another problem with soap and hardness is that when using a bar of soap in hard water, you must work the soap until the lather is built up. When lathering does occur, the water has been "softened" by the soap. The problem is that the precipitate formed by hardness and soap (soap curd) adheres to just about anything (tubs, sinks, dishwashers) and may stain clothing, dishes, and other items. Hardness also affects people in personal ways: the residues of the hardness-soap precipitate may remain in skin pores, causing skin to feel rough and uncomfortable. Today these problems have been largely reduced by the development of synthetic soaps and detergents that do not react with hardness. However, hardness still leads to other problems—scaling and laxative effect. Scaling occurs when carbonate hard water is heated and calcium carbonate and magnesium hydroxide are precipitated out of solution, forming a rock-hard scale that clogs hot-water pipes and reduces the efficiency of boilers, water heaters, and heat exchangers. Hardness, especially with the combined presence of magnesium sulfates, can lead to the development of laxative effect on new consumers.

Rowe and Abdel-Magid (1995) point out that using hard water has some advantages. These include: (1) hard water aids in growth of teeth and bones, (2) hard water reduces toxicity to humans by poisoning with lead oxide from pipelines made of lead, and (3) soft waters are suspected to be associated with cardiovascular diseases.

Fluoride

Fluoride, seldom found in appreciable quantities in surface waters, appears in groundwater in only a few geographical regions, though it is sometimes found in a few types of igneous or sedimentary rocks. Fluoride is toxic to humans in large quantities, and is also toxic to some animals. Some plants used for fodder can store and concentrate fluoride. Animals that consume this forage ingest an enormous overdose of fluoride. Their teeth mottle, they lose weight, give less milk, grow bone spurs, and become so crippled they must be destroyed (Koren, 1991).

Fluoride in small concentrations is beneficial for controlling dental caries. Water containing the proper amount of fluoride can reduce tooth decay by 65% in children between ages twelve and fifteen. Adding fluoride to provide a residual of 1.5 to 2.5 mg/l has become common practice for municipal water plants. Concentrations above 5 mg/l are detrimental and limited by drinking water standards.

How does the fluoridation of a drinking water supply actually work to reduce tooth decay?

Fluoride combines chemically with tooth enamel when permanent teeth are forming. The result, of course, is harder, stronger teeth more resistant to decay. Adult teeth are not affected by fluoride.

The EPA sets the upper limit for fluoride based on ambient temperatures, because people drink more water in warmer climates. Fluoride concentrations should be lower in those areas.

Metals

Iron and manganese are commonly found in groundwater, but surface water, at times, may also contain significant amounts. Metals in water are classified as either toxic or nontoxic. Only those metals that are harmful in relatively small amounts are labeled toxic; other metals fall into the nontoxic group. In natural waters, sources of metals include dissolution from natural deposits and discharges of domestic, agricultural, or industrial wastewater.

Nontoxic metals commonly found in water include the hardness ions (calcium and manganese), iron, aluminum, copper, zinc, and sodium. Sodium (abundant in the earth's crust and highly reactive with other elements) is by far the most common nontoxic metal found in natural waters. Sodium salts (in excessive concentrations) cause a bitter taste in water and are a health hazard for kidney and cardiac patients. The usual low-sodium diets allow for 20 mg/l sodium in drinking water. Sodium, in large concentrations, is toxic to plants.

Although iron and manganese in natural waters (in very small quantities) may cause color problems, they frequently occur together and present no health hazards at normal concentrations. Some bacteria, however, use manganese compounds as an energy source, and the resulting slime growth may produce taste and odor problems. The recommended limit for iron is 0.3 mg/l, and it is 0.05 mg/l for manganese.

Very small quantities of other nontoxic metals are found in natural water systems. Most of these metals cause taste problems well before they reach toxic levels.

Toxic metals are present, fortunately, in only minute quantities in most natural water systems. However, even in small quantities, these toxic metals can be especially harmful to humans and other organisms. Arsenic, barium, cadmium, chromium, lead, mercury, and silver are toxic metals that dissolve in water. Arsenic, cadmium, lead, and mercury—all cumulative toxins—are particularly hazardous to human health. Concentrated in organism bodies, these toxins are passed up the food chain and pose the greatest threat to organisms at the top of the chain.

Organic Matter

Organic matter includes both natural and synthetic molecules containing carbon, and usually hydrogen. All living matter is made up of organic molecules. Some organics are extremely soluble in water (alcohol and sugar are good examples), while some may be quite insoluble (plastics).

Tchobanoglous and Schroeder (1987) point out that the presence of organic matter in water is troublesome for the following reasons: "(1) color formation, (2) taste and odor problems, (3) oxygen depletion in streams, (4) interference with water treatment processes, and (5) the formation of halogenated compounds when chlorine is added to disinfect water" (p. 94).

The main source of organic matter in water—although total amounts in water are low—is from decaying vegetation. The general category of organics in natural waters includes organic matter whose origins could be from both natural sources and from human activities. Distinguishing natural organic compounds from solely man-made compounds (for example, pesticides and other synthetic organic compounds) is critical.

Those natural organic materials soluble in water are generally limited to contamination of surface water only. These dissolved organics are usually divided into two categories: *biodegradable* and *non-biodegradable*.

Biodegradable (tends to break down) material consists of organics that can be utilized for nutrients (food) by naturally occurring microorganisms within a reasonable length of time. These materials usually consist of alcohols, acids, starches, fats, proteins, esters, and aldehydes. They may result from domestic or industrial wastewater discharges, or they may be end products of the initial decomposition of plant or animal tissue. The principal problem associated with biodegradable organics is the effect resulting from the action of microorganisms. Secondary problems include color, taste, and odor problems.

For microorganisms to use dissolved organic material effectively requires *oxidation* and *reduction* processes. In oxidation, oxygen is added to or hydrogen is deleted from elements of the organic molecule. Reduction occurs when hydrogen is added to or oxygen is deleted from elements of the organic molecule. The oxidation process is by far more efficient and is predominant when oxygen is available. In aerobic (oxygen-present) environments, the end products of microbial decomposition of organics are stable and acceptable compounds. On the other hand, anaerobic (oxygen-absent) decomposition results in unstable and objectionable end products.

The quantity of oxygen-consuming organics in water is usually determined by measuring the *biological oxygen demand (BOD)*: the amount of dissolved oxygen needed by aerobic decomposers to break down the organic materials in a given volume of water over a 5-day incubation period at 208°C (68°F).

Nonbiodegradable organics (resistant to biological degradation and thus considered refractory) include constituents of woody plants (tannin and lignic acids, phenols, and cellulose) and are found in natural water systems. Some polysaccharides with exceptionally strong bonds, and benzene with its ringed structure, are essentially nonbiodegradable.

Some organics are toxic to organisms and thus are nonbiodegradable. These include the organic pesticides and compounds that have combined with chlorine.

Pesticides and herbicides have found widespread use in agriculture, forestry, and mosquito control. Surface streams are contaminated via runoff from rainfall. These toxic substances are harmful to some fish, shellfish, predatory birds, and mammals. Some compounds are toxic to humans.

Certain nonbiodegradable chemicals can react with oxygen dissolved in water. The *chemical oxygen demand (COD)*—which is the amount of oxygen needed to chemically oxidize a waste—is a more complete and accurate measurement of the total depletion of dissolved oxygen in water.

Nutrients

Nutrients are elements (carbon, nitrogen, phosphorous, sulfur, calcium, iron, potassium, manganese, cobalt, and boron) essential to the growth and reproduction of plants and animals. Aquatic species depend on their watery environment to provide their nutrients. In water quality terms, however, nutrients can be considered pollutants when their concentrations are sufficient to encourage excessive growth of aquatic plants such as algal blooms. The nutrients required in most abundance by aquatic species are carbon, nitrogen, and phosphorous. Plants, in particular, require large amounts of each of these three nutrients or their growth is limited.

Carbon is readily available from a number of natural sources including alkalinity, decaying products of organic matter, and from dissolved carbon dioxide from the atmosphere. Since carbon is usually readily available, it is seldom a *limiting nutrient* (the nutrient least available relative to the plant's needs) (Masters, 1991). The limiting nutrient concept is important because it suggests that algal growth can be controlled by identifying and reducing the supply of a particular nutrient. In most cases, nitrogen and phosphorous are essential growth factors and are the limiting factors in aquatic plant growth. According to Welch (1980), seawater is most often limited by nitrogen, while freshwater systems are most often limited by phosphorous.

Nitrogen gas (N_2), which is extremely stable, is the primary component of the earth's atmosphere. Major sources of nitrogen in water include runoff from animal feedlots, fertilizer runoff from agricultural fields, from municipal wastewater discharges, and from certain bacteria and blue-green algae that obtains nitrogen directly from the atmosphere. Certain forms of acid rain can also contribute nitrogen to surface waters.

In water, nitrogen is frequently found in the form of nitrate (NO_3). Nitrate in drinking water can lead to serious problems—most notably, nitrate poisoning in infants. Human and animal infants can be affected by nitrate poisoning, which can cause serious illness and even death, if bacteria commonly found in the infant's intestinal tract converts nitrates to highly toxic nitrites (NO_2).

Nitrite can replace oxygen in the bloodstream and result in oxygen starvation, which causes a bluish discoloration of the infant ("blue baby" syndrome).

In aquatic surface water systems, nitrogen (and phosphorous in the form of phosphate) is a chemical that can stimulate biological growth and is classified as a *biostimulent*. As biostimulents, nitrogen and phosphates (derived from fertilizers and detergents) are impurities that can result in greatly increased *eutrophication* (or slow death) of the body of water.

Eutrophication of water systems, especially in lakes, is usually a natural phenomena that occurs over time. Increases in biostimulents (nitrogen and especially phosphate, or any other growth-limiting nutrient) speed up eutrophication, affecting the natural process.

Eutrophication can result when algal blooms get a large supply of nitrogen or phosphorous (or both) and grow out of control. When natural processes falls out of control, nature steps in and lowers the boom. Algal blooms die off at the end of the growing season, then degrade and provide a rich source of organic material for bacteria. With an ample food supply, bacteria may grow exponentially, consuming dissolved oxygen in the process. As food supplies are consumed, waste products accumulate, and the process slows (Chapter 16 discusses this in more detail).

Biological Characteristics of Water

Along with the physical and chemical parameters of water quality, the environmental science practitioner is also concerned with the biological characteristics of water. This concern is well warranted when you consider that the health and well-being of the people who receive and use the product from the "end-of-the-pipe" is at stake. In this context, remember that water may serve as a medium in which thousands of biological species spend part, if not all, of their life cycles. Note that, to some extent, all members of the biological aquatic community serve as parameters of water quality because their presence or absence may indicate in general terms the characteristics of a given body of water.

The biological characteristics of water impact directly on water quality. To a lesser degree this impact includes the development of tastes and odors in surface water and groundwater and the corrosion of and biofouling of heat transfer surfaces in cooling systems and water supply treatment facilities. However, it is the presence or absence, in particular, of certain biological organisms that is of paramount importance to the water specialist. These organisms are the *pathogens*. Pathogens are the organisms that are capable of infecting or transmitting diseases to humans and animals. These organisms are not native to aquatic systems, and usually require an animal host for growth and reproduction. They can be and are transported by natural water systems. These waterborne pathogens include species of bacteria, viruses, protozoa, and parasitic worms (helminths). In the sections to follow, a brief, basic description of each of these is provided, along with a brief description of rotifers, crustaceans, fungi, and algae, which are also microorganisms of concern in water.

Before we begin a discussion of the basics of aquatic microorganism types of concern in environmental science, note that in addition to knowledge of the types of microorganisms that inhabit a watery environment, some knowledge of the environmental factors that affect them is important. For example, the environmental practitioner who specializes in water quality must be familiar with the nutritional requirements of aquatic organisms.

To grow in an aquatic environment, organisms must be able to extract the nutrients they need for cell synthesis, and for the generation of energy from their water environment. Some organisms obtain their energy from photosynthetic light. Others obtain their energy from organic or inorganic matter. Carbon is a critical ingredient for all aquatic microorganisms (actually, carbon is critical to all organisms). Some aquatic organisms (higher plants, algae, and photosynthetic bacteria) get their carbon from carbon dioxide. Others (bacteria, fauna, protozoa, animals) get their carbon from organic matter.

In addition to carbon and energy, oxygen plays a critical role in the growth of cells. Many organisms (aerobes) require molecular oxygen (O_2) for their metabolism. Other organisms (anaerobes) do not require molecular oxygen and derive the oxygen they need for synthesis of cells from chemical compounds.

Along with the nutritional requirements of microorganisms, also consider the affects of environmental factors covered earlier in the text. These factors include chemical composition, pH, temperature, and light.

Bacteria

The word bacteria (singular: bacterium) comes from the Greek word meaning "rod" or "staff," a shape characteristic of many bacteria. Bacteria are single-celled microscopic organisms that multiply by splitting in two (binary fission). To multiply, they need carbon from carbon dioxide if they are autotrophs, which synthesize organic substances from inorganic molecules by using light or chemical energy, or from organic compounds (dead vegetation, meat, sewage) if they are heterotrophs. Their energy comes from sunlight if they are photosynthetic, or from chemical reaction if they are chemosynthetic. Bacteria are present in air, water, earth, rotting vegetation, and the intestines of animals. Gastrointestinal disorders are common symptoms of most diseases transmitted by waterborne pathogenic bacteria.

Virus

A virus is an infectious particle consisting of a core of nucleic acid (DNA and RNA) enclosed in a protein shell. It is an entity that carries the information needed for its replication, but does not possess the machinery to carry out replication. Thus, viruses are obligate parasitic particles that require a host in which to live. They are the smallest biological structures known, and can only be seen with the aid of an electron microscope. Waterborne viral infections, which are known to cause poliomyelitis and infectious hepatitis, are usually indicated by disorders with the nervous system rather than of the gastrointestinal tract.

Protozoa

Protozoa (singular: protozoan) are mobile, single-celled, complete, self-contained organisms that can be free-living or parasitic, pathogenic or nonpathogenic, microscopic or macroscopic. Protozoa range in size from two to several hundred microns in length. They are highly adaptable and widely distributed in natural waters, although only a few are parasitic. Most protozoa are harmless. Only a few cause illness in humans—*Entamoeba histolytica* (amebiasis) and *Giardia lamblia* (giardiasis) being two of the exceptions. Giardiasis (typically contracted by drinking surface water contaminated by wild animals or humans) is the most widespread protozoan disease occurring throughout the world (Tchobanoglous and Schroeder, 1987). Unless properly treated, giardiasis can be chronic. Symptoms of giardiasis usually include diarrhea, nausea, indigestion, flatulence, bloating, fatigue, and appetite and weight loss.

Worms (Helminths)

Worms also are important in water quality assessment from a standpoint of human disease. Normally, worms inhabit organic mud and organic slime. They have aerobic requirements but can metabolize solid organic matter not readily degraded by other microorganisms. Water contamination may result from human and animal waste that contains worms. Worms pose hazards primarily to those persons who come into direct contact with untreated water. Thus swimmers in surface water polluted by sewage or stormwater runoff from cattle feedlots are at particular risk. The *Tubifix* worm is a common organism used as an indicator of pollution in streams.

Rotifers

Rotifers (the name derived from the apparent rotating motion of the cilia located on the head) make up a well-defined group of the smallest, simplest multicellular microorganisms and are found in nearly all aquatic habitats. Strict aerobes, rotifers range in size from about 0.1-0.8 nm. Bacteria are rotifers' primary food source.

Crustaceans

Microscopic crustaceans are important members of freshwater zooplankton and are thus of interest to water specialists. Characterized by a rigid shell structure, they are multicellular, strict aerobes, and primary producers that feed on bacteria and algae. They are important as a source of food for fish.

Fungi

Fungi constitute an extremely important and interesting group of aerobic microbes ranging from the unicellular yeasts to the extensively mycelial molds. Fungi are not considered plants, but are instead a distinctive lifeform of great practical and ecological importance. Like bacteria, fungi metabolize dissolved organic matter. As the principal organisms responsible for decomposition of carbon in the biosphere, the importance of fungi cannot be overstated. Fungi are unique (in a sense, when compared to bacteria) in that they can grow in low moisture areas and in low pH solutions, which aids in the breakdown of organic matter in aquatic environments.

Algae

Algae are a diverse group of autotrophic, photosynthetic, eucaryotic microorganisms containing chlorophyll. The feature that distinguishes algae from fungi is the algae's chlorophyll. Alga microorganisms impact water quality by shifting the balance between oxygen and carbon dioxide in the water, affecting pH levels, and contributing to odor and taste problems.

SUMMARY

One of the blessings of modern life we enjoy (but usually ignore) is readily available, clean, safe, potable water. As anyone who has carried water along for camping or back-packing or has traveled or lived in a country without safe drinking water knows that just being able to turn on the tap to get water is an enormous advantage modern technology has given us. We don't have to carry water in pails from the village well, the river, or our own well or spring. We don't have to boil it or chemically treat it to drink or cook with it. We don't have to brush our teeth with bottled water. We take our safe water for granted, even if we buy springwater to drink because we don't like the taste of the treated water from the faucet.

We also forget that even when our water supplies were relatively unspoiled, unsafe waters killed many because the technology available to test the water was not available. The physical, chemical, and biological characteristics of water and our understanding of the forces at work on groundwater and surface water allow us to control the quality of the water we rely on.

Cited References

ASTM, *Manual of Water*, Philadelphia: American Society for Testing and Materials, 1969.

Clary, D., "What Makes Water Wet," *Geraghty & Miller Newsletter*, 39,4, 1997.

Corbitt, R. A., *Standard Handbook of Environmental Engineering*, New York: McGraw-Hill, 1990.

Kemmer, F. N., *The Nalco Water Handbook*, 2nd ed., New York: McGraw-Hill, 1988.

Kemmer, F. N., *Water: The Universal Solvent*, 2nd ed., Oak Brook, Illinois: Nalco Chemical Company, 1979.

Koren, H., *Handbook of Environmental Health & Safety: Principles and Practices*, Chelsea, Michigan: Lewis Publishers, 1991.

Manahan, S. E., *Environmental Science and Technology*, Boca Raton: Florida, Lewis Publishers, 1997.

Masters, G. M., *Introduction to Environmental Engineering and Science*, Englewood Cliffs, NJ: Prentice-Hall, 1991.

McGhee, T. J., *Water Supply and Sewerage*, New York: McGraw-Hill, 1991.

Moncrief, R. W., *The Chemical Senses*, 3rd ed., London: Leonard Hill, 1967.

Outwater, A., *Water: A Natural History*, New York: Basic Books, 1996.

Peavy, H. S., Rowe, D. R., and Tchobanoglous, G., *Environmental Engineering*, New York: McGraw-Hill, 1985

Rowe, D. R., and Abdel-Magid, I. M., *Handbook of Wastewater Reclamation and Reuse*, Boca Raton, Florida: Lewis Publishers, 1995.

Spellman, F. R., *Microbiology for Water/Wastewater Operators*, Lancaster, PA: Technomic Publishing Company, 1997.

Spellman, F. R., *The Science of Water: Concepts and Applications*, Lancaster, PA: Technomic Publishing Company, 1998.

Tchobanoglous, G., and Schroeder, E. D., *Water Quality*, Reading, Massachusetts: Addison-Wesley Publishing Company, 1987.

U. S Watersheds Have Water Quality Problems, *Environmental Technology*, Nov/Dec., p.10, 1997.

Welch, E. G., *Ecological Effects of Waste Water*, Cambridge, UK: Cambridge University Press, 1980.

Suggested Readings

Arbuckle, J., Randle, G., and Randle, R. V., *Clean Water Handbook*, Rockville, MD: Government Institutes, 1990.

Conservation Foundation, *State of the Environment*, Washington, DC: The Conservation Foundation, 1984.

Consumer Reports Books Edition, and Gabler, R. *Is Your Water Safe to Drink?* New York: Consumer Reports Books, 1987.

Eckenfelder, W. W., *Industrial Water Pollution Control*, New York: McGraw-Hill, 1966

Horne, A. J., and Goldman, C. R., *Limnology*, 2nd ed., New York: McGraw-Hill, 1994

LaMoreaux, P. E., *Environmental Hydrogeology*, Boca Raton, FL: CRC Press/Lewis Publishers, 1997.

McKinney, R. E., *Microbiology for Sanitary Engineers*, New York: McGraw-Hill, 1962.

Mitchell, R., *Environmental Microbiology*, New York: Wiley, 1992.

Randtke, S. J., and Snoeyink, U. L., Evaluating GAC Adsorptive Capacity, *J. Am. Water Works Assoc.*, 75, 406-413, 1983.

Rosenweig, W. D., Minnigh, H., and Pipes, W. O., Fungi in Distribution Systems, *Journal of American Water Works Association*, 78:1:53, 1986.

Salvato, J. A., *Environmental Engineering and Sanitation*, 3rd ed., New York: Wiley Interscience, 1982.

Singh, Vijay, P., *Environmental Hydrology*, Norwell, MA: Kluwer Academic Publishers, 1995.

Spellman, F. R., *Stream Ecology and Self-Purification*, Lancaster, PA: Technomic Publishing Company, 1996.

Thomann, R. V. and Mueller, J. A., *Principles of Surface Water Quality Modeling and Control*, New York: Harper & Row, 1987.

U.S. Environmental Protection Agency, "Ambient Water Quality Criterion for Dissolved Oxygen," *Federal Register*, 50 (76), 1985.

U.S. Environmental Protection Agency, "Drinking water in America: An overview," *EPA Journal*, September, 1986.

U.S. Geological Survey, *Estimated Use of Water in the United States*, Washington, DC: Department of Interior, 1984.

U.S. Water Resource Council, *The Nation's Water Resources, 1975-2000*, Washington, DC: Government Printing Office, 1978.

CHAPTER 16

CHARACTERISTICS OF FRESHWATER BODIES

Early in the spring on a snow and ice-covered high alpine meadow is an example of the continuation of the water cycle. Water, the cycle's main component, has been held in reserve, frozen for the long dark winter months—but with longer, warmer spring days, the sun is higher, more direct, and of longer duration, and the frozen masses of water respond to the increased warmth. The melt begins with a single drop, then two, then more and more. As the snow and ice melts, the drops join a chorus that continues unending; they fall from their ice-bound lip to the bare rock and soil terrain below. The terrain the snowmelt strikes is not like glacial till, the unconsolidated, heterogeneous mixture of clay, sand, gravel, and boulders, dug-out, ground-out, and exposed by the force of a huge, slow, and inexorably moving glacier. Instead, this soil and rock ground is exposed to the falling drops of snowmelt because of a combination of wind and the tiny, enduring force exerted by drops of water as over season after season they collide with the thin soil cover, exposing the intimate bones of the earth.

Gradually the single drops increase to a small rush. They join to form a splashing, rebounding, helter-skelter cascade, many separate rivulets that trickle, then run their way down the face of the granite mountain. At an indented ledge halfway down the mountain slope, a pool forms whose beauty, clarity, and sweet iciness provides the visitor with an incomparable gift—a blessing from the earth.

The mountain pool fills slowly, tranquil under the blue sky, reflecting the pines, snow, and sky around and above it, an open invitation to lie down and drink, and to peer into the glass-clear, deep waters. The pool has no transition from shallow margin to depth; it is simply deep and pure. As the pool fills with more melt water, we wish to hold this place and this pool in its perfect state forever, it is such a rarity to us in our modern world. But this cannot be, and for a brief instant the water laps in the breeze against the outermost edge of the ridge, then a trickle flows over the rim. Gravity reaches out and tips the overflowing melt onward, and it continues the downward journey, following the path of least resistance to its next destination, several thousand feet below.

When the overflow, still high in altitude, but its rock-strewn bed bent downward, toward the sea, meets the angled, broken rocks below, it bounces, bursts, and mists its way against steep, V-shaped walls that form a small valley, carved out over time by water and the forces of the earth.

Within the valley confines, the melt water has grown from drops to rivulets to a small mass of flowing water. It flows through what is at first a narrow opening, gaining strength, speed, and power as the V-shaped valley widens to form a U-shape. The journey continues as the water mass picks up speed and tumbles over massive boulders, then slows again.

At a larger but shallower pool, waters from higher elevations have joined the main body—from the hillsides, crevices, springs, rills, mountain creeks. At the influent poolsides all appears peaceful, quiet, and restful, but not far away, at the effluent end of the pool, gravity takes control again. The overflow is flung over the jagged lip, and cascades downward several hundred feet, where the waterfall again brings its load to a violent, mist-filled meeting.

The water separates and joins again and again, forming a deep, furious, wild stream that calms gradually as it continues to flow over lands that are less steep. The waters widen into pools overhung by vegetation, surrounded by tall trees. The pure, crystalline waters have become progressively discolored on their downward journey, stained brown with humic acid, and literally filled with suspended sediments; the once-pure stream is now muddy.

The mass divides and flows in different directions, over different landscapes. Small streams divert and flow into open country. Different soils work to retain or speed the waters, and in some places the waters spread out into shallow swamps, bogs, marshes, fens, or mires. Other streams pause long enough to fill deep depressions in the land and form lakes. For a time, the water remains and pauses in its journey to the sea. But this is only a short-term pause, because lakes are only a short-term resting place in the water cycle. The water will eventually move on, by evaporation, or seepage into groundwater. Other portions of the water mass stay with the main flow, and the speed of flow changes to form a river, which braids its way through the landscape, heading for the sea. As it changes speed and slows, the river bottom changes from rock and stone to silt and clay. Plants begin to grow, stems thicken, and leaves broaden. The river is now full of life and the nutrients needed to sustain life. But the river courses onward, its destiny met when the flowing rich mass slows its last and finally spills into the sea.

CHAPTER OUTLINE

- Introduction
- Surface Water
- Groundwater

INTRODUCTION

From the space shuttle circling 150 miles above the earth, we could see our world in a way very few people have experienced, viewing it with a sense of unity and completeness, of interconnection difficult to comprehend from our usual position on the ground. What our eyes might be drawn to first is the circulating cloud masses and patterns of the earth's weather as it passes over the globe. Then the intense, deep blues would attract our attention through the cloud cover and thin haze, broken here and there by the brown, green, and tan landmasses on the oceans of blue.

This view from above the earth serves to verify on a visceral, visual level what most of us know only intellectually—that the earth is literally shrouded in water. Still, the sight of so much water drives home another point that most of us don't contemplate, realize, or understand—water not only shrouds the earth, covering roughly 71% of its surface mass, but controls, predominates, dominates—it is everywhere. We occupy landmasses that are nothing more than islands surrounded by water.

Our most abundant resource, water covers almost three-quarters of the earth's surface, yet it is only a thin film, of which almost 97% is salt water. This veil of water helps maintain climate, works to dilute environmental pollution, and, of course, is essential to life. Without fresh water there would be no agriculture, no manufacturing, no transportation, and no life as we know it.

From our space ship high above the earth, we would also notice that water—fresh water—is not evenly distributed. Some areas have too little water and others too much. Human beings have, with varying degrees of success, attempted to correct these imbalances by capturing fresh water in reservoirs, behind dams, transferring fresh water in rivers and streams from one area to another, tapping underground supplies, and endeavoring to reduce water use, waste, and contamination. In some of these efforts we have been successful; in others we are still learning.

Our focus on water in this text, and particularly in this chapter, is on the characteristics of "fresh water" bodies. We pointed out earlier that most of the earth's water supply is salty, and only a small portion (less than 3%) fresh. Three-quarters of the fresh water is found in polar ice caps and glaciers, unavailable for human use, and nearly all of the remaining one fourth is found under the earth's crust in water-bearing rock, or in sand and gravel formations. Only a small proportion (about 0.5%) of all water on the earth is found in lakes, rivers, streams, or in the atmosphere. Obviously, this seems a small amount, relative to the earth's total water supply. But it is more than enough—even this small proportion, if it were kept free of pollution and distributed evenly, could provide for the drinking, food preparation, and agricultural needs of all the earth's people. We simply need to learn how to better manage and conserve the fresh water readily available to us.

Water is classified as either marine or fresh, depending upon its salt content. The salt content of marine waters is fairly consistent, about 35 parts per thousand (ppt). On average the salt content of fresh waters is 0.5 parts per thousand. The salt content of fresh waters tends to vary more than that of marine waters because lakes, rivers, and streams are much more dominated by local environmental conditions, which include the rate of evaporation and the mineral content of soils.

In the following sections, we discuss fresh water found in two basic forms: *surface water* and *groundwater*. Precipitation that does not infiltrate into the ground or return to the atmosphere is known as surface water and becomes *runoff*—water that flows into nearby lakes, wetlands, streams, rivers, and reservoirs.

Precipitation, under the influence of gravity, infiltrates and percolates slowly through porous ground material deep into the earth. It completely saturates water-bearing layers of the earth's crust (*aquifers*) and eventually becomes a part of groundwater stores. Although classified separately, surface water and groundwater are not entirely distinct. Some water from lakes, streams, and rivers may percolate downward to groundwater supplies. Springs feed into surface water bodies, returning to the water cycle. During dry seasons when surface runoff is minimal, groundwater sources help to maintain the flow of rivers and streams, and the water level of lakes.

Before we continue our discussion of surface water and groundwater bodies, you should review the basic concepts of the hydrologic cycle (see Figure 6.6).

Actually a manifestation of an enormous heat engine, the water cycle raises water from the oceans in warmer latitudes by a prodigious transformation of solar energy. Transferred through the atmosphere by the winds, the water is deposited far away over sea or land. Figure 6.6 depicts this ongoing natural circulation of water through the atmosphere. It shows water taken from the earth's surface to the atmosphere (either by evaporation from the surface of lakes, rivers, streams, and oceans or through transpiration of plants), where it forms clouds that condense to deposit moisture on the land and sea as rain or snow. The water that collects on land flows back to the oceans in streams and rivers.

The water that we see is surface water. The EPA defines surface water as all water open to the atmosphere and subject to runoff (1989). Surface water can be divided into five categories: oceans, lakes, rivers and streams, estuaries, and wetlands

Each category of surface water is important. However, the water contained in surface water bodies and available for human consumption is the focus in the next section.

SURFACE WATER

The study of surface or open fresh water bodies is known as *Limnology*. More specifically, limnology is the study of bodies of open fresh water (lakes, rivers, and streams) in terms of their plant and animal biology and their physical properties. Limnology divides fresh water systems into two groups or classes, *lentic* and *lotic*. The lentic (lenis = calm) or standing water systems are the lakes, ponds, impoundments, reservoirs, and swamps. The lotic (lotus = washed) or running water systems are the rivers, streams, brooks, and springs. On occasion, these two different systems are not well differentiated. This can be seen in the case of old, wide, and deep rivers where water velocity is quite low and the system becomes similar to that of a pond.

On its journey over the earth's landscape, surface water produced by melting snow or ice or from rainstorms follows the path of lease resistance. A series of droplets, rills, rivulets, brooks, creeks, streams, and rivers carries water from an area of elevated land surface that slopes down toward one primary watercourse. This *drainage area* is known as a *watershed* or *drainage basin* (see Figure 16.1). A watershed is a basin surrounded by a ridge of high ground, called the *watershed divide*, which separates one drainage area from another.

Lentic (standing or still) Water Systems

Natural lentic water systems consist of lakes, ponds, bogs, marshes, and swamps. The other standing fresh water bodies, such as reservoirs, oxidation ponds, and holding basins, are usually man-made. In this section we focus on lakes and reservoirs, primarily because these two bodies of standing water are widely used as sources of potable water. However, there probably is no better or easier way to gain understanding of lentic water systems than by studying the pond, and thus we begin our discussion of lentic water systems there.

Ponds

Simply defined, a pond is a still body of water smaller than a lake, often of artificial construction. A pond may be ancient ("permanent") or transitory. Those of the shortest duration are known as *vernal ponds* (pertaining to spring) and are generally only flooded in early spring or late winter from melt-water or heavy rains. As summer begins, they begin to shrink, and usually dry up completely before summer's end. Ponds are shallow and the quality of their water is strongly influenced by the soils that form their basins. Pond water temperature is closely related to ambient air temperature.

Though transitory or vernal ponds are usually short-lived, they often exist long enough to support basic

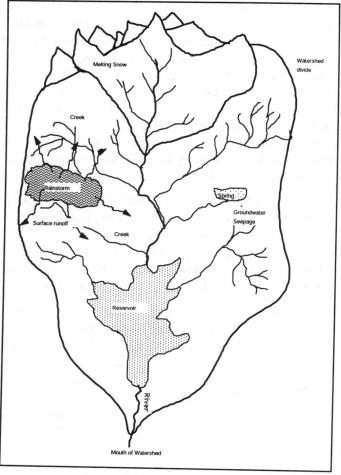

Figure 16.1 Typical watershed

aquatic lifeforms such as branchiopod crustaceans (fairy shrimp), which hatch out and swarm in the cold early season water. Ponds also normally contain various cysts and spores from which rotifers, protozoans, nematodes, and algae emerge. Frogs usually become the next residents of the pool—the perfect setting for mating and egg laying. The eggs hatch quickly and produce rapidly developing tadpoles in short order.

Vernal or transitory ponds are not used for freshwater supplies. Since they lack permanence and stability, they aren't suitable for the permanent potable water supply municipal water systems require. Their transitory nature also causes water problems. Because they are of short duration and subject to contamination from soils and surface runoff, they do not provide water of a quality suitable for human consumption.

Generalizing about so-called *permanent ponds* is difficult because ponds are not really permanent. While the water itself is permanent (we have the same amount of water today as the early Greeks and Roman did, the same amount that has been on earth since the earth was formed), the location is not permanent—water cycles continuously. Generalization is also difficult because if you were to examine a dozen of these ponds, even if they were placed closely in the same geographical location, they would all differ.

A permanent pond is usually defined by the following characteristics: (1) it is shallow enough to permit aquatic plants to penetrate the surface anywhere over its entire mass; (2) its mass is not so great as to allow formation of large waves which could erode the shoreline; (3) it has no temperature layering, rather a gradient of temperatures extending from surface to bottom (Amos, 1969). A pond can also be characterized by its age and productivity. If, for example, it has no planktonic life or rooted plant life, it will not support animal life. Such ponds are nutrient poor— *oligotrophic*. Usually oligotrophic ponds are clear and fairly new. A highly productive pond, one that contains large populations of plants and animals, is nutrient rich—*eutrophic*. Eutrophic ponds can become unhealthy if too much enrichment occurs (usually by phosphorous). The nutrients are decomposed by bacteria and chemical processes, which use large amounts of oxygen from the water, killing aerobic organisms and causing pond stagnation.

Because ponds have a life span (normally ranging from decades to a century or two), which is actually a transformation period from one phase to another, they can also be classified according to their age: young, mature, and senescent (old). Their ages can be measured with a fair amount of accuracy according to their physical appearance and the different organisms resident during each phase of transformation process. This transformation process has been heavily studied, is well-documented, and is known as *pond succession*.

Once a land depression fills with water, a *young pond* forms. Characterized by their lack of sediment, young ponds produce pioneer plants along the shoreline, and their water contains plankton, invertebrates, and fishes (see Figure 16.2 A). As plants grow and decay within the

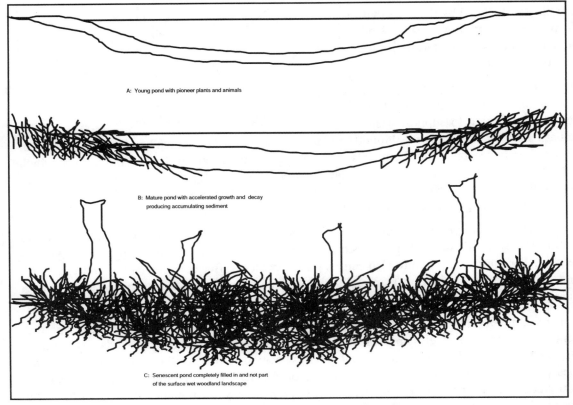

A: Young pond with pioneer plants and animals

B: Mature pond with accelerated growth and decay producing accumulating sediment

C: Senescent pond completely filled in and not part of the surface wet woodland landscape

Figure 16.2 Natural pond succession cycle

pond, along with sediment accumulating from land, the young pond becomes a *mature pond* (see Figure 16.2 B). The mature pond is carpeted with rich sediment, has aquatic vegetation extending out into open water, and has a great diversity of plankton, invertebrates, and fishes. As the pond continues to fill with sediment, it transforms from mature to *senescent* (see Figure 16.2 C). At this point, so much sediment exists within the depression that it is filled and the bottom rises close to the surface. Vegetation grows and covers the entire area. In this environment, the pond can't support fish or plankton, or even many invertebrates. The pond has transformed from a water environment to a land environment.

Pond Habitat

Many of the characteristics and conditions described for ponds also apply to lakes, which we will discuss in detail later.

Ponds contain a number of quite distinct habitats, each inhabited by specifically adapted organisms. Freely swimming organisms are *nektons*. Floating organisms that move with water movement are called *plankton*. Plants and animals living on or near the pond bottom are called *benthic* organisms. Plants and animals living on the surface are *neustons*. Each of these populations can be graded into small subdivisions, often as the result of biotic *zonation* in the pond (Amos, 1969).

Large ponds normally consist of four distinct zones: *littoral, limnetic, profundal,* and *benthic* (see Figure 16.3 A and B). Each zone provides a variety of ecological niches for different species of plant and animal life.

The most obvious (and easiest to observe) zonation is found along the shoreline, the *littoral zone*. In the littoral zone (the outermost shallow-water region), light penetrates to the bottom (see Figure 16.3 A). It provides an interface zone between the land and the open water of ponds. This zone contains rooted vegetation (grasses, sedges, rushes, water lilies and water weeds) that grow in the moist, saturated soil of the pond bank, and a large variety of organisms. The littoral zone is further divided into concentric zones, with one group replacing another as the depth of water changes. Figure 16.3 B shows these concentric zones: *emergent vegetation, floating leaf vegetation*, and *submerged vegetation zones*, proceeding from shallow to deeper water.

The emergents rise through shallow water before producing foliage and flowers, and thus invade the pond a yard or more from the shoreline. Floating-leaf plants begin where the emergents stop. Floating-leaf plants are rooted in the bottom, but have long flexible stems anchoring their buoyant leaves in place. Beyond the floating-leaf plants is the open water of the limnetic zone. Given enough time, however, the floating-leaf plants will literally work their way across the surface of the pond and cover it completely. Below and never quite penetrating the pond surface, the submerged plants grow where light can penetrate and reach them. However, if the pond is covered with floating-leaf plants (thus preventing light penetration), the submerged plants may be totally absent.

The open water portion of the pond that remains shallow enough for effective light penetration is the limnetic zone. The community in this zone is comprised of minute suspended organisms, plankton, and some insects and fish. The species population density is quite low. The rate of photosynthesis is equal to the rate of respiration; thus, the limnetic zone is at compensation level.

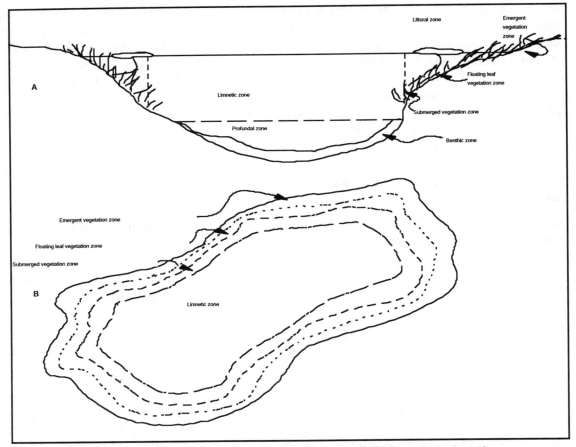

Figure 16.3 A: Vertical section of pond showing major zones; B: View looking down on concentric zones that make up littoral zone

Small shallow ponds do not have this zone; they have a littoral zone only. When all lighted regions of the littoral and limnetic zones are discussed as one, the term *euphotic* is used for both, designating these zones as having sufficient light for photosynthesis and the growth of green plants.

The portion of the pond not penetrated by light is called the profundal zone. Containing darker water with lower oxygen levels, only specially adapted aquatic organisms occupy this zone.

The benthic zone is the bottom of the pond. This zone supports scavengers and decomposers that live on sludge. The decomposers are mostly large numbers of bacteria, fungi, and worms that live on dead plant and animal debris and other wastes that find their way to the bottom.

Lakes

Much of the material covered above about ponds also applies to lakes, especially to their shore-lines and shallows. However, lakes are distinct and separate entities. They are accurately de-scribed as arbitrary flow reactors with long hydraulic residence times. That lakes are an impor-tant source of freshwater is nowhere more apparent and important than in North America, where the United States and Canada share the Great Lakes, considered the largest supply of freshwater in the world. Another huge supply of available freshwater is Lake Baikal in Siberia; it alone contains about the same volume of water as the entire Great Lakes system. Together, the Great Lakes and Lake Baikal contain 40% of the world's available fresh water (NALCO, 1988). Another 15% of the world's available freshwater is contained in several lakes that dot the landscape within Canada.

Lakes are especially sensitive to pollution—much more so than rivers. Rivers, with their moving waters, have the ability, within limits, to self-purify (this process will be explained in detail later). Lakes are still bodies of water that, with the exception of stratification and turnover, do not generally flow and self-purify. Oxygen-demanding wastes are one of the most prevalent lake pollutants. Phosphorus is generally the pollutant that most seriously affects overall water quality in lakes (Davis and Cornwell, 1991). Pathogenic organisms can also thrive in lakes and cause serious health problems for swimmers and others who participate in lake recreational activities.

You must have a basic knowledge of lake systems to understand the role of contaminants in lake pollution. This section is essentially a short course in limnology as it relates to contaminant pollution.

Classification of Lakes

Odum (1971) points out that lakes can be classified in three ways: by eutrophication, by special types of lakes, and by impoundments. *Eutrophication* is a normal aging process that results from recycling and accumulation of organic material over long periods of time. As the sediments continue to accumulate, the lake bed fills until it is transformed into a bog and eventually into a terrestrial ecosystem. In its natural phase, eutrophication is the result of the natural lake succession process. This process is usually slow, and sometimes can take up to tens of thousands of years to complete. However, lake succession via accelerated eutrophication can occur when massive amounts of organic material are dumped into the lake over time. This accelerated process is called *cultural eutrophication.*

Classification of Lakes Based on Eutrophication

Lakes can be classified into three types based on their eutrophication state.

1) *Oligotrophic lakes* (few foods). These are young, deep, nutrient-poor lakes with crystal-clear water, usually more aesthetically pleasing, but with little biomass productivity. The water quality is usually more suitable for a wider range of uses. Lake Superior is an oligotrophic lake, as is Lake Tahoe. Both of these lakes will eventually turn eutrophic; this process is inevitable.

2) *Mesotrophic lakes.* To draw a distinct line between oligotrophic and eutrophic lakes is hard, and often the term *mesotrophic* is used to describe a lake that falls somewhere between the two extremes. Mesotrophic lakes develop with the passage of time. Nutrients and sediments are added through runoff, and the lake becomes more biologically productive. Mesotrophic lakes hold a greater diversity of species with very low populations at first, but a shift toward higher and higher populations with fewer and fewer species occurs. Sediments and solids contributed by runoff and organisms make the lake shallower. At an advanced mesotrophic stage, lakes may have undesirable odors and colors in certain areas. Turbidity increases and organic deposits accumulate on the bottom. Lake Ontario has reached this stage.

3) *Eutrophic lakes* (good foods). These are lakes with large or excessive supplies of nutrients. As nutrients continue to enter the lake system, large, unsightly algal blooms de-

Figure 16.4 A Secchi disk, used to measure water clarity. The disk is lowered into the water until it can no longer be seen. The depth of visual disappearance becomes the Secchi Disk Transparency Extention Coefficient, which will range from a few centimeters in very turbid waters to 35 m in a very clear lake.

velop, fish populations increase, types of fish change from sensitive to more pollution-tolerant species, and biomass productivity becomes very high. The lake takes on undesirable characteristics such as offensive odors, very high turbidity, and a blackish color. This high level of turbidity can be seen in studies of Lake Washington in Seattle. Laws (1993) reports that secchi (a device used to measure turbidity; shown in Figure 16.4) depth measurements made in Lake Washington from 1950 to 1979 show an almost fourfold reduction in water clarity. Along with an increase in turbidity, a eutrophic lake becomes very shallow. Lake Erie and Green Lake (also in Seattle) are at this stage.

Special Types of Lakes

Odum (1971) refers to several special lake types.

1) *Dystrophic* (bog) *lakes* have low pH, and the water color ranges from yellow to brown. Dissolved solids, nitrogen, phosphorus, and calcium are low and humic matter is high. These lakes are sometimes void of fish fauna; other organisms are limited. When fish are present, production is poor (Welch, 1983).

2) *Deep, ancient lakes* are specialized environments found in only a few places—for example, Lake Baikal in Russia.

3) *Desert salt lakes* are specialized environments like the Great Salt Lake, Utah, where evaporation rates exceed precipitation rates, resulting in salt accumulation.

4) *Volcanic lakes* form on volcanic mountain peaks, as in Japan and the Philippines.

5) *Chemically stratified lakes* include the Big Soda Lake in Nevada. These lakes are stratified because different dissolved chemicals have altered water supplies. They are *meromictic*, which means partly mixed.

6) *Polar lakes* exist in polar regions, with surface water temperature mostly below 4°C.

Classification of Lakes by Impoundments (Shut-ins)

Impoundments or *shut-ins* are artificial man-made lakes made by trapping water from rivers and watersheds. They vary in characteristics according to the region and nature of drainage. They have high turbidity and a fluctuating water level. The biomass productivity, particularly of benthos (bottom-dwellers), is generally lower than that of natural lakes.

Impoundments also include man-made enclosed pits and earth-embankments filled artificially to form artificial lakes, reservoirs, or holding basins. Some of these structures may be spring fed, and all are subject to infiltration by runoff.

A *reservoir* is a special type of impoundment. Reservoirs used for storage of raw water are generally lakes, ponds, or basins that are either naturally formed or constructed. The term reservoir also applies to aboveground and underground storage, and ground-level tanks designed to store treated or finished water. Some reservoirs (both raw and finished water types) are large and deep enough that they exhibit many of the characteristics of natural lakes systems, in particular, the characteristics of stratification and turnover, which only occur in lake-type water bodies.

Lake Thermal Stratification and Turnover

Water's temperature-density relationship is unique, and is especially important for lakes (only those located in some tropical and all temperate regions, because of their depth and long-term residence time). The temperature-density relationship of water in temperate lakes (of more than 25 ft. in depth) leads to lake *stratification* and subsequent *turnover*.

Water has its greatest density at 4°C (39.2°F); in the spring, a temperate lake's water may hold at this temperature throughout its entire depth. One result of this density characteristic (which is specific to the liquid water only) is that ice floats because the surrounding water is slightly warmer and denser. For the biota occupying a lake, this fact of nature is good—otherwise they and their watery environment would freeze solid.

During summer, the *epilimnion* (upper levels—see Figure 16.5) of the lake warm up, while the *hypolimnion* (deepest portions) retain their 4°C temperature. The epilimnion is heated directly by the sun, and indirectly by contact with warm air, forming a narrow band of warm water. When water is warm, it is less dense than colder water, and so tends to remain near the surface until mixed downward by turbulence. The epilimnion layer of warm water floats on the lower, colder water of the hypolimnion. These two layers are separated by a rather narrow horizontal zone of abrupt temperature change known as the *thermocline*. If you swim in a temperate lake,

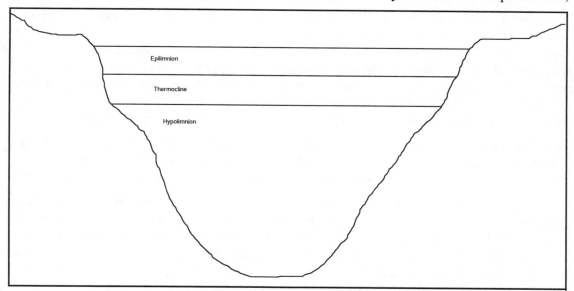

Figure 16.5 Thermal stratification of a temperate lake

you may experience the thermocline. When you swim horizontally on the surface, the water is warm. But tread water with your body vertical, and you experience the colder layers underneath.

In the summer, most of the plankton and nekton are found in the epilimnion, where sufficient dissolved oxygen exists, rather than below the thermocline (in the hypolimnion) where oxygen levels are very low. In each of these layers are differences in viscosity. Because the water viscosity changes between each level, the epilimnion, on top of the colder water, is circulated by wind action (wind-induced lake water circulation), while the hypolimnion is quiet and almost lifeless. These conditions remain fairly stable; stability actually increases with increased temperatures in the epilimnion throughout the summer season, while the hypolimnion remains at a fairly constant temperature. This leads to a very stable layering effect known as *thermal stratification*. When thermal stratification occurs in a lake, it causes turnover.

During summer, stratification occurs because the top layer, warmer than the bottom, results in layers of different density—the top light, the bottom heavy. With increased rise in temperature, the top layer becomes even lighter, and the thermocline (medium density) forms. From top to bottom, we now have the lightest and warmest on top, medium weight and relatively warm in the middle, and the heaviest and coldest below, with a sharp drop in temperature at the thermocline. There is no circulation of water in these three layers. If the thermocline is below the range of effective light penetration, which is quite common, the oxygen supply becomes depleted in the hypolimnion, since both photosynthesis and the surface source of oxygen are cut off. This state is known as *summer stagnation*.

In the fall, as air temperature drops, so does the temperature of the epilimnion. The marked stratification of summer begins to disappear, until the epilimnion temperature is the same as that of the thermocline. At this point, these two layers mix. The temperature of the whole lake is now the same, with complete mixing. As the temperature of the surface water reaches 4°C, it becomes denser than the water below, which is not in direct contact with the air and does not cool as rapidly at the lower levels. The denser, oxygen-rich surface layer stirs up organic matter as the water sinks, causing overturning; this is known as *fall turnover*.

During winter, the epilimnion, which may be icebound, is at the lowest temperature and is thus lightest; the thermocline is at medium temperature and medium weight; and the hypolimnion is at about 4°C and is heaviest. This is *winter stratification*. In winter, the oxygen supply is usually not greatly reduced, as the low temperature solubility of oxygen is higher and bacterial decomposition (along with other life activities) is operating at a low rate. When there is too much ice with heavy snow accumulation, light penetration is reduced. This reduces the rate of photosynthesis, which, in turn, causes oxygen depletion in hypolimnion, resulting in *winter kill* of fish.

As the ambient temperatures rise in spring, the ice of the epilimnion melts and the top two layers mix. As the epilimnion reaches 4°C, it sinks down, causing *spring overturn*. Odum (1971) describes this spring overturn phenomenon as being analogous to the lake taking a "deep breath."

Lake turnover is important to water quality in two ways: (1) it causes changes in distribution of nutrients and temperature, and (2) it causes movement of bottom sediments throughout the volume. Normally, nutrient materials accumulate in the lower depths, either as sediments or because biological activity is lower. When turnover occurs, these sediments are brought to the surface and are exposed to sunlight and higher temperatures and oxygen concentrations, and eutrophication rates increase (Tchobanoglous and Schroeder, 1987).

Lotic (flowing) Water Systems

In Chapter 15 we compared the human circulatory system to the earth's water circulation system, pointing out that water (analogous to blood) is pumped by the hydraulic cycle and continuously circulated through air, water bodies, and various vessels. Like vessels in the human circulatory system, water vessels (rivers) are essential to the earth's circulatory system. The earth's rivers are fed by capillary creeks, brooks, streams, rills, and rivulets. In this section we discuss how this part of the earth's circulatory system works with the complete system—how rivers/streams (for this text's purposes, the terms are interchangeable) carry water through the terrestrial part of the hydrological cycle.

Rivers

What exactly is a river? Several standard characteristics help define exactly what elements and parameters fit—massive flows of moving water, braided flows of moving water, white-water rapids, riffles, river pools, riparian areas—but definitions are slippery things, and for something as elemental as a river, unsatisfactory. But most people would have no difficulty in recognizing a river when they see one, even though rivers can be described in many different ways. Maybe this description by Watson (1988) will help to illustrate the degree and depth of this dilemma.

> *A river, almost by definition, is a body of moving water large enough to occupy one's mind. Something redolent of distant and legendary origins, filled with news of the sources from which it takes its strength. Rivers are metaphors for change, dreams of history. They move by, tangling the threads of time, braiding their waters into mixtures of the moment, open systems hinting at a single destiny—a predetermined and overwhelming need for the sea.*

> *On the same slope, large rivers flow faster than smaller ones, because there is less friction with their banks and beds. And, because even the air has a braking effect, the speed of any river's flow is greatest at a point somewhere between one-tenth and four-tenths of its depth below the surface. In addition, every bend and shift in direction, every rock and ledge below, produces whirls and upwellings which create random turbulence. The result is an environment where nothing is certain but change.*

For our purposes, we simplify the process by defining a river as a large, natural body of water, the movement of which erodes land surfaces and transports and deposits materials along its course, eventually emptying into an ocean, lake, or other body of water, and usually fed along its course by converging tributaries. The origination of this converging of tributaries (it looks like the branching of a tree) is best seen by examining a network analysis for a typical river system. In 1930, Robert E. Horton, a famous engineer, devised such a network analysis model (see Figure 16.6 for an example).

Rivers are fed from precipitation that does not infiltrate into the ground or evaporate. Rivers, like lakes, pass through various stages. The life span can be divided into four stages.

1)   *Establishment of a river.* Beginning as an outlet of lakes or ponds, arising from seepage areas or springs, or from runoff in a watershed, a river may be a dryrun or a headwater riverbed before it is eroded to the level of groundwater.

2)   *Young rivers.* A youthful river becomes permanent as its bed is eroded below the ground-water level and it begins to receive runoff and spring water.

3)   *Mature rivers.* A river reaches maturity as it becomes wider, deeper, and more turbid. Its velocity slows and water temperature rises. Sand, silt, mud, and/or clay form the bottom.

4)   *Old rivers.* Old rivers have reached their geologic base level. The flood plain may be very broad and flat. During normal flow periods, the channel refills and many shifting bars develop.

Remember that runoff affects rivers—varying amounts of runoff affect the rate and volume of flow—and that runoff is dependent, in part, upon the condition of the watershed soil. River channels are designed to handle normal river flow; contrary to popular belief, they are not designed to handle flow during all weather event conditions, such as flooding (see Case Study 16.1).

Case Study 16.1
Why Do Rivers Flood?

If we were to have asked this question in 1998, the answer we would have received would probably have had something to do with El Niño.

But is El Niño responsible for the flooding events that occurred in the United States in 1973 and 1993, when gauging stations on the Mississippi River at St. Louis, Missouri peaked at 43 feet and 49 feet above flood level, respectively? If you ask this question of a scientist who studies weather-related flooding events, or a scientist who knows weather (climatology), geology, and geophysics and who is an expert on flooding events, he or she might reply "maybe."

Right now, no one is certain what the exact impact El Niño, or its lesser-known associated phenomenon, Southern Oscillation (together ENSO), is on global weather patterns. However, enough historical evidence is available to accurately state why most rivers flood their basins.

Based on actual observation and measurement, we know for certain that all river systems naturally experience high discharge at a time of heavy precipitation. Flow alternates with time and is a consequence of the fluctuation of low and high precipitation periods. In short, when it rains, the river rises.

We are also certain that no river system in the world forms a channel that would convey without overflow all possible overflow events. You might pause for a moment and think to yourself that this last statement is strange or maybe inaccurate. That deep incomparable canyon is on your mind; namely, the Grand Canyon in Arizona (see Figure 16.7). "The Colorado River certainly has a steep enough channel to convey just about any type flood event that is imaginable, right?" The next question would be: "Isn't the Grand Canyon itself evidence that during geologic time some great event carved out what we see today?"

The Grand Canyon is over one mile deep in places, dwarfing the Colorado River with walls that tell of at least two billion years of history, and is probably deep enough to convey just about any flood event short of Noah's Ark. But what about the river system itself? The Grand Canyon area is just a small section of landscape the river courses through in a large river system that flows through hundreds and thousands of miles of widely diverse landscape.

Some may have a misconception that some mammoth event must have occurred in the distant geologic past to have carved the Grand Canyon as we know it today. However, geologic evidence indicates that the Grand Canyon and other similar formations were actually carved out by modest, frequent flows that carved, deepened, shaped, and altered the channel structure we see today.

A river system usually contains a channel that can contain within its banks only a discharge of modest size—its normal flow. But nature has not overlooked the fact that rivers do overflow. It solves such significant events by providing room for greater overflow discharges onto the valley floor or floodplain. Humans tend to forget that a river system consists not only of its channel, but also its floodplain. When humans use this part of the river for agriculture or construction, they encroach on the river. The result is that when floods occur, roads, homes, buildings, crops, and whatever else is in the river's path may be destroyed.

At this point we understand that a flood is (1) an event of such magnitude that the channel cannot handle peak discharge, (2) a flow in excess of channel capacity, and (3) a normal and expected characteristic of rivers (on the average, most rivers experience discharges in excess of channel capacity about twice a year). We understand all of these things about floods, but we still have not answered the main question of what causes a river to flood.

When a river discharges enough water to overtop its banks and flow over the floodplain, many factors are involved. Precipitation combined with snowmelt is often enough to cause a river to flood, but other factors can easily enhance the runoff. If the landscape is already wet (saturated), infiltration is prevented or decreased and runoff increases. Reduced or zero infiltration can also result if the landscape is frozen.

Human activities also have an affect on runoff, and subsequently on flooding. Small-scale floods are affected by such activities as grazing by animals, surface paving, agricultural practices, or deforestation, which decrease infiltration. Manmade alterations in surface conditions that come with urbanization of an area, including street gutters, storm water drains, and even roof downspouts speed the movement of surface water downslope and prevent soil infiltration as well. The river channel alterations communities put into place to prevent flooding in town also alter river flow, forcing greater masses of water further downstream by preventing local infiltration. This can cause flood levels to increase for communities down river as the water collects and moves onward. Large-scale floods (those caused by extraordinary total amounts of rainfall or widespread rapid snow/ice-melt) are not normally significantly influenced by urbanization, forests, and agricultural practices.

In rivers, temperature and dissolved oxygen (DO) levels follow different rules than those of still bodies of water. Temperature and DO in rivers are generally evenly distributed, though some variations in DO exist between rapidly flowing, turbulent areas and deeper, quiet river pools because of physical aeration. The amount of oxygen in aquatic systems is controlled by the solubility of gaseous oxygen in water. Since the DO is usually high and evenly distributed (the amount of DO in rivers and lakes is generally eight to ten parts per million), river organisms are adapted to this environment and have a narrow range of tolerance for DO. Organisms residing in aquatic systems are dependent upon the exact amount of oxygen present, which is controlled by the solubility of gaseous oxygen in water. The best example is trout, which are adapted to high oxygen rivers and cannot survive in rivers with DO levels below 5 mg/l—the fishkill level. River systems that receive large amounts of organic pollution are especially susceptible to fishkill because of the corresponding reduction in oxygen levels.

Rivers exhibit a large area for land-water interchange (this can be clearly seen in Figure 16.6). Most rivers are primarily detritus-based food chains—their primary source of energy comes not from green plants, as in most ecosystems, but from organic matter from the surrounding land, which is used as food by decomposers. Nutrients and waste products are transported by the flowing water to and away from many aquatic organisms, which helps to maintain a productivity level many times greater than that in standing waters (Smith, 1974).

Current is the outstanding feature of rivers and the major factor limiting the distribution of organisms. The current is determined by the steepness of the bottom gradient, the roughness of the streambed, and the depth and width of the streambed. Current in rivers has promoted many special biological adaptations by stream organisms.

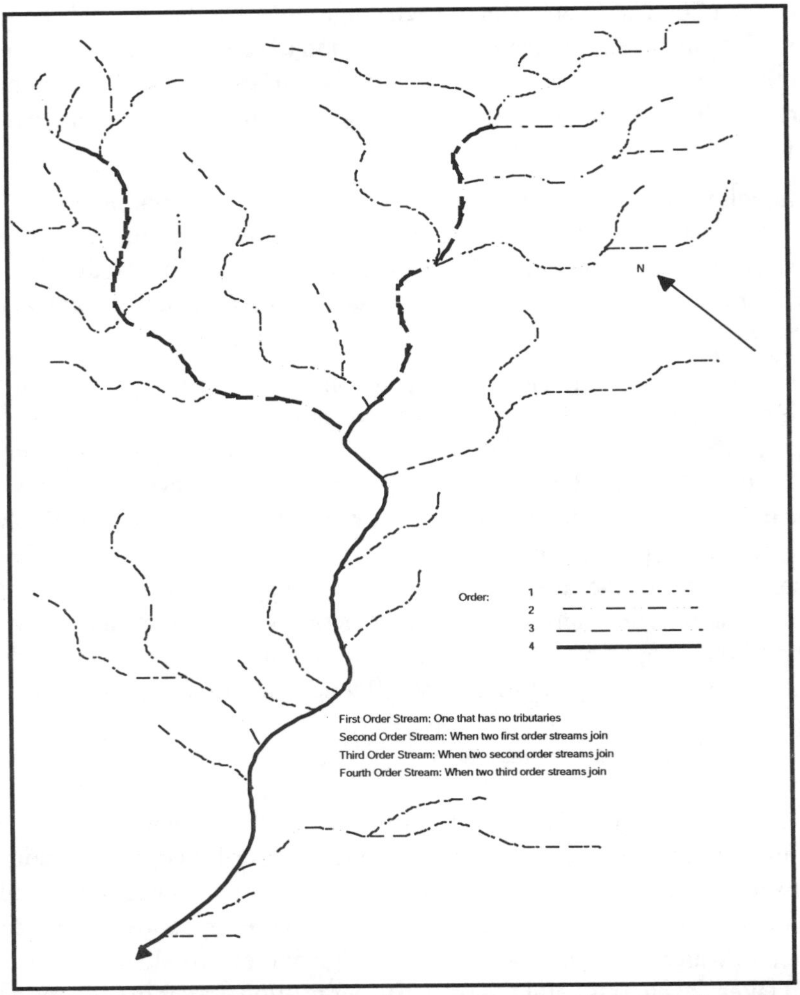

River Habitat

Rivers typically have two types of zones: rapids and pools. In the *rapids zone*, the current velocity is great enough to keep the bottom clear of silt and sludge, providing a firm bottom for organism growth. Organisms living in the rapids zone are adapted for life in running water. Trout, for example, have streamlined bodies to help in respiration and obtaining food. River organisms that live under rocks to avoid the strong current have flat or streamlined bodies. Others have hooks or suckers to cling or attach to a firm substrate to avoid the washing-away effect of the strong current.

Order:
1
2
3
4

First Order Stream: One that has no tributaries
Second Order Stream: When two first order streams join
Third Order Stream: When two second order streams join
Fourth Order Stream: When two third order streams join

Figure 16.6 Hypothetical channel network system. Shows small creeks joining to form larger streams, and the highly organized successive merging pattern they follow in the process. As tributaries meet the main stream, the discharge, width, and depth of the main river increase.

Figure 16.7 The Colorado River, braiding its way through part of the Grand Canyon (photo by John Goeke)

River *pool zones* (see Figure 16.8) are usually deeper water regions where reduced water velocity allows silt and other settling solids to provide a soft bottom, unfavorable for sensitive bottom dwellers. Decomposition of these solids causes lower levels of dissolved oxygen (DO). Some river organisms spend part of their time in the rapids and other times in the pool zone.

Figure 16.8 Pool zone in a river

River Water Quality

Lakes are separate, isolated bodies of water, typically landlocked, and they usually lack a separate freshwater feed system (with the exception of precipitation, infiltration, and runoff) to replenish total water supply. When contaminants run off land into such a body of water, or when people dispose rubbish, garbage, or other unwanted items into the water body, the contaminants must remain within it. They have no place to go, and no transport system to take them elsewhere. Water body systems like lakes do not have self-purifying capabilities. While lake water in temperate lakes may turnover twice each year, mixing the water and replacing surface water with bottom water, this process only works to exacerbate a contaminated lake's pollution problem. If highly toxic pesticides are dumped into a lake system, they will eventually sink to the bottom sediments and accumulate. During lake turnover, these toxins and any others present in the lake will be brought up from the bottom and redistributed throughout the lake system. Lakes have a distinct disadvantage—they lack the physical capability to maintain acceptable water quality on their own.

River systems don't have this problem. Rivers, because of their main characteristic—current— have the built-in capability to purify themselves, to a point. Since we began to settle near rivers, we have used them for transportation, water power, and drinking and irrigation water, and we have used the fish and other organisms in the river as a food source. We have also used the river as the local garbage dump.

When civilization was in its infancy, our river disposal practices were not terribly damaging to the local area we inhabited. We could simply walk over to the river, throw our unwanted items in, and stand there and watch them sink or move off. The unwanted items disappeared, and this is what we wanted.

As civilization advanced and settlements along river systems turned into cities, we disposed an increasing number of unwanted items into the river, but now the unwanted items included human sewage, organic waste, and chemical runoff. The river continued to carry away what it could.

We seemed to think that because rivers flow and because water in the river seemed to be only momentarily filthy, they could continue endlessly to carry unwanted items away.

But more and more people moved in, and along with more people, the production of unwanted items accelerated and dumping in the river increased. The river did its best to clean itself and move our unwanted items out of sight. But these unwanted items were really only out of the sight of those who threw them away. To others who settled downstream they were not.

For instance, in Great Britain during the Industrial Revolution, industry brought in more people and more people accumulated more unwanted items to throw into the legendary Thames River. But the Thames could handle only so much of the increased input of waste it received. A limit was eventually reached. The Thames had been turned by human actions from a river to an open cesspool. It was still used as a depository for unwanted items, but these items were no longer out of sight. Instead, the Thames was a floating mess, a water supply deadly with disease and rank with a horrendous, unbearable stench. Eventually, Londoners had enough; something had to be done to clean up the Thames.

This river contamination scenario has been repeated many times in many places throughout the world, and in some places it continues. Rivers have the physical capability to clean themselves and to maintain water quality, but only to a point. We always seem to find that point, one way or another.

Rivers and Self-Purification

Normally, rivers maintain a balance between plant and animal life, with considerable interdependence among the various life forms. In a healthy river, when organic matter enters, bacteria metabolizes it and converts it into carbon dioxide, ammonia, sulfates, nitrates, and so forth, which are used, in turn, by algae and plants to produce oxygen and carbohydrates. Microscopic animals (rotifers and protozoa) feed on the plant life and, in turn, provide food for insects, worms, crustaceans, and fish. Even the wastes produced by organisms living in the river environment provide a source of food (which assists the bacterial degradation process) for some river organisms.

However, when excessive quantities of pollutants are dumped into the river, they can upset this natural balance in a number of ways, and may eventually lead to the river's death. Changes in pH or excessive quantities of organic material may cause rapid bacterial growth and depletion of the DO resources of the river. Generally, polluted rivers are characterized by very large numbers of relatively few species and the absence of higher forms. If they could not naturally self-clean, they would turn into nothing more than masses of stagnating filth.

How do rivers self-clean? When pollutants enter a river system, the system works to reduce the concentration of pollutants by dilution, precipitation, bacterial oxidation, and/or other natural processes. Given enough time, these processes work to reestablish the normal cycle and distribution of life forms within the river system. In river systems, water quality standards are based on the maintenance of minimum dissolved oxygen concentrations, nontoxic concentrations of specific chemical species, and a near-neutral pH (McGhee, 1991). A healthy river has a natural assimilative capacity to assist in waste treatment without adversely affecting downstream users.

Before we begin our discussion of river self-purification, you must understand two terms important for their impact on river pollution.

1)   *Dissolved Oxygen (DO)* is the amount of oxygen dissolved in a river. It indicates the river's degree of health and its ability to support a balanced aquatic ecosystem. As DO drops below 5 mg/l, the forms of life that can survive begin to be reduced. In extreme cases, when anaerobic conditions exist, most higher forms of life are killed or driven off and noxious conditions prevail. The deoxygenation caused by microbial decomposition of wastes and oxygenation by reaeration are competing processes that are simultaneously removing and adding oxygen to a river (Masters, 1991). Oxygen comes from the atmosphere by solution, and from photosynthesis of water plants. In fast rivers, oxygen is added primarily through reaeration from the atmosphere in rapids, waterfalls, and cascades (see Figure 16.9). Dissolved oxygen concentrations are usually higher and more uniform from surface to bottom in streams than in lakes.

2)   *Biochemical Oxygen Demand (BOD)* is the amount of oxygen required to biologically oxidize the organic waste over a stated period of time. BOD is important in the self-purification process, because to estimate the rate of deoxygenation in the river, the 5-day and the ultimate BOD must be known.

NOTE: See Figure 16.10 A and B for more information on self-purification.

Figure 16.9 Waterfall in a river system, aiding in re-aeration of water

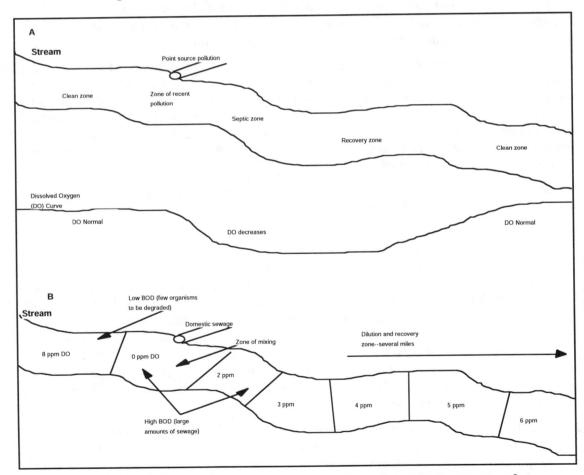

Figure 16.10 A: Changes that occur in a river after it receives excessive amounts of raw sewage; B: Effects of waste on DO (adapted from Enger et al., Environmental Science: The Study of Interrelationships, *1989, p. 411)*

When a river receives an excessive amount of organic wastes, it exhibits various changes, which can be differentiated and classified into zones. Upstream, before a single point of pollution discharge, the river is defined by having a *clean zone* (Figure 16.10 A). At the point of waste discharge, the water becomes turbid—the *zone of recent pollution*. Not far below the discharge point, the level of dissolved oxygen falls sharply, and in some cases may fall to zero; this is called the *septic zone*.

After the organic waste has been largely decomposed, the dissolved oxygen level begins to rise in the *recovery zone*. Eventually, given enough time and no further waste discharges, the river will return to conditions similar to those in the clean zone.

Figure 16.10 B shows the effect of organic wastes on dissolved oxygen in a river and the result of organic waste being attacked by organisms that use oxygen in the degradation process. An inverse relationship exists between oxygen and organic matter in the river. The greater the BOD, the less desirable the river is for human use.

Aquatic Organisms and Their Role in Self-Purification

The self-purification process in rivers is similar to the purification process of secondary sewage treatment (see Chapter 18), which employs biological and chemical processes to remove most organic matter. In this discussion, we address the biological process.

In the biological self-purification process, certain factors indicate water quality. Four important ones include coliform bacteria count, concentration of DO, BOD, and the Biotic Index (the biota that exist at various stages in the self-purification of a river are direct indicators—a biotic index—of the water conditions).

Aquatic organisms degrade or decompose organic wastes. The river exhibits a change in the type of organisms present as the strength of the waste decreases. As the organic wastes are received by the river, a very large number of bacteria predominate because they thrive on the energy they receive from the organic waste. Some of these bacteria are normally found in rivers. Others, such as enteric microbes (coliform bacteria, found in great numbers in the intestines and thus in the feces of humans and other animals), are not normally found in the stream environment. While the growth of normal stream bacteria is greatly enhanced by organic nutrients, coliforms and pathogens generally die out within a few days, perhaps because of predation and unfavorable conditions. The bacteria predominate during the recent pollution zone to near the end of the septic zone. If the organic load is high, the bacterial type changes from aerobic to anaerobic because of similar changes in conditions that affect bacteria.

As stabilization continues, bacterial food declines because of consumption by high bacterial populations. Protozoans start to increase and eventually predominate. The one-celled protozoans (amoeba, paramecium, and other ciliates) feed on bacteria. As their food supply diminishes, protozoans decrease in population, and are in turn consumed by rotifers and crustaceans in the recovery zone. During this period, turbidity decreases and algal growth increases.

Aquatic insects are also affected in a polluted river. In the septic zone, for example, intolerant insects such as the mayfly nymph disappear. Only air-breathing or specially adapted insects such as mosquito larvae can survive in the low dissolved oxygen levels present in the septic zone. When the river has completely purified the organic waste, algae returns. Higher life organisms such as insects eat the algae, and they serve as food for fish. This entire process, known as *general biological succession*, is critical to the river self-purification process.

GROUNDWATER

On average, approximately three feet of water falls each year on every square foot of the earth. Approximately six inches of this goes back to the sea. Another two feet are lost to evaporation. The remaining six inches seeps through interstices, voids, hollows, and cavities into the sponge-like soil. In its journey downward through the soil, water may go only a few feet or several hundred feet before it joins the subterranean stores of liquid that make up the earth's groundwater supply—literally, an ocean below our feet.

The groundwater supply is a colossal reservoir that feeds all the natural fountains and springs of the earth. At times and in certain places, it bubbles up in cool, blue pools, and in other places it heats up, forms steam, and bursts back to the surface in geysers and hot springs.

Most of the earth's groundwater supply lies just beneath the surface and can easily be reached by drilling a borehole or well to the level of the water table. This practice has gone on for millennia, and accelerates as more and more people inhabit the earth. In this way groundwater has served as a reliable source of potable water for millions of the earth's inhabitants, and when used with moderation, groundwater should remain a viable source for years to come.

Groundwater Uses and Sources

Large cities are supplied primarily by surface water, while most small communities use groundwater. This situation helps to explain why a larger portion of the U.S. population is supplied by surface water, but the total number of communities supplied by groundwater is four times that supplied from surface water.

Groundwater has several characteristics that make it desirable as a water supply source: (1) a groundwater system provides natural storage, eliminating the need for man-made impoundments; (2) since the groundwater supply is usually available at point of demand, the cost of transmission is reduced significantly; and (3) because groundwater is filtered by natural geologic strata, it usually appears clearer to the eye than surface water (McGhee, 1991). For these reasons, groundwater is generally preferred as a source of municipal and industrial water supplies.

Historically, because groundwater has been considered to be safe to drink, many water utilities delivered it untreated to their customers. However, we are quickly learning that using groundwater does have some disadvantages. These include the possibility of contamination by toxic or hazardous materials leaking from waste treatment facilities, natural sources, or landfills, which may not be evident to either the public or regulatory agencies. Also, when groundwater becomes contaminated, restoring it is difficult, if not possible. Recent discoveries of contaminated groundwater in some areas has led to the shutdown of thousands of potable water wells across the U.S.

Aquifers

In the simplest terms, the charging of the subsurface with water that then becomes groundwater occurs when surface water seeps down from the rain-soaked surface, sinks to a certain level, collects above an impermeable layer, and fills all the pores and cracks of the permeable portions. The top of this *saturated zone* is called the *water table*.

Figure 16.11 illustrates that the groundwater system is a bit more complex and complicated than just described. Groundwater occurs in unconfined aquifers in two different zones, distinguished by whether or not water fills all the cracks and pores between particles of soil and rock. The *unsaturated zone* lies just beneath the land surface and is characterized by crevices that contain both air and water. The unsaturated zone contains water (*vadose water*) that is essentially unavailable for use. When an aquifer is not overlain by an impermeable layer it is said to be *unconfined* (see Figure 16.11, which shows an unconfined aquifer situated above a confined bed). A *confined aquifer* consists of a water-bearing layer sandwiched between two less-permeable layers (Figure 16.12). Water flow in a confined aquifer is restricted to vertical movement only. In contrast, water flow in an unconfined aquifer has more freedom and is similar to flow in an open channel.

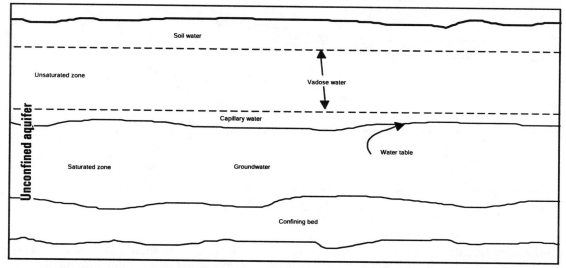

Figure 16.11 An unconfined aquifer with its saturated and unsaturated zones. To remove water from the water table, a well would have to penetrate the saturated zone.

Groundwater Flow

Whether groundwater flow occurs in the open channel-like flow of an unconfined aquifer or the vertical-only (pipe-like) flow of a confined aquifer, to have any flow at all, a hydraulic gradient must exist. Simply stated, the *hydraulic gradient* is the difference in *hydraulic head* divided by the distance along the fluid flow path. For our purposes, what you need to understand is that groundwater moves through an aquifer in the direction of the hydraulic gradient and at a rate proportional to the gradient and inversely related to the permeability of the aquifer. The steeper the slope and the more permeable the substrate, the more rapidly the water flows.

Groundwater does not flow like a river. Instead it percolates downward, moving from high elevations to lower elevations, at a variable rate. It sometimes moves slowly and sometimes surprisingly fast, from less than an inch to a few feet a day.

Groundwater aquifers are important sources of water. As we have said previously, they supply a large portion of the U.S. population—almost all of the rural population. Demand for groundwater use continues to increase, and with this increase both the quantity and quality of this vast

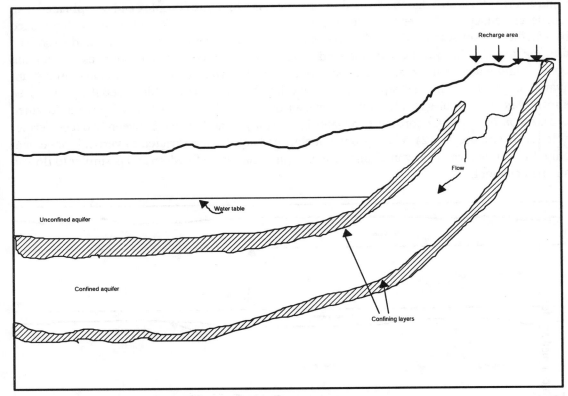

Figure 16.12 A confined aquifer

resource are threatened. Two important points you should remember about groundwater: (1) we do not have an inexhaustible groundwater supply; and (2) groundwater is not completely purified as it percolates through ground—it is not exempt from surface contamination. We'll discuss this second point more fully in Chapter 17.

SUMMARY

The interconnectedness of the hydrological cycle works to our advantage in self-purification, but with persistent pollutants, the processes integral to the water cycle can trap toxins, complicating efforts to clean them up. While the natural processes that clean our water as it travels through the hydrological cycle worked well for centuries, in many places humans have overloaded the capacity of the water cycle to self-purify. We are now cleaning up problems created by past environmental abuse and ignorance. Inevitably we will create problems that future generations will have to clean up due to unforeseen problems with the solutions we try now, but we have no excuse for re-creating past mistakes. Our water system is too valuable—in fact it is priceless.

Cited References

Amos, W. H., *Limnology: An Introduction to the Fresh Water Environment*, Chestertown, Maryland: LaMotte Company, 1969.

Davis, M. L. and Cornwell, D. A., *Introduction to Environmental Engineering*. New York: McGraw-Hill, Inc., 1991.

Laws, E. A., *Aquatic Pollution: An Introductory Text*. New York: John Wiley & Sons, Inc., 1993.

Masters, G. M., *Introduction to Environmental Engineering and Science*. Englewood Cliffs, NJ: Prentice-Hall, 1991.

McGhee, T. J., *Water Supply and Sewerage*. New York: McGraw-Hill, Inc., 1991.

NALCO, *The NALCO Water Handbook*. 2nd ed., New York: McGraw-Hill, Inc. 1988.

Odum, E. P., *Fundamentals of Ecology*. Philadelphia: Saunders College Publishing, 1971.

Smith, R. L., *Ecology and Field Biology*. New York: Harper & Row Publishers, 1974.

Spellman, F. R., *Stream Ecology and Self-Purification: An Introduction for Wastewater and Water Specialists*. Lancaster, PA: Technomic Publishing Company, 1996.

Tchobanoglous, G. and Schroeder, E. D., *Water Quality*. Reading, Massachusetts, Addison-Wesley Publishing Company, 1987.

U.S. Environmental Protection Agency, Surface Water Treatment Regulations, *Federal Register*, title 54, part 124, June 29, 1989, p. 27486.

Watson, L., *The Water Planet: Celebration of the Wonder of Water*. New York: Crown Publishers, Inc., 1988.

Wetzel, R. G., *Limnology*. New York: Harcourt Brace Jovanovich College Publishers, 1983.

Suggested Readings

AWWA Manual M21, *Groundwater*, Denver, Colorado: American Water Works Association, 1989.

Bensen, M. A., *Factors influencing the occurrence of floods in a humid region of diverse terrain*. U.S. Geological Survey Water Supply Paper 1580-B, 1962.

Berner, E. K., and Berner, R. A., *The Global Water Cycle, Geochemistry and Environment*. Englewood Cliffs, NJ: Prentice-Hall, 1987.

Cooper, M. H., *World Simmers over Water*, San Francisco Examiner, p. 48, Jan. 14, 1996.

Egna, H. and Boyd, C., *Dynamics of Pond Aquaculture,* Boca Raton, FL: CRC/Lewis Publishers, 1997.

Emerson, H. W., Channelization: A Case Study. *Science*, vol. 173, pp. 325-326, 1971.

Horne, A. J., and Goldman, C. R., *Limnology*. 2nd ed., New York: McGraw-Hill, Inc., 1994.

Laenen, A. and Dunnette, D. A., *River Quality: Dynamics and Restoration*. Boca Raton, FL: CRC/Lewis Publishers, 1997.

Leopold, L. B., *Water, Rivers and Creeks*, Sausalito, California: University Science Books, 1997.

Outwater, A., *Water: A Natural History*. New York: Basic Books, 1996.

Postel, S., "Where Have all the Rivers Gone?" *World Watch*, vol. 8, no. 3, pp. 9-19, 1995.

Rogers, P., "The Future of Water," *The Atlantic*, p. 80, July 1983.

Thomann, R. V., and Mueller, J. A., *Principles of Surface Water Quality Modeling and Control*. New York: Harper & Row, 1987.

U.S. Environmental Protection Agency, "Ambient Water Quality Criterion for Dissolved Oxygen," *Federal Register*, 50 (76), 1985.

U.S. Geological Survey, *Estimated Use of Water in the United States, 1980*. Washington, DC: Department of the Interior, 1984.

White, G. F., and Myers, M. F, *Coping with the Flood: The Next Phase,* Water Resources Update, issue 95, 1994.

Willis, R. and Yeh, W. W-G., *Groundwater Systems Planning & Management*. Englewood Cliffs, NJ: Prentice-Hall, 1987.

World Resources Institute, *World Resources 1994-1995*, New York: Oxford University Press, 1994.

CHAPTER 17

WATER POLLUTION

Is it not enough for you to drink clean water?
Must you also muddy the rest with your feet? —Ezekiel 34:18

CHAPTER OUTLINE

- Introduction
- Point and Nonpoint Sources of Pollution
- Industrial Sources of Water Pollution
- Hazardous Waste Disposal
- Acid Mine Drainage
- Agricultural Sources of Surface Water Pollution
- Acid Rain
- Groundwater Pollution

INTRODUCTION

Unless you swim in a water body best described as a cesspool, and/or drink water that smells foul, tastes worse, and ultimately makes you ill, you may think that water pollution is relative, and find it hard to define. Once you come up with a definition (it might have something to do with physical characteristics and negative impact) you may also consider the idea that freshwater pollution is not a new phenomenon. Only the issue of freshwater pollution as a major public concern is relatively new.

Natural forms of pollutants have always been present in surface waters. Many of the pollutants we have discussed in previous chapters were being washed from the air, eroded from land surfaces, leached from the soil, and ultimately found their way into surface water bodies long before people walked on the earth. After all, floods and dead animals pollute, but their effects are local and generally temporary. In prehistoric times, and even in more recent times, natural disasters have contributed to surface water pollution. Cataclysmic events—earthquakes, volcanic eruptions, meteor impact, transition from ice age to interglacial to ice age—have all contributed to surface water pollution. Natural purification processes over time were able to self-clean surface water bodies. We can accurately say that without these self-purifying processes, the water-dependent life on earth could not have developed as it did (Peavy et al., 1985).

For our purposes, we define *water pollution* as the presence of unwanted substances in water beyond levels acceptable for health or aesthetics. Water pollutants may include organic matter (living or dead), heavy metals, minerals, sediment, bacteria, viruses, toxic chemicals, and volatile organic compounds. In this chapter we discuss the sources of water pollutants, point and nonpoint sources, which include industrial sources, hazardous waste disposal, acid mine drainage, agricultural sources, and acid rain. We also discuss groundwater pollution. Finally, we address the major concern of water pollution—the health effects.

POINT AND NONPOINT SOURCES OF POLLUTION

Because of the need to control and regulate sources of water pollution, environmental scientists established a means to distinguish between point sources and nonpoint sources of water pollution from human activities. A *point source* (usually easy to identify) discharges pollution (or any effluent) from an identifiable, specific source or point. Pipes that discharge waste into streams and smokestacks that emit smoke and fumes into the air are point sources of pollution. Industries or facilities usually identified as point sources include factories, electric power plants, sewage treatment plants, coal mines, offshore oil wells, and oil tankers.

Nonpoint-source pollution (much more difficult to identify) cannot be traced to a specific source, but rather comes from multiple generalized sources, contributed from throughout an area, and can include runoff into surface water and seepage into groundwater from croplands, livestock feedlots, logged forests, construction areas, roadways, parking lots, and urban and suburban lands. Because it is so much harder to identify than point source pollution, nonpoint-source pollution is much harder to control.

INDUSTRIAL SOURCES OF WATER POLLUTION

The disposal of noxious industrial wastes into rivers and streams has been a common industrial practice for many years and ranks as one of the most serious forms of water pollution. In the past, industrial pollutants were dumped into rivers and streams in smaller quantities, and streamflows were adequate to dilute and carry away the wastes with minimal environment damage. But as industry and population increased, streamflows could no longer handle the waste load, and industrial pollution became serious.

Industry is the largest U.S. water user, with nearly 250,000 industries in the United States using more than 260 billion gallons of water per day. The organic chemical and plastics industries are the largest sources of toxic chemical pollution in the U.S. The iron and steel and metal finishing industries are next. Other significant sources of pollutants include metal foundries, petroleum refineries, and the pulp and paper industry.

Spills provide one of the most troublesome sources of industrial contamination. Chemical and oil spills are particularly troublesome because they can never be completely removed or cleaned up. Consider, for example, oil dispersed by several routes into a large surface water body (a lake, for example). The largest percentage, the insoluble fraction, is lighter than water, and gradually spreads and thins to form an ever-widening oil slick on the water's surface. Approximately 30% of the components in crude oil immediately begin to evaporate into the atmosphere. In areas that experience wind action that produces waves, the oil is whipped into an oil-water emulsion. If the spill occurs near the lakeshore, the wind and waves may transport the oil to recreational beach areas, which become coated with oil. With time the emulsion is distributed throughout the upper layer of the water. A portion of the emulsion is broken down by bacteria. That portion not degraded by bacterial actions collects to form floating tar balls. Crude oil, heavier than water, sinks to the bottom, where it coats and kills the organisms there.

Unfortunately, such toxic discharges are fairly commonplace. In the U.S. alone, more than 20 major oil spills occur every year. Collins (1988) reports that a recent study by the EPA documents nearly 7,000 accidents involving hazardous substances between 1980 and 1985.

Certain chemical contaminants are highly dangerous environmental threats, because they *bioaccumulate* through food webs—a serious matter. Persistent toxic substances like nutrients pass from one trophic level to the next. If such substances are not excreted or broken down by the organism, they remain in its tissues. If the organism continues to ingest contaminated materials, the chemical concentrations rise and are passed up to the next trophic level when that organism itself is ingested. In this way, persistent pollutants enter the food webs and become considerably more concentrated and dangerous by the time they reach the highest trophic level. As a result, organisms feeding at the highest trophic levels are exposed to the highest doses of toxic substances. In aquatic webs, which usually consist of four to six trophic levels, accumulation is especially pronounced. The ultimate upper trophic level occupant—human beings—run the risk of ingesting the largest concentrations of toxic substances.

Another form of industrial pollution occurs because many industries use water for cooling. *Thermal pollution* occurs when an industry returns heated water to its source. Generation of electricity, for example, requires enormous quantities of water for cooling, and that water is usually drawn from major rivers and large lakes, then circulated through hot electricity-generating machinery. The heat then transfers to the water, raising its temperature. When heated water discharges into a water body, it can adversely affect aquatic ecosystems by causing organisms to respond by elevating respiration rates, while simultaneously, the elevated water temperatures reduces DO content by decreasing the solubility of oxygen in water.

HAZARDOUS WASTE DISPOSAL

Because only a small fraction of industrial waste is recycled, detoxified, or destroyed, the rest must go somewhere. Unfortunately, in many cases, *hazardous wastestreams* end up in surface water bodies and in groundwater. Almost 80,000 disposal sites exist in the United States alone. Many wastestreams are produced in industrial, municipal, agricultural, and mining activities and from brines from oil and gas extraction. Other toxic pollutants enter water supplies from nonpoint sources, including runoff from mining activities, farms, and urban areas. Urban runoff from streets, for example, carries heavy metals (cadmium and lead) as well as other pollutants. Chemicals containing cyanide are used to de-ice streets, and herbicides used in agriculture are also potential sources of pollution (Concern, 1986).

ACID MINE DRAINAGE

Several environmental and human health costs are associated with extracting minerals. Miners are at risk for diseases caused by the substances with which they work (for example, black lung disease from coal dust). Miners are also at risk from mine collapse, underground explosions and fires, and other safety hazards. Mining disturbs soil and overlying vegetation, which disrupts ecosystems, reduces productivity, and leads to soil erosion. Mining causes siltation of streams, lakes, and rivers by eroding soil. Mining can diminish or reduce productivity of land. Mining degrades water used in mining processes. Finally, mining produces tailings and mine drainage that can contaminate soil and water. This last environmental cost is the focus of this discussion.

Mining operations to uncover deposits of coal, copper, nickel, zinc, and lead can lead to surface water and groundwater pollution. When any of these ores are uncovered, they are exposed to air. Since these ores contain sulfides, the air + sulfide combination leads to oxidation, and that process is often catalyzed by certain strains of bacteria. When water filters through those oxidized minerals, it becomes more acidic (water + oxidized sulfides = sulfuric acid), which increases the solubility of metals in water that filters through (is leached from) the newly aerated deposits and runs off into the streams below the mine. Such water often has harmful levels of toxic materials. This pollution, known as *acid mine drainage*, kills fish and disrupts normal aquatic life cycles. Quigg (1976) pointed out that drainage from mining operations pollutes several thousand miles of streams in the United States, and mine drainage can be expected to grow with the world's increasing reliance on coal as an energy source.

AGRICULTURAL SOURCES OF SURFACE WATER POLLUTION

Surface water pollution from agricultural sources, both organic and inorganic, usually is produced by pesticides, fertilizers, and animal wastes. All these enter water bodies via runoff and groundwater absorption in areas of agricultural activity.

Pesticides include insecticides, herbicides, and fungicides. Their purpose is to kill unwanted plant species and insects to protect the crop. Because of the ever-increasing worldwide population, pesticides have become essential to food production and other crop products to produce the yield required to feed the increasing population.

Natural pesticides are biodegradable and are thus less harmful to the environment. Nonbiodegradable pesticides (DDT, for example) can accumulate in the environment and have been proven harmful to human health.

When pesticide pollution is in appreciable concentrations in aquatic environments, they pose a serious health hazard for organisms that inhabit the water body and for those who use the water as a drinking supply. Although pesticides are normally sold for a specific pest, they often kill non-pest species and have side effects on the growth and reproduction of birds and fish.

Fertilizers, like pesticides, are used to increase food production. Agricultural water pollution results primarily from the phosphates and nitrates present in the fertilizers, which enter the water supply through erosion of topsoil and water runoff.

Nitrates can be toxic to animals and humans in high enough concentrations. Nitrates can be reduced to nitrites, which interfere with the transport of oxygen by hemoglobin in blood. Nitrates and phosphates both contribute to the excessive growth of microscopic plant algae in lakes, a condition that affects the local ecosystem.

Animal wastes are a source of water pollution with potential health hazards. High waste concentration areas (barnyards, land treated with animal wastes, feedlots) are a major threat to water supplies when rainfall runoff carries them into streams, lakes, and rivers. The problem of animal waste disposal is magnified when animals are raised in feedlots. To gain an appreciation for how serious a problem disposal problems can be for animal feedlots, consider that the U.S. Soil Conservation Service points out that while a human being produces 0.33 pounds per day of human waste, a cow produces 52.0 pounds per day. The human's waste would probably be processed by a sewage treatment plant, but the cow's probably would not be.

In the past, animal waste was a benefit to farmers and others because it was used as a fertilizer. However, compared to modern chemical fertilizers, the benefits of soil fertilization from animal wastes often do not justify the costs of hauling and application. Thus, the livestock owner may have to contend with a daily supply of animal waste that cannot be sold, burned, or given away, and that may cause water pollution in water runoff. When such runoff enters rivers and streams (carrying with it nitrates, phosphates, and ammonia) the organic matter is broken down by microorganisms, which use up oxygen in the process. As we have said, streams have the potential to clean themselves, but a high degree of organic pollution cannot be handled by the stream and creates problems for most aquatic life. The most critical effect of agricultural water pollution on human beings is the possibility of waterborne bacterial diseases—cholera, typhoid fever, and dysentery.

Another agricultural practice that contributes to water pollution is irrigation. Irrigating crops to ensure their growth is important, but a problem arises when water used to irrigate crops becomes contaminated with salt, which, through runoff, contaminates streams, lakes, and rivers. In the western United States, where water from the Colorado River is withdrawn for irrigation and is later returned, the water leaches large quantities of salts from the irrigated land and adds them to the river. Pimentel (1989) points out that during dry periods in other western U.S. locations where the same irrigation practices are used, like the Red River in Oklahoma and Texas, the river becomes more salty than seawater.

ACID RAIN

Acid rain or *acidic deposition* was briefly discussed in Chapter 8. This section concentrates on acid rain effects on lakes, streams, and river bodies. To understand how acid rain affects surface water bodies, you need to recall a few important facts. Acid rain is normally formed when atmospheric water picks up acidic particles from the air, reacts with them to form acids, then falls to the earth as rain, snow, or other precipitation. Acid rain, more precisely defined, is rain with a pH below 5.6. It can produce pH values as low as vinegar or lemon juice (pH of 2.8 and 2.3, respectively).

One of the most obvious effects of acid rain is dead or crippled lakes. The damage from acid rain in lakes begins when the smaller organisms die off first, leaving the larger organisms to starve. As a lake becomes more acidic, sometimes the larger organisms (fish, for example) are killed directly. High acid levels in lakes dissolve heavy metals like mercury, lead, zinc, and especially aluminum, leading to concentrations at toxic and often lethal levels.

Acid rain can also impact human health. It can poison reservoirs and water supply systems by dissolving toxic metals from soils and bedrock in watersheds. In 1988, the New York Times, in an article based on an EPA survey, reported that more than 4% of the streams surveyed in the Middle Atlantic states were acidic, and roughly half had a low capacity to neutralize acidic rain.

GROUNDWATER POLLUTION

Groundwater, the ocean of fresh water beneath the earth's surface, is a precious and little-protected resource. In the United States, groundwater is the drinking water source for half the population—most of it used directly from the ground without any treatment. Groundwater pollution can be a very serious problem.

Through experience and study, we have determined that any pollutant that comes in contact with the ground may contaminate groundwater. As water enters the ground, it filters naturally through the soil, and in some soils, that process quite effectively removes many substances, including suspended solids and bacteria. Some chemicals bind themselves to the surface of soil particles (phosphates) and thus are removed. In some areas, though, industrial and municipal wastes are sprayed on the ground surface so that they will filter through the soil, become purified in the process, and recharge the groundwater reservoir. Though natural purification of water as it passes through the soil is beneficial, it is a slow process because the water is not readily diluted and has no access to air.

The same drainage-basin activities that pollute surface waters can also contaminate groundwater. Septic tanks, agriculture, industrial waste lagoons, underground injection wells, underground storage tanks, and landfills can all lead to groundwater contamination (see Figure 17.1). A major

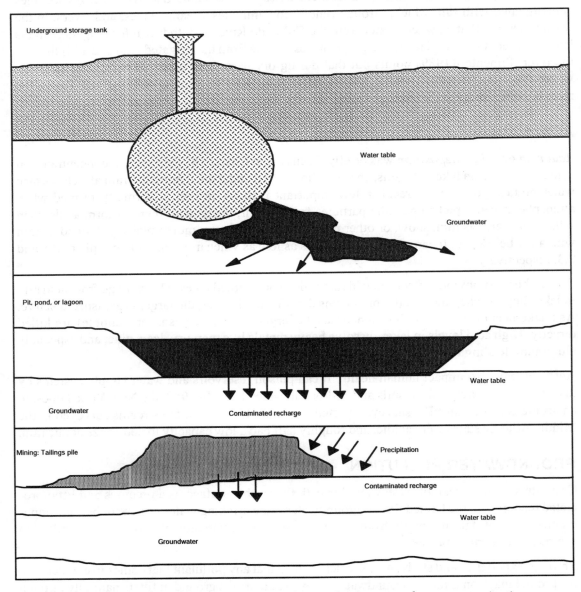

Figure 17.1 Examples of common activities that can lead to groundwater contamination: leaky underground storage tank; surface pits, ponds, or lagoons; and mining tailings contributing to acid mine drainage.

problem occurs when waste disposal sites are situated in unsuitable soils and even directly over fractured dolomites and limestones. When located directly on top of such rock, polluted water finds its way into wells. Unfortunately, at least 25% of the usable groundwater (from wells) is already contaminated in some areas (Draper, 1987).

Groundwater contamination occurs in several different ways. Increasing occurrences of groundwater contamination from saltwater, microbiological contaminants, and toxic organic and inorganic chemicals are being observed. The disposal of toxic industrial wastes is the major source of groundwater contamination in the U.S. This type of contamination is magnified any time waste disposal sites are not protected by some type of lining, when disposal sites are located in permeable materials lying above usable water aquifers, and when these sites are located in proximity to a water supply well. The Conservation Foundation (1982) reported that groundwater contamination was responsible for the closing of hundreds of wells in the U.S.

Groundwater supplies are routinely replenished and purified by wetlands. Wetland plants absorb excess nutrients and immobilize pesticides, heavy metals, and other toxins, preventing them from moving up the food chain. In some locations wetlands have been used to treat sewage. However, the capacity of wetlands to cleanse polluted water is limited, and many have been overwhelmed by pollution. Wetland areas are becoming rare.The United States, for example, has already lost half its wetlands to urban and agricultural development (Goldsmith and Hildyard, 1988).

SUMMARY

As the world population grows, we are forced by circumstances we have created to face the realization of our resources' limitations. Many of us in the U.S. have been fortunate and have always had enough of whatever we needed, and even whatever we wanted. When something we like breaks or wears out we throw it away and buy a new one, and often we don't even make an attempt to fix the problem. We neglect basic maintenance until we damage our belongings beyond repair, and we expect that we will always have enough. But some things are beyond our control, and beyond our power or financial ability to replace or repair. Our water supply is one of these. Without our concern, attention, and preventive maintenance and reclamation, our water supply will not be able to support the needs of the future.

Cited References

Concern Inc., *Drinking Water: A Community Action Guide*, Washington, DC, pp. 1-4, 1986.

Draper, E., Groundwater Protection, *Clean Water Action News*, p. 4, Fall 1987.

Goldsmith, E. and Hildyard, N. (eds.), *The Earth Report: The Essential Guide to Global Ecological Issues*, Los Angeles: Price Stern Sloan, 1988.

Peavy, S., Rowe, D. R., and Tchobanoglous, G., *Environmental Engineering*. New York: McGraw-Hill, Inc., 1985.

Pimentel, D., *Waste in Agriculture and Food Sectors*, Cornell University, College of Agriculture and Life Sciences, unpublished paper, 1989.

Quigg, P. W., *Water: The Essential Resource*, New York: National Audubon Society, 1976.

The Conservation Foundation, *State of the Environment 1982*, Washington: DC, p. 110, 1982.

The New York Times, p. A16, May 22, 1988.

Suggested Readings

Council on Environmental Quality, *Contamination of Groundwater by Toxic Organic Chemicals*, Washington, DC: U.S. Government Printing Office, 1981.

Conservation Foundation, *Groundwater Pollution*, Washington, DC: Conservation Foundation, 1987.

Hunt, C. A. and Garrels, R. M., *Water—The Web of Life*, New York: W.W. Norton, 1972.

King, Jonathan, *Troubled Water*, Emmaus, PA: Rodale Press, 1985.

Rice, R. G., *Safe Drinking Water. The Impact of Chemicals on a Limited Resource*. Chelsea, MI: Lewis Publishers, 1985.

National Research Council, *Groundwater Quality Protection*, Washington, DC: National Academy Press, 1986.

World Resources Institute, Natural Resources Consumption, *World Resources 1994-95*. New York: Oxford University Press, 1994.

World Resources Institute, Water, *World Resources 1994-95*, New York: Oxford University Press, 1994.

Worldwatch Institute, *State of the World Resources 1994-95*. New York: W.W. Norton and Company, 1995.

CHAPTER 18

WATER POLLUTION CONTROL

Using the state's freedom-of-information law, we quickly obtained the sludge hauling and alum [aluminum sulfate] purchasing records for the town of Newburgh's Chadwick Lake filtration plant. Those records showed that while the plant had produced over a million gallons of alum sludge during 1984, none had been hauled away. It could only have been dumped into the creek. Records from the previous five years also showed giant discrepancies between alum sludge produced and the amount hauled away. We immediately informed the DEC [Department of Environmental Conservation] about the illegal dumping, hoping for a criminal prosecution against the town.

It was disappointing to me [Robert F. Kennedy, Jr.] that the DEC responded instead by writing the town a temporary permit to discharge alum backwash into the creek. I had read just enough environmental law by then to recognize that the purported permit was utterly illegal under the Clean Water Act. The permit had not been subject to public notice or comment, as the act requires, and it allowed discharges of pollutants that the act flatly forbade. There were no penalties and no requirement that the town clean up the creek.... I was still naive enough to be shocked that a government agency, charged with protecting the public from pollution, would so blatantly intervene to protect a polluter from the public and the law. —Cronin and Kennedy

CHAPTER OUTLINE

- Introduction
- The Safe Drinking Water Act (SWDA)
- The Federal Water Pollution Control Act
- Affect of Regulations on Preventing Water Pollution
- Water Treatment
- Wastewater Treatment
- Thermal Pollution Treatment
- Pollution Control Technology: Underground Storage Tanks
- Pollution Control Technology: Groundwater Remediation

INTRODUCTION

This chapter's opening statement (from Cronin and Kennedy's book *The Riverkeepers*) illustrates a prime example of the issues involved when regulation mandates specific actions by covered parties to prevent an environmental problem such as water pollution. Several years ago, in the early 1970s, the public, environmentalists, and legislators came to the realization that something had to be done to protect and clean up our water resources, specifically our traditional fresh water supplies: lakes, streams, and rivers. Actually, environmentalists were well aware of our water resource problems several years before legislation was enacted. However, in the 1950s and 1960s, when environmentalists were voicing their environmental concerns, most people ignored them; polluters simply played their standard trump card and declared them to be bleeding heart liberals. Fortunately for the rest of us, we woke up and put pressure on Congress to enact two important regulations intended to protect our water resources. Let's briefly look at these two major regulations specifically designed to protect the nation's water resources: the Safe Drinking Water Act of 1974 and the Federal Waster Pollution Control Act of 1972.

SAFE DRINKING WATER ACT (SDWA)

The *Safe Drinking Water Act (SWDA)* of 1974 (a U.S. federal law) came about when federal legislators became aware of the unhealthy condition of many local drinking water supplies, and the reluctance of local and state officials to remove pollutants from their wastewater. The act set national drinking water standards, called *maximum contaminant levels (MCLs)*, for pollutants that might adversely affect public health and welfare. The first standards went into effect three years later and specifically covered every public water supply in the country serving at least 15 service connections or 25 or more people. Because over 200 contaminants from hazardous wastes injected into the soil were identified in groundwater, SDWA also established standards to protect groundwater from such practices. Specifically, SDWA requires establishment of programs to protect critical groundwater sources of drinking water and areas around wells that supply public drinking water systems, and regulates the underground injection of wastes above and below drinking water sources.

Later, in 1982 and 1983, the EPA established a priority list for setting regulations for over 70 substances. These substances were listed because they are toxic and likely to be found in drinking water. In 1986, when Congress reauthorized the SWDA, it amended the Act and directed the EPA to monitor drinking water for unregulated contaminants and to inform public water suppliers about which substances to look for. The 1986 reauthorization also instructed the EPA to set standards within 30 years for all 70 substances on its priority list. By the end of 1994, the priority list had been expanded, and the EPA had set standards for more than 80 substances.

Under the Act, local public water systems are required to monitor their drinking water periodically for contaminants with MCLs and a broad range of other contaminants, as specified by the EPA. Enforcement of the standards, monitoring, and reporting are the responsibility of the individual states, but the 1986 amendments require the EPA to act when the state fails or is too slow to do act, and authorizes substantial civil penalties against the worst violators.

SWDA and its amendments also authorize the EPA to set *secondary drinking water standards* regarding public welfare, by providing guidelines regarding the taste, odor, color, and aesthetic aspects of drinking water that do not present a health risk. These guidelines are non-enforceable and are called *suggested levels*. The EPA recommends these levels to the states as reasonable goals, but federal law does not require water systems to comply with them, though some individual states have enforceable regulations regarding these concerns.

The 1996 amendments also ban all future use of lead pipe and lead solder in public drinking water systems, and require public water systems to tell their users of the potential sources of lead contamination, its health effects, and the steps they can reasonably take to mitigate lead contamination.

FEDERAL WATER POLLUTION CONTROL ACT

In 1972 Congress enacted the *Federal Water Pollution Control Act*, commonly called the *Clean Water Act (CWA)*. CWA stems originally from a much-amended 1948 law aiding communities in building sewage treatment plants. CWA is the keystone of environmental law (the law Kennedy was referring to in the chapter's opening) and is credited with significantly cutting the amount of municipal and industrial pollution fed into the nation's water bodies.

Through the 1970s and 1980s, the primary aim of the Clean Water Act was to make national waters fishable and swimmable. Specifically, it sought to eliminate discharge of untreated municipal and industrial wastewater into waterways (many of which are used as sources of drinking water), providing billions of dollars to finance building of sewage treatment plants.

In the 1987 amendments (The Water Quality Act, which reauthorized the original Clean Water Act), the Act focused on updating standards for dealing with toxic chemicals, since much of the toxic pollution still fouling the nation's surface water bodies came from companies that had installed 1970s-era pollution control technologies. For the first time it also attempted to deal with water pollution stemming from nonpoint sources (city streets and croplands, for example), requiring states to identify waters that do not meet quality standards and developing programs to deal with the problem. The 1987 amendments also granted the U.S. Army Corp of Engineers the authority to regulate the dredging and filling of wetlands.

EFFECT OF REGULATIONS ON PREVENTING WATER POLLUTION

On the surface, we might assume that the laws passed since the early 1970s, designed to protect our water resources, are adequate in themselves to ensure water quality. We would be wrong, of course. Kennedy's expressed shock that "a government agency, charged with protecting the public from pollution, could so blatantly intervene to protect a polluter from the public and the law" makes a valid point that always holds true: you can make all the laws you want, but requirement does not necessarily mean enforcement. Many factions have criticized the EPA for failing to properly enforce the Safe Drinking Water Act and the Clean Water Act. Another problem has developed in protecting groundwater supplies—those programs have been underfunded or unfunded.

In addition, many environmentalists have criticized the EPA for exempting from the SDWA requirements over 100,000 public water systems as not serving year-round residents, although these systems include schools, factories, seasonal resorts, summer camps, roadside restaurants, and hospitals.

Simply stated, regulations designed to protect the environment, and at the same time protect public health and welfare, are only the first step across the bridge between pollution and prevention. Once in place, regulations must be complied with and enforced. Then the effort shifts from one of determining direction, objectives, and goals to one of implementation. In this implementation phase, technology comes into play—in the key role. A society can have all kinds of plans, objectives, goals, and regulations to stipulate what needs to be done to correct or mitigate an environmental problem, but none of these will bring about positive results unless the means (technology) is available to accomplish the requirements. With the means must also come a certain amount of common sense. Only solid, legitimate, careful scientific analysis can provide the answers and the solutions to environmental problems.

Obviously we need to stop polluting our surface water and groundwater. But we shouldn't do anything rash to our environment until science supports our investment. The problem we consistently face, however, is how to compel politicians to think like scientists.

We are fortunate in one way. The technology needed to protect and to clean up our water resources is available. Advances in water pollution technology are on-going. In a number of cases, to facilitate installation, much of the technology needed to protect and/or treat water and wastewater has been funded by the Federal Government. In the sections that follow, we briefly discuss traditional treatment technologies (water, wastewater, thermal, underground storage tanks (USTs) and groundwater) currently available and widely used.

WATER TREATMENT

Municipalities normally control contaminants in drinking water supplies by following established treatment procedures. Before distribution to local destinations, water withdrawn from its source undergoes treatment. Initial water quality determines the needed degree of treatment. Most water systems, large or small, include certain basic treatment steps. In this chapter's opening statement, we mentioned alum (aluminum sulfate). Alum (or lime) is added to water supplies to create floc (small gelatinous particles—insoluble precipitates), which gather dirt and other solids. Gentle mixing of the water causes floc particles to join and form larger particles; floc and sediment fall to the bottom and are eventually removed as sludge. The water is then filtered through a granular material (sand or carbon—crushed anthracite coal). Chlorine, sodium hypochlorite, or ultra-violet light is then added to kill bacteria and other organisms, and a chemical (usually lime) is added to raise pH (reduce acidity) in water and prevent corrosion in city and household pipes. Many municipal plants also add fluoride to water supplies to prevent tooth decay.

Treated water is sent through a network of pipes to customers. To ensure its quality, water must be monitored and tested by licensed operators throughout the treatment and delivery process. Generally, surface water is more complicated to treat than groundwater, because contamination is more likely. However, sometimes groundwater is hard (containing calcium or magnesium), which means an additional step is added to the treatment process (softening by using alum and lime) to remove hardness.

WASTEWATER TREATMENT

What is wastewater and how is it generated and treated? Five major sources generate wastewater, and each source's wastewater presents specific characteristics.

1) *Human and Animal Wastes*: Considered by many to the most dangerous from a human health viewpoint. Contains the solid and liquid discharges of humans and animals, and millions of bacteria, viruses, and other organisms, some pathogenic.

2) *Household Wastes*: The wastes, other than human and animal wastes, discharged from the home. Contains paper, household cleaners, detergents, trash, garbage, and any other substance that users may discharge into the sewer system.

3) *Industrial Wastes*: All materials that can be discharged from industrial processes into the collection system are included in this category. May contain chemicals, dyes, acids, alkalies, grit, detergents, and highly toxic materials. Characteristics are industry specific and cannot be determined without detailed information on the specific industry and processes used.

4) *Stormwater Runoff*: If the collection system is designed to carry both the wastes of the community and the stormwater runoff, wastewater can, during and after storms, contain large amounts of sand, gravel, road-salt, and other grit, and the amount of water may be excessive.

5) *Groundwater Infiltration*: If the collection system is old or not sealed properly, groundwater may enter the system through cracks, breaks or unsealed joints. This can add large amounts of water to the wastewater flows, as well as additional grit.

Wastewater can be classified according to the sources of the flow.

1) *Domestic Wastewater (sewage)*: Domestic wastewater consists mainly of human and animal wastes, household wastes, small amounts of groundwater infiltration, and perhaps small amounts of industrial wastes.

2) *Sanitary Wastewater*: Consists of domestic wastes and significant amounts of industrial wastes. In many cases, the industrial wastes can be treated without special precautions. In some cases the industrial wastes will require special precautions, or a pretreatment program to ensure the wastes do not cause compliance problems for the plant.

3) *Industrial Wastewater:* Often an industry will determine that treating its wastes independent of the domestic wastes is more economical.

4) *Combined Wastewater*: A combination of sanitary wastewater and stormwater runoff. All the wastewater and stormwater of the community is transported through one system and enters the treatment system.

5) *Stormwater*: Many communities have installed separate collection systems to carry stormwater runoff. Stormwater flow should contain grit and street debris, but no domestic or sanitary wastes.

Wastewater contains many different substances that can be used to characterize it. Depending on the source, the specific substances present will vary, as will the amounts or concentration of each. For this reason, wastewater characteristics are normally described for *average domestic wastewater*. Other sources and types of wastewater can dramatically change the characteristics.

The physical characteristics of wastewater are listed as follows:

Color: Typical wastewater is gray and cloudy. Wastewater color will change significantly if allowed to go septic. Typical septic wastewater will be black.

Odor: Fresh domestic wastewater has a musty odor. This odor will change significantly if septic. Septic wastewater develops the rotten egg odor associated with hydrogen sulfide production.

Temperature: Wastewater temperature will normally be close to that of the water supply. Significant amounts of infiltration or stormwater flow can cause major temperature changes.

Flow: The volume of wastewater is normally expressed in terms of gallons per person per day. Most treatment plants are designed using an expected flow of 100 to 200 gallons per person per day. This figure may have to be revised to reflect the degree of infiltration or stormwater flow the plant receives. Flow rates will vary throughout the day. This variation, which can be as much as 50 to 200% of the average daily flow, is known as the *diurnal flow variation.*

The chemical characteristics of wastewater are:

Alkalinity: A measure of the wastewater capability to neutralize acids. Measured in terms of bicarbonate, carbonate, and hydroxide alkalinity, alkalinity is essential to buffer (hold the neutral pH) wastewater during the biological treatment process.

Biochemical Oxygen Demand (BOD): A measure of the amount of biodegradable matter in the wastewater. Normally measured by a five-day test conducted at 20°C. The BOD5 domestic waste is normally in the range of 100 to 300 mg/l.

Chemical Oxygen Demand (COD): A measure of the amount of oxidizable matter present in the sample. The COD is normally in the range of 200 to 500 mg/l. The presence of industrial wastes can increase this significantly.

Dissolved Gases: Gases that are dissolved in wastewater. The specific gases and normal concentrations are based on the composition of the wastewater. Typical domestic wastewater contains oxygen (relatively low concentrations), carbon dioxide, and hydrogen sulfide (if septic conditions exist).

Nitrogen Compounds: The type and amount of nitrogen present varies from raw wastewater to treated effluent. Nitrogen follows a cycle of oxidation and reduction. Most of the nitrogen in untreated wastewater will be in the forms of organic nitrogen and ammonia nitrogen, presence and levels determined by laboratory testing. The sum of these two forms of nitrogen is also measured and is known as *Total Kjeldahl Nitrogen (TKN).* Wastewater will normally contain 20 to 85 mg/l of nitrogen. Organic nitrogen will normally be in the range of 8 to 35 mg/l and ammonia nitrogen will be in the range of 12 to 50 mg/l.

pH: A method of expressing the acid condition of wastewater. For proper treatment, wastewater pH should normally be in the range of 6.5 to 9.0.

Phosphorus: Essential to biological activity, phosphorus must be present in at least minimum quantities or secondary treatment processes will not perform. Excessive amounts can cause stream damage and excessive algal growth. Phosphorus will normally be in the range of 6 to 20 mg/l. The removal of phosphate compounds from detergents has had a significant impact of the amounts of phosphorus in wastewater.

Solids: Most pollutants found in wastewater can be classified as solids. Wastewater treatment is generally designed to remove solids, or to convert solids to a form that is more stable or can be removed. Solids can be classified by their chemical composition (organic or inorganic), or by their physical characteristics (settleable, floatable, colloidal). Concentration of total solids in wastewater is normally in the range of 350 to 1,200 mg/l.

Water: Always the major component of the wastewater. In most cases water makes up 99.5 to 99.9% of wastewater. Even in the strongest wastewater, the total amount of contamination present is less than 0.5% of the total, and in average-strength wastes it is normally less than 0.1%.

As a process, wastewater treatment is designed to use the natural purification processes to the maximum level possible, and to complete these processes in a controlled environment rather than over many miles of stream. Removing contaminants not addressed by natural processes and treating the solids generated by the treatment steps are further tasks of wastewater treatment. The specific goals wastewater treatment plants are designed to accomplish include:

- Protecting public health
- Protecting public water supplies
- Protecting aquatic life
- Preserving the best uses of the waters
- Protecting adjacent lands

Wastewater treatment is accomplished by applying up to seven principal treatment steps to the incoming wastestream. The processes and equipment for each step are specific to the task. The major categories of treatment steps used in many treatment plants include preliminary treatment, primary treatment, secondary treatment, advanced waste treatment, disinfection, and biosolids treatment.

Preliminary treatment removes materials (wood, rocks, and other forms of debris) that could damage treatment plant equipment or would occupy treatment capacity without being treated.

Primary treatment removes larger particles by filtering through screens and settleables and floatable solids in ponds or lagoons. Water is removed from the top of the settling lagoon and released. Water that has been treated in this manner has had its sand and grit removed, but it still carries a heavy load of organic matter, dissolved salts, bacteria, and other microorganisms. Primary treatment removes up to about 60% of suspended solids. In larger cities, where several cities a few miles or less from each other take water and return it to a stream, primary wastewater treatment is not adequate.

Secondary treatment usually follows primary treatment and is designed to remove BOD5 and dissolved and colloidal suspended organic matter by biological action. Organics are converted to stable solids, carbon dioxide, and more organisms by holding the wastewater until the organic material has been degraded by the bacteria and other microorganisms. It removes up to 90% of the oxygen-demanding wastes by using either *trickling filters*, where aerobic bacteria degrade sewage as it seeps through a large vat bed filled with media (rocks, plastic media, etc.) covered with bacterial growth, or an *activated sludge process*, in which the sewage is pumped into a large tank and mixed for several hours with bacteria-rich biosolids and air to increase bacterial degradation. To optimize this action, large quantities of highly oxygenated water for aerating water are added directly by a blower system.

Advanced Wastewater Treatment (Tertiary Sewage Treatment) uses physical, chemical and biological processes to remove additional BOD5, solids, and nutrients. Advanced wastewater treatment is normally used in facilities that have unusually high amounts of phosphorus and nitrogen present.

Disinfection is used to kill pathogenic microorganisms to eliminate the possibility of disease when the flow is discharged.

Biosolids treatment works to stabilize the solids removed from the wastewater during treatment, inactivates pathogenic organisms, and/or reduces the volume of the biosolids by removing water (*dewatering*).

THERMAL POLLUTION TREATMENT

Approximately half of the water withdrawn in the United States is used for cooling large power-producing plants. The most common method (because it is easiest and cheapest) is to withdraw cold water from a lake or river, pass it through heat exchangers in the facility, and return the heated water to the same body of water. The warm water discharge raises the receiving body's temperature, lowers DO content, and causes aquatic organisms to increase their respiration rates and consume the already depleted oxygen faster.

We can minimize the harmful effects of excess heat on aquatic ecosystems in a number of ways. Two of the most commonly used methods are the cooling tower and dry tower methods.

In the *cooling tower method*, the heated water is sprayed into the air and cooled by evaporation. The obvious disadvantage of this treatment method is the loss of large amounts of water to evaporation. Production of localized fogs is another disadvantage.

The *dry tower method* does not release water into the atmosphere. Instead, the heated water is pumped through tubes and the heat is released into the air, which is similar to the action performed by an automobile's radiator. The disadvantage to using the dry tower method is its high cost, both to construct and to operate.

POLLUTION CONTROL TECHNOLOGY: UNDERGROUND STORAGE TANKS

Recent estimates have ranged from five to six million, but no one is quite sure just how many *underground storage tanks (USTs)* containing hazardous substances or petroleum products are in use in the United States. Compounding the issue, no one can even guess how many USTs are no longer being used, the contents of which have been escaping, fouling water, land, and air. Another potential problem is just biding its time; USTs that are not leaking today will probably leak soon. Environmental contamination from leaking USTs poses a significant threat to human health and the environment.

Besides the obvious problem of fouling environmental mediums (water, soil, and air), many of these leaking USTs ironically pose serious fire and explosion hazards. The irony is in the fact that USTs came into common use primarily as a fire and explosion prevention measure (the hazard was buried under the ground). Today, however, the hazards we worked to protect ourselves from are finding other ways of presenting themselves.

The problem with leaking USTs goes beyond fouling the environment (especially groundwater, which 50% of the U.S. population relies on for drinking water) and presenting fire and explosion hazards. Products released from these leaking tanks can damage sewer lines and buried cables, and can poison our crops.

The EPA, under its *Resource Conservation and Recovery Act (RCRA)*, defines USTs as tanks with 10% or more of their volume (including piping) underground. The largest portion of the USTs regulated by the EPA are petroleum storage tanks owned by gas stations; another significant percentage are petroleum storage tanks owned by a group of other industries (airports, trucking fleets, farms, manufacturing operations, and golf courses) that store petroleum products for their own use.

In 1986 the U.S. Congress established a UST cleanup fund known as *Leaking Underground Storage Tank (LUST)* trust fund. The EPA, tasked with the responsibility of exploring, developing, and disseminating new cleanup technologies and funding mechanism, must still leave the primary job of cleaning up LUST sites to the various state and local governments. Owners and operators of tank facilities are liable for clean-up costs and damage caused by their tanks—not a small matter in any way. The average current cost for remediating a site containing petroleum contamination of the soil and groundwater is on the order of $200,000 to $350,000, depending on the lateral and vertical extent of the contamination and the required clean-up target levels. Experience has shown that in some cases the cost of cleanup may exceed the value of the property.

When the EPA and other investigators initially investigated the problems with leaking USTs in 1985, they found that many of the existing USTs were more than 20 years old were or of unknown age. Compounding the problem of tank age, these older tanks were found to be constructed of bare steel, not protected against corrosion, and nearing the end of their useful lives (Holmes et al., 1993). Exacerbating the problem, many of the old tanks systems were found to have already leaked, or were right on the edge of leaking. Many of these old tanks were found in abandoned gas stations (shut down because of the oil crisis in the 1970s).

Because of the findings of the EPA and others on the scope of the problem with USTs, regulatory requirements were put into place. The regulatory requirements for USTs depend on whether the system is an existing or a new installation. An existing installation is defined as one that was installed prior to 1988.

Certain requirements for USTs must now be met. All existing USTs (in current use) must have overfill and spill protection. In addition, corrosion protection and leak detection systems will have to be installed in accordance with the schedule mandated by the Federal regulations. The compliance schedule ensures that the oldest tanks (those with the greatest potential for failure) are addressed first.

Under Federal Regulations (40 CFR Part 280), all existing tanks must have corrosion protection and spill and overfill prevention devices installed. Both pressured and suction piping installed prior to December 1988 is required to have corrosion protection by December 1998.

To evaluate the integrity of an installed UST, all owners must abide by certain regulatory requirements (minimum requirements). The specific requirements are listed and explained in Table 18.1.

Under Federal Law, facilities having USTs containing petroleum and hazardous substances must respond to a leak or spill within 24 hours of release, or within another reasonable period of time as determined by the implementing agency. Responses to releases from USTs are site-specific, and depend on several different factors. Corrective action usually involves two stages. Stage one (initial response) is directed toward containment and collection of spilled material. Stage two (permanent corrective response) involves technical improvements designed to ensure that the incident does not occur again. Preventive-action technology usually includes employing either containment, diversion, removal, or treatment protocols. The choice of which technology to employ in spill prevention and correction depends on its suitability, life-span, ease of implementation, and ease of performing required maintenance checks.

Table 18.1: UST Requirements—What You Have to Do

Leak Detection

For New Tanks	Must be monitored monthly. These checks may be made by using automatic tank gauging, vapor monitoring, interstitial (secondary containment) monitoring, groundwater monitoring, and other approved methods.
or you may	Perform monthly inventory control and tank tightness testing every 5 years (only for up to 10 years after installation).
For Existing Tanks	Monthly monitoring, or Monthly inventory control and annual tank tightness testing (could only be used until December 1998), or Monthly inventory control and annual tank tightness testing every 5 years (this choice can only be used for 10 years after adding corrosion protection and spill/overfill prevention, or until December 1998, whichever date is later.

New and Existing Pressurized Piping

Choice of one of the following:

- automatic flow restrictor
- annual line testing
- automatic shutoff device and monthly monitoring (except automatic tank gauging)
- continuous alarm system

New and Existing Suction Piping

Choice of monthly monitoring (except automatic tank gauging) or
Line testing every 3 years.

Corrosion Protection

New Tanks	Choices: Coated and cathodically protected steel Fiberglass Steel tank clad with fiberglass
Existing Tanks	Choices: Same options as for New Tanks Add cathodic protection system Interior lining and cathodic protection
New Piping	Choices: Coated and cathodically protected steel Fiberglass
Existing Piping	Choices: Same options as for new piping Cathodically protected steel

Spill/Overfill Protection

All Tanks	Catchment Basins and ▪ Automatic shutoff devices or ▪ Overfill alarms or ▪ ball float valves

Source: 40 CFR Part 280

The EPA issued *Cleanup of Releases from Petroleum USTs: Selected Technologies* (1988), which has become the standard reference in making a decision of which technology to employ for use in cleanup of releases from petroleum USTs. Although only a limited number of technologies are available to clean environmental mediums of the contaminants associated with gasoline, their practicality, removal efficiencies, limitations, and costs are well-documented.

In recovering free product from the water table, two technologies are presently used to limit the migration of floating gasoline across the water table: The Trench Method and the Pumping Well Method. In recovering the free product using the trench method, a variety of equipment is available for use, including skimmers, filter separators, and oil/water separators. In the pumping well method, both single- and dual-pump systems are available for use.

When the water table is no deeper than 10-15 ft. below the ground surface, the *trench method* is most effective. The advantages of this method include the ease in which the trench can be excavated and the ability to capture the entire leading edge of the plume. The disadvantage in using the trench method is that it does not reverse groundwater flow, which means it may not be appropriate for use when a potable well supply is threatened. The cost of this system is about $150 per cubic yard of soil excavated.

When a spill is deep (water table depth exceeds 20 ft. below the ground surface), a *pumping well system* is the preferred method used to recover free product from the water table. The major advantage of using this system is that it can reverse the direction of groundwater flow. Including the cost of labor and engineering, this system ranges from about $150 to $300 per foot for 4- to 10-in. gravel-packed galvanized steel wells (EPA, 1988).

Because gasoline spilled onto the soil may eventually find its way to groundwater, removal of gasoline from unsaturated soils is an essential component of any corrective action plan. A number of removal techniques are available, but they all vary in effectiveness and cost. The most widely used corrective action is excavation and disposal. Other methods include volatilization, incineration, venting, soil washing/extraction, and microbial degradation.

The advantage of using excavation and disposal is that it can be 100% effective. The main disadvantages are that usually only a small portion of the contaminated soil can be removed because of high cost, limitations of excavation equipment (backhoes normally only reach down to about 16 ft.), the fact that landfills may not accept the contaminated soil, and a lack of uniform guidelines for the proper disposal of contaminated soil.

Volatilization will effectively remove about 99% of volatile organic compounds (VOCs), but the process does not have an extensive track record (because it is little used) to make definitive statements as to its efficiency and/or effectiveness in the field.

Incineration, like volatilization, will remove approximately 99% of gasoline constituents in soil. Having proven itself highly reliable, incineration of gasoline contaminated soil is widely practiced. The practice does have a few limitations, however: (1) the soil must be brought to the surface, increasing the risk of exposure, (2) it is usually appropriate only when toxics other than volatiles are present, (3) there may be time delays in the permitting process.

The big advantage of using *venting*, which can be up to 99% effective, is that it allows for the removal of gasoline without excavation. However, because critical parameters have not yet been defined, venting is not widely used in the field. Venting is relatively easy to implement, but its effectiveness is uncertain because soil characteristics may impede free movement of vapors, and could even lead to an explosion hazard.

Soil washing and extraction works to leach contaminants from the soil into a leaching medium, after which the extracted contaminants are removed by conventional methods. Under ideal conditions, up to 99% of VOCs can be removed. If the contaminated soil contains high levels of clay and silt, they may impede the separation of the solid and liquid after the washing phase. The soil's suitability to be decontaminated using this method should be verified before this procedure is implemented.

Microbial degradation, theoretically, can remove up to 99% of the contaminants. This method is still in the research mode with field testing still in progress; thus, its cost effectiveness and overall effectiveness have not been verified. If further testing supports its viability for use in the field, the advantage will be that *in situ* treatment is possible with volatiles completely destroyed.

POLLUTION CONTROL TECHNOLOGY: GROUNDWATER REMEDIATION

As you should recall, groundwater is an important water source, supplying a significant percentage of the water used for drinking. For many years groundwater was not only the only source of potable water available in certain areas, it was the source of choice, even when other sources were available, because of people's perception of groundwater as pure. For many years people held the perception that groundwater was safe and that it only needed disinfectant before being sent to the household tap. Most people knew groundwater was supplied from precipitation in the form of rain, sleet, and snow. Once on the surface, precipitation entered rock and soil and filtered its way through the earth's strata to the water table, where it was held in a "clean" state, and by the nature of its confinement there, was protected from surface contamination.

This view of groundwater has problems, though. In the first place, since groundwater is so widely used, and because the populations using it have increased at a steady pace, many groundwater supplies have either been depleted, or lowered to the point in coastal areas where saltwater intrusion takes place. Secondly, groundwater supplies may become polluted.

Both groundwater depletion and groundwater pollution may be irreversible. Depletion can cause an aquifer to consolidate, diminishing its storage capacity. Groundwater may be contaminated by both naturally occurring and artificial materials. Just about anything water comes into contact with will be dissolved in or mixed with the flow. If contaminated, the water is likely to remain that way.

Of particular concern in groundwater pollution are nonaqueous-phase liquids (NAPLs). NAPLs are classified as either *light (LNAPLs)* or *dense (DNAPLs)*. LNAPLs include such products as gasoline, heating oil, and kerosene. Because of the widespread use of underground storage tanks, these products are common in many soils. Because LNAPLs are light, they tend to float on the groundwater, penetrating the capillary fringe and depressing the water surface. Even when the source of the spill is controlled, the soil will remain contaminated and the floating layer will serve as a long-term source of contamination (McGhee, 1991). From a health standpoint, DNAPLs are a much more serious problem. They include trichloroethane, carbon tetrachloride, creosote, dichlorobenzene, and others. Because these compounds are toxic and have low viscosity, great density, and low solubility, they are not only health hazards but are also very mobile in groundwater, spreading quickly throughout a localized aquifer.

In contaminated groundwater mitigation and treatment, usually only localized areas of an aquifer need reclamation and restoration, because the spread of contaminants is usually confined to the plume. Experience has shown, however, that even after the original source of contamination is removed, cleanup of a contaminated aquifer is often costly, time-consuming, and troublesome. Problems with cleanup include difficulty in identifying the type of subsurface environment, locating potential contamination sources, defining potential contaminant transport pathways, determining contaminant extent and concentration, and choosing and implementing an effective remedial process (Davis and Cornwell, p. 712, 1991).

Cleanup is possible, but not simple. Certain methods have in some cases (especially where groundwater has been pumped from the subsurface) proven successful. These efforts have been refined from processes used to treat industrial wastes. However, attempting to treat site-contaminated groundwater using these methods is often confusing. The contaminants themselves may also dictate what methodologies should work for mitigation. When the contaminant is a single chemical, the treatment system employed may be simple—but in cases involving multiple contaminants, treatment can be extremely complex. To determine which treatment should be employed, only representative samples and laboratory analysis will provide the needed information. Cleanup technologies commonly used for groundwater containing organic contamination include air stripping and activated carbon. The chemical precipitation process is used for inorganics in groundwater. Each of these treatment processes is briefly described in the following.

In *air stripping* (a relatively simple mass transfer process), a substance in solution in water is transferred to solution in a gas. Air stripping uses four basic equipment configurations, including diffused aeration, countercurrent packed columns, cross-flow towers, and coke tray aerators. The countercurrent packed tower system has significant advantages (provides the most liquid interfacial area and high air-to-water volume ratios) over the other systems, and is most often used in removing volatile organics from contaminated groundwater.

Carbon adsorption occurs when an organic molecule is brought to the activated carbon surface and held there by physical and/or chemical forces. When activated carbon particles are placed in water containing organic chemicals and mixed to give adequate contact, adsorption of the organic chemicals occurs. Activated carbon adsorption has been successfully employed for removing organics from contaminated groundwater.

Biological treatment (a new technology still under evaluation by pilot studies) works to remove or reduce the concentration of organic and inorganic compounds. To undergo biological treatment, contaminated groundwater must first be pretreated to remove toxins that could destroy microorganisms needed to metabolize and remove the contaminants.

In removing inorganic contaminants, the established and commonly used methodology is *chemical precipitation*. Accomplished by the addition of carbonate, hydroxide, or sulfide chemicals, chemical precipitation has successfully removed heavy metals from groundwater.

When groundwater near a potable water system (well) is contaminated, the most common way to protect the water from an approaching plume of contaminated groundwater is to use some combination of *extraction wells* and *injection wells*. Extraction wells are used to lower the water table, creating a hydraulic gradient that draws the plume to the wells. Injection wells raise the water table and push the plume away. Working in combination, extraction well and injection well pumping rates can be adjusted in such a way to manipulate the hydraulic gradient, which helps keep the plume away from the potable water well, drawing it toward the extraction well. Once extracted, the contaminated water is treated, and either re-injected back into the aquifer, reused, or released into the local surface water system (Masters, 1991).

SUMMARY

Cleanup of contaminated water supplies presents us with a complex problem. The contaminated areas are often out of sight and hard to define clearly. The technologies, while becoming more effective, are also costly. Without the EPA as guardian and enforcer, without the Safe Drinking Water Act and the Federal Water Pollution Control Act, without pressure from concerned citizens over the condition and quality of our resources, an attitude related to "not in my backyard" creeps in—"Ignore it. It's someone else's problem." Fortunately, people are becoming more aware that contaminated water supplies will affect us all.

Cited References

Cronin, J., and Kennedy, R. K., Jr., *The Riverkeepers*, New York: Scribner, 1997.

Davis, M. L., and Cornwell, D. A., *Introduction to Environmental Engineering*, 2nd ed., New York: McGraw-Hill, 1991.

Holmes, G., Singh, B. R., and Theodore, L., *Handbook of Environmental Management and Technology*, New York: John Wiley *and* Sons, 1993.

Masters, G. M., *Introduction to Environmental Engineering and Science*, Englewood Cliffs, NJ: Prentice-Hall, 1991.

McGhee, T. J., *Water Supply and Sewerage*, 6th ed., New York: McGraw-Hill, 1991.

U.S. Environmental Protection Agency, *Cleanup of Releases from Petroleum USTs: Selected Technologies*, April 1988.

Suggested Readings

American Society of Civil Engineers, *Management of Water Treatment Plant Residuals*, New York: American Society of Civil Engineers, 1996.

Canter, L. W., and Knox, R. C., *Ground Water: Pollution Control*, Chelsea, Michigan: Lewis Publishers, 1990.

Carmichael, J., (ed.), *Industrial Water Use and Treatment*, New York: Tayler and Francis, Inc., 1986.

Comella, P. A., "Waste Minimization/Pollution Prevention," *Pollution Engineering*, 1990.

Conservation Foundation, *Groundwater Pollution*, Washington, DC: Conservation Foundation, 1987.

Consumer Reports Books Editors, and Gabler, R., *Is Your Water Safe to Drink?* New York: Consumer Reports Books, 1987.

Driscoll, F. G., *Groundwater and Wells*, 2nd ed., St. Paul, Minnesota: Johnson Division, 1986.

Freeze, R. A., and Cherry, J. A., *Groundwater*, Englewood Cliffs, NJ: Prentice-Hall, 1979.

Griffin, R. D., *Principles of Hazardous Materials Management*, Ann Arbor, Michigan: Lewis Publishers, 1988.

HDR Engineering, *Handbook of Public Water Systems*, New York: Van Nostrand Reinhold, 1997.

Horan, N. J., *Biological Wastewater Treatment Systems: Theory and Operation*, New York: Wiley, 1990.

Josephson, J., Restoration of Aquifers, *Environmental Science and Technology*, 17:347A - 350A, 1983.

King, J., *Troubled Water*, Emmaus, PA: Rodale Press, 1985.

National Academy of Sciences, *Groundwater Contamination*, Washington, DC: National Academy Press, 1984.

Office of Technology Assessment, *Protecting the Nation's groundwater from Contamination*, Washington, DC: Government Printing Office, 1984.

Pettyjohn, W. A., (ed.), *Protection of Public Water Supplies from Ground-Water Contamination*, Park Ridge, NJ: Noyes Data Corporation, 1987.

Roques, H. (ed.), *Chemical Water Treatment: Principles and Practice*, New York: VCH Publishers, 1996.

Rules and Regulations, *Federal Register*, 52(185) September 1988.

U.S. Environmental Protection Agency, *Handbook—Remedial Action at Waste Disposal Sites* (Revised), 1985.

U.S. Environmental Protection Agency, Drinking water in America: An overview, *EPA Journal*, September 1986.

U.S. Environmental Protection Agency, *Underground Storage Tank Corrective Action Technologies*, January 1987.

U.S. Environmental Protection Agency, *Environmental Progress and Challenges: EPA's Update*, August 1988.

U.S. Environmental Protection Agency, *Basics of Pump-and-Treat Groundwater Remediation Technology*, 1990.

Viessman, W., Jr., and Hammer, M. J. M., *Water Supply and Pollution Control*, 4th ed., New York: Harper and Row, 1985.

Welch, E. B., *Groundwater Systems Planning and Management*, Cambridge, UK: Cambridge University Press, 1980.

Wentz, C. S., *Hazardous Waste Management*, New York: McGraw-Hill, 1989.

PART IV
SOIL QUALITY

CHAPTER 19

SOIL CHARACTERISTICS

Soils are crucial to life on earth...soil quality determines the nature of plant ecosystems and the capacity of land to support animal life and society. As human societies become increasingly urbanized, fewer people have intimate contact with the soil, and individuals tend to lose sight of the many ways in which they depend upon soils for their prosperity and survival. The degree to which we are dependent on soils is likely to increase, not decrease, in the future. Of course, soils will continue to supply us with nearly all of our food and much of our fiber. On a hot day, would you rather wear a cotton shirt or one made of polyester? In addition, biomass grown on soils is likely to become an increasingly important source of energy and industrial feedstocks, as the world's finite supplies of petroleum are depleted over the coming century. The early signs of this trend can be seen in the soybean oil-based inks, the cornstarch plastics, and the wood alcohol fuels that are becoming increasingly important on the market. —Brady and Weil

CHAPTER OUTLINE

- Introduction
- Soil: What Is It?
- Soil Basics

INTRODUCTION

If we were transported back in time, we would instantly recognize the massive structure before us, even though we might be taken aback at what we saw: A youthful mountain range with considerable mass, steep sides, and a height that certainly reached beyond any cloud. We would instantly relate to one particular peak—the tallest, most massive one. The polyhedron-shaped object, with its polygonal base and triangular faces culminating in a single sharp-tipped apex, would have looked familiar—comparable in shape, though larger in size, to the largest of the Great Egyptian Pyramids, though the Pyramids were originally covered in a sheet of limestone, not the thick, perpetual sheet of solid ice and snow that covered the mountain peak.

But if we walked this same site in modern times, if we knew what had once stood upon this site, the changes would be obvious and startling—and entirely relative to time. What stood as an incomparable mountain peak eons ago, we cannot today see in its ancient majesty. In fact, we wouldn't give it a second thought as we walked across its remnants and through the vegetation that grows from its pulverized and amended remains.

The pyramid-shaped mountain peak 300 million years ago stood in full, unchallenged splendor above the clouds, wrapped in a cloak of ice, a mighty fortress of stone, seemingly invulnerable, standing higher than any mountain on earth ever stood or will ever stand.

And so it remained, for millions upon millions of passings of the Earth around the Sun. Born when Mother Earth took a deep breath, the pyramid-shaped peak stood tall and undisturbed until millions of years later, when Mother Earth stretched. Today we would call this stretch a massive earthquake—one of such a magnitude that we've never witnessed. But when this massive earthquake shattered the earth's surface, nothing we would call intelligent life lived here—and it's a good thing.

During this massive upheaval, the peak shook to its very foundations, and after the initial shockwave and the hundred plus aftershocks, the solid granite structure had fractured. This immense fracture was so massive that each aftershock widened it and loosened the base foundation of the pyramid-shaped peak itself. Only 10,000 years later (a few seconds relative to geologic time), the fracture's effects totally altered the shape of the peak forever. During a horrendous windstorm, one of an intensity known only in earth's earliest days, a sharp tremor (emanating from deep within the earth and shooting up the spine of the mountain itself, up to the very peak) widened the gaping wound still more.

Decades of continued tremors and terrible windstorms passed (no present day structure could withstand a blasting from such a wind), and finally the highest peak of that time, of all time, fell. It broke off completely at its base, and following the laws of gravity, tumbled from its pinnacle position and fell more than 20,000 feet. It collided with the expanding base of the mountain range, the earth-shattering impact destroying several thousand acres. It finally came to rest on a precipitous ledge, at 15,000 feet in elevation. The pyramid-shaped peak, much smaller now, sat precariously on the precipitous ledge for about 5 million years.

Nothing is safe from time. The most inexorable natural law is that of entropy. Time and entropy mean change and decay—harsh, sometimes brutal, but always inevitable. The bruised, scarred, truncated, but still massive rock form, once a majestic peak, was now a victim of nature's way. Nature, with its chief ally, Time, at its side, works to degrade anything and everything that has substance and form. For better or worse, in doing so, Nature is harsh, sometimes brutal, and always inevitable—but never without purpose.

While resting on the ledge, the giant rock, over the course of that 5 million years, was exposed to constantly changing conditions. For several thousand years, the earth's climate was unusually warm—almost tropical—everywhere. Throughout this warm era, the rock was not covered with ice and snow, but instead baked in intense heat, steamed in hot rain, and seared in the gritty, heavy windstorms that arose and released their abrasive fury, sculpting the rock's surface each day for more than ten thousand years.

Then came a pause in the endless windstorms and upheavals of the young planet—a span of time when the weather wasn't furnace-hot or arctic-cold, but moderate. The rock was still exposed to sunlight, but at lower temperatures, to rainfall at increased levels, and to fewer windstorms. The climate remained so for some years. Then the cycle repeated itself—arctic cold, moderately warm, furnace hot—and the cycle continued.

During the last of these cycles, the rock, considerably affected by physical and chemical exposure, was reduced in size and shape. Considerably smaller now than when it landed on the ledge, and a mere pebble compared to its former size, it fell again, 8,000 feet to the base of the mountain range, coming to rest on a bed of talus. Further reduced, it remained on its sloping talus bed for many more thousand years.

Somewhere around 15,000 B. C., the rock form, continuously exposed to chemical and mechanical weathering, its physical structure weakened by its long-ago falls, fractured, split, broken into smaller rocks, until the largest intact fragment left from the original rock was no bigger than a four-bedroom house. But change did not stop, and neither did time, rolling on until (about the time the Egyptians were building their pyramids) the rock was reduced, by this long, slow decaying process, to roughly ten feet square.

Over the next thousand years, the rock continued to decrease in size, wearing, crumbling, flaking away, surrounded by fragments of its former self, until it was about the size of a beach ball. Covered with moss and lichen, a web of fissures, tiny crevices and fractures were now woven through the entire mass.

Over the next thousand or so years, via bare rock succession, what had once been the mother of all mountain peaks, the highest point on earth, had been reduced to nothing more than a handful of soil.

If a layer of soil is completely stripped off land by natural means (water, wind, etc), by anthropogenic means (tillage plus erosion), or by cataclysmic occurrence (a massive landslide or earthquake), only after many years can a soil-denuded area return to something approaching its original state, or can a bare rock be converted to soil. But given enough time—perhaps a millennium—the scars will heal and a new, virgin layer of soil will form where only bare rock once existed. We call the series of events that take place in this restoration process *bare rock succession*. It is indeed a true "succession," with identifiable stages. Each stage in the pattern dooms the existing community as it succeeds the previous state.

Bare rock, however it is laid open to view, is exposed to the atmosphere. The geologic processes of weathering begins, breaking down the surface into smaller and smaller fragments. Many forms of weathering exist, and all effectively reduce the bare rock surface to smaller particles or chemicals in solution.

Lichens cover the bare rock first. These hardy plants grow on the rock itself (see Figure 19.1). They produce weak acids that assist in the slow weathering of the rock surface. The lichens also trap wind-carried soil particles, which eventually produce a very thin soil layer—a change in environmental conditions that gives rise to the next stage in bare rock succession.

Mosses replace lichens, growing in the meager soil the lichens and weathering provide. They produce a larger growing area, and trap even more soil particles, providing a moister

Figure 19.1 The first stage of bare rock succession: lichens growing on bare rock

bare rock surface. The combination of more soil and moisture establishes abiotic conditions that favor the next succession stage.

Now the seeds of herbaceous plants invade what was once bare rock. Grasses and other flowering plants take hold. Organic matter provided by the dead plant tissue is added to the thin soil, while the rock still weathers from below. More and more organisms join the community as it becomes larger and more complex.

By this time the plant and animal community is fairly complicated. The next major invasion is by weedy shrubs that can survive in the amount of soil and moisture present. As time passes, the process of building soil speeds up as more and more plants and animals invade the area. Soon trees take root and forest succession is evident. Many years are required, of course, before a climax forest will grow here, but the scene is set for that to occur (Tomera, 1989).

SOIL: WHAT IS IT?

In any discussion about soil (the third environmental medium), we must initially describe, explain, and define exactly what soil is, and why soil is so important to us. We must also clear up a major misconception about soil. As the chapter's introduction indicates, people often confuse soil with dirt. Soil is not dirt. Dirt is misplaced soil—soil where we don't want it, contaminating our hands or clothes, tracked in on the floor. Dirt we try to clean up and keep out of our environment.

But soil is special. It is almost mysterious, critical to our survival, and whether we realize it or not, it is essential to our existence. We have relegated soil to an ignoble position. We commonly degrade it, considering only feces as a worse substance. But soil deserves better, and in this chapter and the other chapters in this section, we'll give soil the attention it rightly deserves.

In Chapter 6 we discussed the basics of soil. In this chapter, we review many of those points and develop others to form a foundation for advanced information on using technology to reuse or recycle contaminated soil presented in Chapter 21.

Before we move on, let's take another look at that handful of "dirt" after the mountain peak was crafted into soil by the sure hand of Mother Nature over millions and millions of years. What does someone really have in hand when he or she reaches down and grabs a handful of "dirt?" We make the point that it isn't dirt, its soil. But what is soil?

Perhaps no question causes more confusion in communication between various groups of "lay persons" and "professionals" (environmental scientists, environmental engineers, specialized groups of earth scientists, and engineers in general) than does the word "soil." From the professional's perspective, the problem lies in the reasons why different groups study soils.

Pedologists (soil scientists) constitute a group interested in soils as a medium for plant growth. A corresponding branch of engineering soils specialists (*soil engineers*) look at soil as a medium that can be excavated with tools. A *geologist's* view of soil falls between pedologists and soil engineers—they are interested in soils and the weathering processes as past indicators of climatic conditions, and in relation to the geologic formation of useful materials ranging from clay deposits to metallic ores.

Consider the following descriptions of soil to better understand what it is and why it is critically important to us.

1)   A handful of soil is alive. It is as lively as an army of migrating caribou and as fascinating as a flock of egrets. Literally teeming with life of incomparable forms, soil deserves to be classified as an independent ecosystem or, more correctly, as many ecosystems.

2)   When we reach down and pick up an handful of soil, exposing the earth's stark bedrock surface, it should remind us that without its thin living soil layer, the earth would be a planet as lifeless as our own moon.

If you still prefer to call soil dirt, that's okay. Maybe you view dirt in the same way as 1996 Newbery Award Winner E. L. Konigsburg's character Ethan does:

> *The way I see it, the difference between farmers and suburbanites is the difference in the way we feel about dirt. To them, the earth is something to be respected and preserved, but dirt gets no respect. A farmer likes dirt. Suburbanites like to get rid of it. Dirt is the working layer of the earth, and dealing with dirt is as much a part of farm life as dealing with manure: neither is user-friendly, but both are necessary.*

SOIL BASICS

Soil is the layer of bonded particles of sand, silt, and clay that covers the land surface of the earth. Most soils develop multiple layers. The topmost layer (*topsoil*) is the layer in which plants grow. This topmost layer is actually an ecosystem composed of both biotic and abiotic components—inorganic chemicals, air, water, and decaying organic material that provides vital nutrients for plant photosynthesis and living organisms. Below the topmost layer (usually no more than a meter in thickness) is the *subsoil*, much less productive, partly because it contains much less organic matter. Below that is the *parent material*, the bedrock or other geologic material from which the soil is ultimately formed. The general rule of thumb is that it takes about 30 years to form one inch of topsoil from subsoil; it takes much longer than that for subsoil to be formed from parent material, the length of time depending on the nature of the underlying matter (Franck and Brownstone, 1992).

Soil Properties

From the environmental scientist's view (regarding land conservation and remediation methodologies for contaminated soil remediation through reuse and recycling), four major properties of soil are of interest. They are soil texture, slope, structure, and organic matter. *Soil texture* (refer to Figure 19.2) is a given and cannot be easily or practically changed in a significant way. It is determined by the size of the rock particles (sand, silt, and clay particles) within the soil. The largest soil particles are gravel, which consists of fragments larger than 2.0 mm in diameter. Particles between 0.05 and 2.0 mm are classified as sand. Silt particles range from 0.002 to 0.05 mm in diameter, and clay particles, the smallest, are less than 0.002 mm in diameter. Though clays are composed of the smallest particles, the particles have stronger bonds than silt or sand, although once broken apart they erode more readily. Particle size has a direct impact on erodibility. Rarely does a soil consist of only one single size of particle—most are a mixture of various sizes.

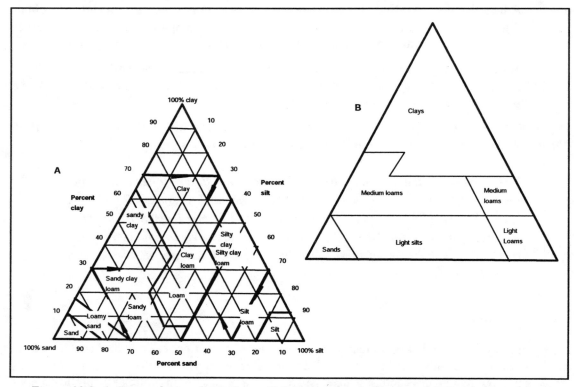

Figure 19.2 A: Textural triangle similar to U.S. Dept. of Agriculture model; B: Broad groups of textural classes (adapted from Briggs et al., Fundamentals of the Physical Environment, *p. 323)*

The *slope* (or steepness) of the soil layer is another given, important because the erosive power of runoff increases with the steepness of the slope. Slope also allows runoff to exert increased force on soil particles, which breaks them apart more readily and carries them farther away.

Soil structure (*tilth*) should not be confused with soil texture—they are different. In fact, in the field, the properties determined by soil texture may be considerably modified by soil structure. Soil structure refers to the way various soil particles clump together. The size, shape, and arrangement of clusters of soil particles called *aggregates* form larger clumps called *peds*. Sand particles do not clump—sandy soils lack structure. Clay soils tend to stick together in large clumps. Good soil develops small *friable* (easily crumbled) clumps. Soil develops a unique, fairly stable structure in undisturbed landscapes, but agricultural practices break down the aggregates and peds, lessening erosion resistance.

The presence of decomposed or decomposing remains of plants and animals (*organic matter*) in soil helps not only fertility, but also soil structure, and especially the soil's ability to store water. Live organisms—protozoa, nematodes, earthworms, insects, fungi, and bacteria—are typical inhabitants of soil. These organisms work to either control the population of organisms in the soil or to aid in the recycling of dead organic matter. All soil organisms, in one way or another, work to release nutrients from the organic matter, changing complex organic materials into products that can be used by plants.

Soil Formation

Soil is formed as a result of physical, chemical, and biological interactions in specific locations. Just as vegetation varies among biomes, so do the soil types that support that vegetation. The vegetation of the tundra and rain forest differ vastly from each other and from vegetation of the prairie and coniferous forest; soils differ in a like manner.

In the *soil forming process*, two related but fundamentally different processes are occurring simultaneously. The first is the *formation of soil parent materials* by weathering of rocks, rock fragments, and sediments. This set of processes is carried out in the *zone of weathering*. The end point, to produce parent material for the soil to develop in, is referred to as C horizon material (see Figure 19.3) and applies in the same way for glacial deposits as for rocks. The second set of processes is the *formation of the soil profile* by *soil forming processes*, which change the C horizon material into A, E and B horizons. Figure 19.3 illustrates two soil profiles, one on a hard granite, and one on a glacial deposit.

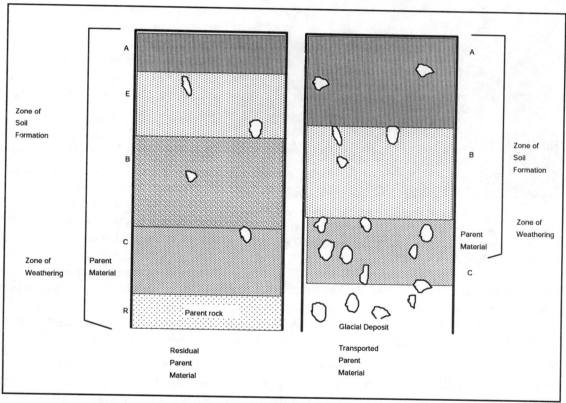

Figure 19.3 Soil profiles on residual and transported parent materials

Soil development takes time and is the result of two major processes: weathering and morphogenesis. *Weathering*, the breaking down of bedrock and other sediments that have been deposited on the bedrock by wind, water, volcanic eruptions, or melting glaciers, happens physically, chemically, or as a combination of both.

Physical weathering involves the breaking down of rock primarily by temperature changes and the physical action of water, ice, and wind. When a geographical location is characterized as having an arid desert biome, the repeated exposure to very high temperatures during the day, followed by low temperatures at night, causes rocks to expand and contract, and eventually to crack and shatter. At the other extreme, in cold climates, rock can crack and break as a result of repeated cycles of expansion of water in rock cracks and pores during freezing, and contraction during thawing. Figure 19.4 shows another example of physical weathering in which various vegetation types can not only spread their roots and grow, but the roots can exert enough pressure to enlarge cracks in solid rock, eventually splitting the rock. Plants such as mosses and lichens also penetrate rock and loosen particles.

In addition to physical weathering, bare rocks are subjected to *chemical weathering*, which involves chemical attack and dissolution of rock. Accomplished primarily through oxidation via exposure to oxygen gas in the atmosphere, acidic precipitation (after having dissolved small amounts of carbon dioxide gas from the atmosphere), and acidic secretions of microorganisms (bacteria, fungi, and lichens), chemical weathering speeds up in warm climates and slows down in cold ones.

Figure 19.4 An example of a bare rock area where various vegetation types are spreading their roots and exerting pressure to enlarge cracks, which will eventually split the rocks and reduce them to soil.

Physical and chemical weathering do not always (if ever) occur independently of each other. Instead, they normally work in combination, and the results can be striking. A classic example of the effect, the power of their simultaneous actions can be seen in an ecological process known as bare rock succession, explained in the chapter opening. To show just how effective and dramatic this combination weathering process can be, we present an example that can actually be seen today. The Natural Bridge in Virginia (see Case Study 19.1) illustrates the power of physical and chemical processes working in tandem, reshaping the earth and producing and transporting the particle material upon which and from which soil will eventually form.

The final stages of soil formation consist of the processes of *morphogenesis,* or the production of a distinctive *soil profile* with its constituent layers or *horizons* (see Figure 19.3). The soil profile (the vertical section of the soil from the surface through all its horizons, including C horizon) gives the environmental scientist critical information. When properly interpreted, soil horizons can provide warning on potential problems in using the land, and tell much about the environment and history of a region. The soil profile allows us to describe, sample, and map soils.

Soil horizons are distinct layers, roughly parallel to the surface, which differ in color, texture, structure and content of organic matter (see Figure 19.3). The clarity with which horizons can be recognized depends upon the relative balance of the migration, stratification, aggregation, and mixing processes that take place in the soil during morphogenesis. In *podzol-type soils,* striking horizonation is quite apparent; in *vertisol-type soils*, the horizons are less distinct. When horizons are studied, they are each given a letter symbol to reflect the genesis of the horizon (see Figure 19.3).

Certain processes work to create and destroy clear soil horizons. Formation of soil horizons that tend to create clear horizons by vertical redistribution of soil materials includes the leaching of ions in the soil solutions, movement of clay-sized particles, upward movement of water by capillary action, and surface deposition of dust and aerosols. Clear soil horizons are destroyed by mixing processes that occur because of organisms, cultivation practices, creep processes on slopes, frost heave, and by swelling and shrinkage of clays—all part of the natural soil formation process.

Case Study 19.1
The Natural Bridge of Virginia

Thomas Jefferson stated that the Natural Bride of Virginia is "the most sublime of Nature's Works." The great stone causeway, situated a few miles west of the Blue Ridge Mountains in the heart of the great Appalachian Valley of Western Virginia, has been proclaimed one of the natural wonders of the world.

The proportions of the Natural Bridge are enormous. It has been measured and is described as 90 feet long, while the width varies from 150 feet at one end to 50 feet at the other, and it is higher than Niagara Falls. The span contains approximately 450,00 cubic feet of rock. If it could be weighed, the mass would balance about 36,000 tons (72,000,000 pounds). At its feet flows Cedar Creek, now only a small trickle in comparison to the massive, roaring flow of water it once was.

The normal question asked about the Natural Bridge is: "How was it formed?" Many theories have been proffered on the exact origin of the Natural Bridge. Thomas Jefferson held the theory that it was formed by some sort of cataclysmic event (what he called "some great convulsion") and that its formation was relatively recent (at the time the earth was believed to be only several thousand years old. To Jefferson, that such an event could have occurred over the course of millions of years was inconceivable).

Today we know that ideas about natural features such as the Natural Bridge change as we gain more knowledge. Spencer (1985) makes clear that in talking about the exact age of the Natural Bridge, we must be careful to distinguish several important events. He points out that the rocks which compose the bridge are early Ordovician (about 500 million years old). Toward the end of the Paleozoic Era (about 200+ million years ago) the internal forms of these rocks (the folds and breaks in the layers) were imposed during the Appalachian Mountain building process. Probably no more than a few million years ago, the formation of the stream drainage and the carving out of the bridge started.

Jefferson, with input from others, later modified his "great convulsion" theory as the cause of the formation of the bridge. He was an astute student of science and was aware that other natural bridges had been formed by the work of water—the wearing action of the water running through it, rather than a convulsion of nature, might have formed the Natural Bridge.

One of Jefferson's friends, Francis W. Gilmer, put forward a detailed description of the origin of the Natural Bridge in 1816. His thinking on the subject is outlined in a paper he presented to the American Philosophical Society:

> *...instead of its being the effect of a sudden convulsion, or an extraordinary deviation from the ordinary laws of nature, it will be found to have been pro-duced by the very slow operation of causes which have always, and must ever continue, to act in the same manner...the country above the bridge...is calcar-eous.... This rock is soluble in water to such a degree, as to be found in solution with all the waters of the country, and is so soft as to yield not only to its chemi-cal agency, but also to its mechanical attrition.... Here, as in calcareous coun-tries generally, there are frequent and large fissures in the earth, which are sometimes conduits for subterraneous streams, called 'sinking rivers.'*

> *...It is probable, then, that the water of Cedar Creek originally found a subterraneous passage beneath the arch of the present bridge.... The stream has gradually widened, and deepened this ravine to its present situation. Frag-ments of its sides also yielding to the expansion and contraction of heat and cold, tumbled down even above the height of the water...The stone and earth composing the arch of the bridge, remained there and no where else; because, the hill being of rock, the depth of rock was greatest above the surface of the water where the hill was highest, and this part being very thick, and the strata horizontal, the arch was strong enough to rest on such a base.... Indeed, the very process by which the natural bridge was formed, is still visibly going on; the water...is excavating the rock, and widening the channel, which, after a long lapse of time, may become too wide to support the arch, and this wonder of our country will disappear.*

Since the time of Gilmer's original theory about the origin of the Natural Bridge, all geologists who subsequently studied the Natural Bridge agree with his view that the bridge was formed by the action of running water diverted from the surface of the ground into a subterranean passage beneath the arch. They differ only in the details of this diversion. Spencer (1985) points out that C. D. Walcott (1893), for example, "suggested the bridge was once the site of a waterfall and at that time the valley floor of Cedar Creek was at a level close to the top of the canyon. Walcott postulated that the water somehow diverted just upstream from the waterfall into an under-ground passage that emptied out of the base of the falls. This would have left the span of dolo-mite between the diversion and edge of the waterfalls intact" (p. 44).

Most of the others who have studied the origin of the Natural Bridge (Kain, 1818; Ashburner, 1884; Mallot *and* Shrock, 1930; and Woodward, 1936) favor ideas much closer to Gilmer's— that a surface stream was diverted into an opening in the earth (a cave) from which water issued farther downstream. This underground flow formed a long, natural tunnel. Over time, the roof of this tunnel collapsed, leaving only the span of Natural Bridge (see Figure 19.5).

Today, Cedar Creek still flows beneath the Natural Bridge (see Figure 19.6). It originates in the Allegheny Mountains and empties into the James River. The structure of the rock of which the

Natural Bridge is made determined its location. The arch of the bridge, which is massive and compact, is formed from dolomite; its structural integrity seems sound (see Figure 19.7). However, in time the bridge will fall into Cedar Creek; it will be gone, but just as Nature works to modify and eventually destroy the Natural Bridge, she also works to form other natural wonders of the world.

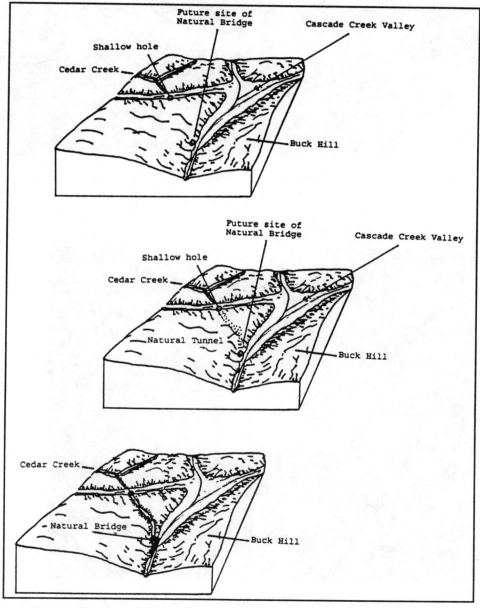

Figure 19.5 How the Natural Bridge was formed. Beginning with a shallow hole, Cedar Creek formed a natural tunnel; later the tunnel roof caved in. Because the Natural Bridge was formed of a hard dolomite, it remained in place.

Figure 19.6 Cedar Creek as it appears today. Notice the inclined layers of limestone in the creek bed and rising to the right, up the slope.

Figure 19.7 Natural Bridge

SUMMARY

When we talk about learning something from practical experience, we say we need to get in there and get our hands dirty. Increasing urbanization means increasing separation from practical knowledge in many areas of land use—literally from getting our hands dirty. American society has reached the point where "Fresh Air Kids" are sent from New York to what amounts to smaller cities for a week in the "country." Many members of our society have never grown anything but houseplants potted in soil that comes packaged in plastic from a store—a packaged, sterile experience of what soil is and does.

But soil doesn't begin wrapped in plastic, any more than does the meat and produce people buy at the supermarket. When we forget that we are reliant on what our earth produces out of the fertility of that thin, fine layer of topsoil, we become wasteful—and we put ourselves at risk. We underestimate the value of our soil. We clear-cut it, we pave over it, we expand our communities needlessly on it, through carelessness and unconcern we poison it—in short, we waste it.

As awareness of the serious soil pollution problems that we must now mitigate and remediate grows, as we work to develop effective methods to reuse and recycle contaminated soil, we still have a tendency to think of soil pollution only as it affects our water supply. Again, we undervalue soil's worth. We should not lose sight of the mountains of stone and eternity of time that went into making the soil under our feet.

Cited References

Brady, N. C., and Weil, R. R., *The Nature and Properties of Soils*, 11th ed., New York, Prentice-Hall, 1996.

Franck, I., and Brownstone, D., *The Green Encyclopedia*, New York: Prentice-Hall, 1992.

Konigsburg, E. L., *The View From Saturday,* New York: Scholastic Books, 1996.

Malot, C. A. and Shrock, R. R., *Origin and Development of Natural Bridge*, Virginia: American Journal of Science 5th ser., vol. 19, pp. 257-273.

Spencer, E. W., *Guidebook: Natural Bridge and Natural Bridge Caverns*, Lexington, VA: Poorhouse Mountain Studios, 1985

Tomera, A. N., *Understanding Basic Ecological Concepts*, Portland, Maine: J. Weston Walch, Publisher, 1989.

Woodward, H. P., *Natural Bridge and Natural Tunnel*, Virginia: Journal of Geology, vol. 44, no. 5, pp. 604-616, 1936.

Suggested Readings

Batie, S. S., *Soil Erosion: Crisis in America's Croplands?* Washington, DC: Conservation Foundation, 1983.

Birkeland, P. W., *Soils and Geomorphology*, New York: Oxford University Press, 1984.

Bohn, H. L., McNeal, B. L., and O'Connor, G. A., *Soil Chemistry*, end ed., New York: John Wiley and Sons, 1985.

Briggs, D., Smithson, P., Addison, K., and Atkinson, K., *Fundamentals of the Physical Environment*, 2nd ed., New York: Routledge.

FitzPatrick, E. A., *Soils: Formation, Classification and Distribution*, London: Longman.

Rowell, D. L., *Soil Science: Methods and Applications*, London: Longman.

Wilde, A., ed., *Russell's Soil Conditions and Plant Growth*, 11th ed., London: Longman.

CHAPTER 20

SOIL POLLUTION

The Global Assessment of Soil Degradation study conducted for the United Nations Environment Programme found that in recent decades, nearly 11 percent of the Earth's fertile soil has been so eroded, chemically altered, or physically compacted as to damage its original biotic function (its ability to process nutrients into a form usable by plants); about 3 percent of soil has been degraded virtually to the point where it can no longer perform that function.
—World Resources Institute

CHAPTER OUTLINE

- Introduction
- Surface Origins of Soil Contaminants
- Industrial Practices and Soil Contamination

INTRODUCTION

Soil fertility is a major concern, not only throughout the United States, but worldwide. The impact on soil fertility from agricultural practices (erosion, salination, and waterlogging) is well known, well studied, and well documented. Remediation practices are also known, and actually are in place in many locations throughout the world. Indeed, solving problems related to soil fertility has received considerable attention, driven not only by a growing and hungry worldwide population, but also by pocketbook issues—economics. However, one major problem related to soil fertility has only recently become apparent, important, and critical in our continuing fight to maintain soil for its primary purpose (as pointedly and correctly stated by The World Resource Institute—"its ability to process nutrients into a form usable by plants"). This problem is soil contamination or pollution.

Soil pollution generated by industrial contamination, management of Superfund sites, exploration and production, mining and nuclear industrial practices, among others, is having an impact on soil quality that we have only recently begun to comprehend. Complicating the problem is that soil pollution remains difficult to assess. However, some evidence clearly indicates the impact of a few industrial practices related to soil pollution. For example, we know that petroleum-contaminated soil affects the largest number of sites and the largest total volume of contaminated material. However, the volume of petroleum-contaminated soil that is either discovered or generated each year is not consistently tracked on a local basis, so this total is unknown. From the evidence, we also know (e.g., in Oklahoma, contaminated soil accounts for about 90% of the waste generated as a one-time occurrence) that the overall amount of contaminated soil generated can be staggering (Testa, 1997).

In this chapter, we focus both on the surface origins of soil contaminants and the industrial practices that can contaminate soil. The concepts of remediation and resource recovery are covered in Chapter 21.

SURFACE ORIGINS OF SOIL CONTAMINANTS

Pollution of soil and water is a problem common to all human societies. Throughout the history of civilization, we have probably had little problem recognizing surface-water contamination. Treatment of surface water for drinking became common in the late nineteenth century, and health problems linked to impure drinking water in developed countries are now rare. Underdeveloped countries, however, are still faced with a lack of safe drinking water.

Only in the past several decades has a new problem come to light: contamination of the soil and its underground environment. In developed countries, this problem is much more serious because of their history of industrialization and the wide range of hazardous materials and other chemicals that have been introduced, either by design or accident, to the underground environment. Ignorance, more than intent, is the culprit. We were ignorant in the sense that we did not comprehend the degree to which contaminants could migrate through the soil, the damage they could do to the soil medium and the groundwater under its protective surface, or the difficulty we would encounter in tracing and removing most contaminants after discovery.

Starting with the most contaminated sites, the response (in developed countries) to underground contamination has been a massive effort to define the extent of contamination, and to remediate the subsurface. This response has been driven by governmental regulations dealing with waste handling and disposal, and many other potentially contaminating activities.

The range of activities that cause underground contamination is much larger than most environmental scientists would have guessed even a few years ago. These activities are discussed briefly in the sections that follow.

Soil quality problems originating on the surface include natural atmospheric deposition of gaseous and airborne particulate pollutants; infiltration of contaminated surface water; land disposal of solid and liquid waste materials; stockpiles, tailing, and spoil; dumps; salt spreading on roads; animal feedlots; fertilizers and pesticides; accidental spills; and composting of leaves and other yard wastes.

Though we do not discuss them in detail in this text, note that other sources of soil contamination relate to petroleum products. These other sources include direct disposal of used oils on the ground by individuals or industries; seepage from landfills, illegal dumps, unlined pits, ponds, and lagoons; and spills from transport accidents. Even auto accidents make contributions to the soil burden (Tucker, 1989)

> NOTE: The following discussion focuses on contamination originating on the land surface. However, note that soil and subsurface contamination may also originate below ground but above the water table from septic tanks, landfills, sumps and dry wells, graveyards, USTs, leakage from underground pipelines, and other sources. In addition, soil, subsurface, and groundwater contamination may also originate below the water table from mines, test holes, agricultural drainage wells and canals, etc.

Gaseous and Airborne Particulate Pollutants

We don't commonly associate soil as being a prominent member of the biogeochemical cycles (carbon, nitrogen, and sulfur cycles) but we should, because they are. Not only is soil a prominent part of the rapid, natural cycles of carbon, nitrogen, and sulfur, but along with these cycles, it has a strong and important interface with the atmosphere. Consider the nitrogen cycle, where nitrates and ammonium ions in rainwater are absorbed by plant roots and soil microorganisms, and converted to amino acids or gaseous N_2 and N_2O, which diffuse back to the atmosphere. N2 uptake and conversion to amino acids (*nitrogen fixation*) by symbiotic and free-living soil microorganisms balances this loss of gaseous nitrogen. NO, NO_2, and NH_3 (other nitrogen gases) are also emitted and absorbed by soils. Soil reactions are major determinants of trace gas concentrations in the atmosphere.

Air pollutants (sulfur dioxide, hydrogen sulfide, hydrocarbons, carbon monoxide, ozone, and atmospheric nitrogen gases) are absorbed by soil. Because the reactions are subtle, they have often been disregarded in considerations of air pollution. Sulfur dioxide, in arid regions, is probably the most obvious example of direct soil absorption. The basicity of arid soils makes them an active sink for sulfur dioxide and other acidic compounds from the atmosphere.

Two classic examples of *airborne particulate soil contamination* can be seen in the accumulation of heavy metals around smelters, and in soils in urban areas, contaminated by exhaust fumes associated with auto emissions. These two soil polluters are serious in localized areas but otherwise are generally thought to be minor.

Infiltration of Contaminated Surface Water

Often wells are intentionally installed near streams and rivers to induce recharge from the water body and to provide high yield with low drawdowns. Occasionally, if the stream or river is polluted, contamination of the soil-water well field can result. This process normally occurs when a shallow water-supply well draws water from the alluvial aquifer adjacent to the stream. The cone of depression imposed by pumping the well or well field creates a gradient on the water table directed toward the well, pulling or drawing the polluted water through and contaminating the well field and well.

Land Disposal of Solid and Liquid Waste Materials

Land disposal, stockpiling, or land-applying wastes or materials, including liquid and sludge (biosolids) wastes from sewage treatment plants (nearly half of the municipal sewage biosolids produced in the U.S. is applied to the soil, either for agricultural purposes or to remediate land disturbed by mining and other industrial activities), food processing companies, and other sources, has become common practice. The purpose of this practice is twofold: it serves as a means of disposal and provides beneficial use/reuse of such materials as fertilizers for agricultural lands, golf courses, city parks, and other areas. The objective is to allow biological and chemical processes in the soil, along with plant uptake, to break down the waste products into harmless substances. In many cases such practices are successful. However, a contamination problem may arise if any of the wastes are water-soluble and mobile, which allows them to be carried deep into the subsurface. If the drainage or seepage area is underlain by shallow aquifers, a groundwater contamination problem may arise.

Stockpiles, Tailings, and Spoils

Stockpiles of certain chemical products can contribute to soil and subsurface contamination. Stockpiling road salt, for example, is a common practice used by many local highway departments and some large industries as a precautionary measure to treat snow- and ice-covered surfaces in winter. Tailings are usually produced in mining activities and commonly contain materials (asbestos, arsenic, lead, and radioactive substances) that are a health threat to humans and other living organisms. Recall that tailings from mining operations may contain contaminants such as sulfide, which, when mixed with precipitation, form sulfuric acid. As the chemically altered precipitation runs off or is leached from the tailing piles, it infiltrates the surface layer, contaminates soil, and ultimately may reach groundwater. Spoil is generally the result of excavations such as road-building operations, where huge amounts of surface cover are removed, excavated, and piled, and then moved somewhere else. Problems with spoil are similar to the tailing problems: precipitation removes materials in solution from the spoil by percolating waters (leaching) any contaminants from the spoil. Those contaminants find their way into the soil and ultimately into shallow aquifers.

Dumps

Until recently, it was common practice to take whatever we didn't want and dump it somewhere out of sight. Today, uncontrolled dumping is prohibited in most industrialized countries, but the "old" dumping sites can contain just about anything, and may still constitute a threat of subsurface contamination. Another problem that is still with us is "midnight dumping." Because dumping today is controlled and regulated, many disposers attempt to find ways to "get rid of junk." Unfortunately, much of this "junk" consists of hazardous materials and toxins that end up finding their way into and through soil to aquifers. In addition to the "midnight dumping" problem, another illegal disposal practice has developed in some industries because of the high cost involved with proper disposal. This practice, commonly called "immaculate conception," occurs when workers in industrial facilities discover unmarked drums or other containers of unknown wastes that suddenly appear on loading docks or elsewhere in the facility. Then, of course, these immaculately conceived vessels of toxic junk end up being thrown out with the common trash, and their contents eventually percolate through the soil to an aquifer.

Salt Spreading on Roads

In northern climates, especially in urban areas, spreading deicing salts on highways is widespread. In addition to causing deterioration of automobiles, bridges, and the roadway itself, and adversely affecting plants growing alongside a treated highway or sidewalk, salt contamination quickly leaches below the land surface. Because most plants cannot grow in salty soils, the productivity of the land decreases. Continued use can lead to contamination of wells used for drinking water.

Animal Feedlots

Animal feedlots are a primary source of nonpoint surface water pollution. Animal feedlots are also significant contributors to groundwater pollution. Because animal waste in feedlots literally piles up and is stationary (sometimes for extended periods), runoff containing contaminants may not only enter the nearest surface water body, but may also seep into the soil, contaminating it. If the contaminated flow continues unblocked through the subsurface, the flow may eventually make its way into a shallow aquifer.

Fertilizers and Pesticides

Fertilizers and pesticides have become the mainstays of high yield agriculture. They have also had significant impact on the environment, with each yielding different types of contaminants.

When we apply fertilizers and pesticides to our soil, are we treating the soil or poisoning it? This question is relatively new to us, and is one we are still trying to definitively answer. One thing is certain; with fertilizer and pesticide application, and the long-term effects of such practices, the real quandary is that we do not know what we do not know. We are only now starting to see and understand the impact of using these chemicals. We have a lot to learn. Let's take a look at a few of the known problems with using chemical fertilizers and pesticides.

Nitrogen fertilizers are applied to stimulate plant growth, but often in greater quantities than plants can utilize at the time of application. Nitrate, the most common chemical form of these fertilizers, can easily leach below the plant root zone by rainfall or irrigation. Once it moves below the root zone, it usually continues downward to the water table.

Another serious problem is that when we put total dependency upon chemical fertilizers, they can change the physical, chemical, and biotic properties of the soil.

Pesticides (any chemical used to kill or control populations of "unwanted" animals, fungi, or plants) are not as mobile as nitrate, but they are toxic at much lower concentrations. If we were to design the perfect pesticide, it would be inexpensive, affect only the target organism, have a short half-life, and break down into harmless substances. However, we have not yet developed the perfect pesticide, and herein lies a multi-faceted problem. Pesticides used in the past—and some of those being used today—are very stable, persistent, and can become long-term problems. They may also be transported from the place of original application to other parts of the world by wind or ocean currents.

Another problem associated with persistence in pesticide use is that they may accumulate in the bodies of organisms in the lower trophic levels. Recall that when a lower trophic level animal receives small quantities of certain pesticides in its food and is unable to eliminate them, the concentration within the organism will increase. You should also recall that when lower trophic level organisms accumulate higher and higher amounts of materials within their bodies, *bioaccumulation* takes place. Eventually, this organism may pass on its accumulation to higher trophic level organisms.

Accidental Spills

Accidental spills of chemical products can be extremely damaging to any of the three environmental mediums—air, water, and soil. Disturbingly common, chemical spills in the soil media that are not discovered right away may allow the contaminant to migrate into and through the soil (contaminating it) to the water table. As an oversimplified, general rule of thumb, we can say that the impact of a chemical spill in soil (or any other medium) is directly related to the concentration present at the point and time of release, the extent to which the concentration increases or decreases during exposure, and the time over which the exposure continues. Much more will be said about this important topic in Part Five of this text.

Composting of Leaves and Other Wastes

Composting, a common practice for many homeowners (especially gardeners), has proven its worth as an environmentally friendly way to dispose of or beneficially reuse common waste products. However, when the feed material (leaves, twigs, and other organics) has been treated with chemical pesticides and some fertilizers, composting this material may be harmful to the soil. In the composting process, the organic material is degraded via a curing process that occurs over time. When water is intentionally added with a garden hose or by precipitation, any chemicals present can be washed or leached from the decaying organic material and drain into the soil, contaminating it.

INDUSTRIAL PRACTICES AND SOIL CONTAMINATION

Industrial practices that can contaminate soil include use of underground storage tanks (USTs), contamination from oil field sites, contamination from chemical sites, contamination from geothermal sites, contamination from manufactured gas plants, contamination from mining sites, from many other industrial activities, and contamination from environmental terrorism (for example in the Persian Gulf War).

In this section we focus on these industrial practices that so consistently contaminate the soil (with the exception of USTs, which we covered in Chapter 18). We address these specific areas because they give us a well-documented representative sample of the types of industrial practices that contaminate soil.

Contamination from Oil Field Sites

People often repeat two cliches: "the past has a way of catching up with us," and "if we don't learn from past mistakes we are doomed to repeat them." But cliches often become cliches by being true. And these are certainly true when we consider the problems with soil contamination from oil field sites, which are a source of large volumes of hydrocarbon-contaminated soil, resulting from past and existing oil fields. The extent of the impact of this problem is location-specific. For example, past and present day oil exploration and production activities located in remote parts of Oklahoma and Texas are not highly visible and thus not subject to public scrutiny. In these remote locations, disposing of hydrocarbon-contaminated soil is easy and inexpensive.

However, in highly urbanized locations (Los Angeles County, CA, for example, where more than 3,000 acres of prime real estate is being or has been exploited for petroleum), most developers sit back and eagerly wait until existing fields reach their productive ends. When this occurs, the developers move in and redevelop the real estate. However the ill-informed developer may quickly find out that just because the well runs dry does not mean that he or she can begin developing the land. In these areas, disposal of contaminated soils has emerged as a serious and expensive undertaking.

On or near petroleum-producing properties, the primary sources of soil contamination include oil wells, sumps, pits, dumps, leakage from above-ground storage tanks, and leakage and or spillage. Secondary sources include USTs, transformers, piping ratholes, well cellars, and pumping stations. In addition, the large stationary facilities used for the refining of petroleum have the potential to cause chronic pollution by the discharge of hydrocarbon-laden wastewater frequent small spills. The primary hazardous constituents associated with old field properties include drilling mud and constituents, methane, and crude oil. When crude contains certain constituents above maximum contaminant levels—arsenic, chloride, chromium, lead, polychlorinated biphenols (PCBs)—and has a flash point less than the minimum standard as set by American Society for Testing and Materials (ASTM) for recycled products, it may be considered a hazardous waste.

Close to many oil fields exist complete handling and processing ancillaries—refineries, terminals, and pipelines—which also contribute to the overall volume of contaminated soil generated. Of primary concern are contaminants such as crude oil, refined products, and volatile organic compounds.

Contamination from Chemical Sites

The 1979 PEDCO-Eckhardt Survey (commonly referred to as the Eckhardt Survey) of more than 50 of the largest manufacturing companies in the U.S. reported 16,843 tons of organic generated waste were disposed. Of this total, more than 10 million tons were untreated (residing in landfills, ponds, lagoons, and injection wells). Approximately 0.5 million tons were incinerated, and approximately 0.5 million tons were either recycled or reused. The volume of contaminated soil as a result of one-time occurrence was not addressed in this survey.

Soil contamination from organic chemicals is a serious matter. Some of these organic compounds are biologically damaging even in small concentrations. When they do find their way into the soil, certain organic chemicals may kill or inhibit certain soil organisms, which can undermine the balance of the soil community. Once in the soil, the contaminant may be transported from the soil to the air, water, or vegetation where it may be inhaled, ingested, or contacted over a wide area by a number of organisms. Because of their potential harm, controlling the release of organic chemicals and understanding their fate and effects in the soil is imperative.

Contamination from Geothermal Sites

Geothermal energy is natural heat generated beneath the earth's surface. The earth's mantle, 15 to 30 miles below the earth's crust, is composed of a semi-molten rock layer. Beneath the mantle, intense pressure, caused by molten rock of iron and nickel and decaying radioactive elements, helps warm the earth's surface. Geothermal energy generally lies too deep to be harnessed, but in certain areas, where the molten rock has risen closer to the earth's surface through massive fractures in the crust, underground reservoirs of dry steam, wet steam, and hot water are formed. As with oil deposits, these deposits can be drilled and their energy used to heat water, drive industrial processes, and generate electricity.

Generally, geothermal resources are more environmentally friendly than nuclear energy or fossil fuels. However, several drawbacks to geothermal energy use adversely impact the environment. As with oil field operations, geothermal operations provide another example of the close relationship between site usage and the potential for adverse environmental impact. The two constituents associated with geothermal plants that may be considered hazardous are brine and lead-mine scale (Testa, 1997).

Disposal of wastewater from geothermal wells containing brine is a major problem (brine are geothermal mineralizing fluids composed of warm to hot saline waters, containing sodium, potassium, chloride, calcium, and minor amounts of other elements that may be harmful to plants and animals). The problem with *lead-mine scale* is more directly related to process equipment failures—developing from scale-buildup in pipes and other process equipment—than environmental problems. However, scale buildup may lead to equipment failure (pipe rupture, for example), which in turn may lead to geothermal fluid spills, with the ultimate result being soil, air, and/or water contamination.

Contamination from Manufactured Gas Plants

The manufacture of gas is not a new process. Since the late 1890s, manufactured gas plants (approximately 3000 of them located in the U.S.) have been in operation, have been upgraded, or have been completely redeveloped in one way or another. The environmental soil pollution problem associated with manufactured gas plants is the production and disposal of tarry substances, primarily produced in the coal gasification processes—coal carbonization, carburetted waste gas, natural gas, or combination processes.

Other than the obvious mess that the production of any tar-like substance can produce, the main environmental problem with tars is that they are known to contain organic and inorganic compounds that are known or suspected carcinogens. The volume of tar-contaminated soil averages 10,000 yd^3 per site (Testa, 1997).

Contamination from Mining Sites

According to the U.S. Departments of Interior and Agriculture, since the mid 1860s, more than 3 million acres of land in the United States have been surface-mined for various commodities. Leading the list of commodities related to acreage is mined coal, followed by sand and gravel, stone, gold, phosphate rock, iron ore, and clay, respectively.

Mining operations can give rise to land and water pollution. Sediment pollution via erosion is the most obvious problem associated with surface mining. Sediment pollution to natural surface water bodies is well-documented. The Chesapeake Bay, for example, is not the fertile oyster-producing environment that it once was. Many environmentalists initially blamed the Bay's decline on nutrient-rich substances and chemical pollutants. However, recent studies of the Bay and its tributaries indicate that oysters may be suffering from sedimentation rather than nutrient contamination.

Less known because less studied is the effect mining sediments and mining wastes (from mining, milling, smelting, and leftovers) have and are having on soil. Typical mining wastes include acid produced by oxidation of naturally occurring sulfides in mining waste, asbestos produced in asbestos mining and milling operations, cyanide produced in precious-metal heap-leaching operations, leach liquors produced during copper-dump leaching operations, metals from mining and milling operations, and radionuclides (radium) from uranium and phosphate mining operations.

One soil-contaminant source that is well known and well documented is acid mine drainage (see Chapter 17). Acid formations occur when oxygen from the air and water react with sulfur-bearing minerals to form sulfuric acid and iron compounds. These compounds may directly affect the plant life that absorbs it or have an indirect effect on the flora of a region by affecting the soil minerals and microorganisms.

Another problem with mining is solid wastes. Metals are always mixed with material removed from a mine. These materials usually have little commercial value and thus must be disposed somewhere. The piles of rock and rubble are not only unsightly, but they are also prone to erosion, and leaching releases environmental poisons into the soil.

Contamination from Environmental Terrorism

Many human activities that have resulted in environmental contamination have been due to accidents, poor planning, poor decision making, inferior design, shoddy workmanship, ignorance, or faulty equipment. In the past, at least prior to 1991, whenever the public heard about or witnessed environmental contamination, they could assume that one or more of those factors were behind the contamination.

The Persian Gulf War has changed this perception. After the Gulf War, almost half of Kuwait's 1500 oil wells were releasing oil into the environment. Each day an estimated 11 million barrels of oil were either being spilled, or burned in 600 wells that were aflame. After the well-capping operation got underway and more than 200 wells had been capped, this amount was reduced to approximately 6 million barrels. The harmful effects to the atmosphere and Persian Gulf were only part of the problem, however. Numerous pools of oil formed, some of them up to 4 ft. deep, collectively containing an estimated 20 million barrels of oil (Andrews, 1992).

The larger long-term problem of this act of terrorism is twofold: the presence of the oil pools and the huge volumes of petroleum-contaminated soil.

SUMMARY

In many respects, soil pollution is to environmental science in the 1990s what water and air pollution were to earlier decades—the pressing environmental problem at hand. While developing methods to control air and water pollution was difficult, socially and politically, once the regulations were in place and working effectively, the problems confronting the environmentalists were relatively easy to locate.

Soil pollution has come to our attention in a time when regulation to begin remediation is in place, and when the public knows the importance of mitigating the contaminated areas. Soil pollution, however, presents us with a new problem. These contamination sites, especially those from old underground sources (USTs, for example), create a difficult game of contamination "hide and seek." While new techniques for handling the contamination show signs of promise, and while we have no shortage of sites in need of remediation, we must also remember that we are still being affected by hidden sites beneath our feet.

Cited References

Andrews, J. S., Jr., "The Cleanup of Kuwait," *Hydrocarbon Contaminated Soils*, Vol. II, (Edited by P. T. Kostecki et al.), Boca Raton, FL: CRC/Lewis Publishers, 1992.

PEDCO, "PEDCO analysis of Eckhardt Committee Survey for Chemical Manufacturer's Association," Washington, DC: PEDCO Environmental Inc., 1979

Testa, S. M., *The Reuse and Recycling of Contaminated Soil*, Boca Raton, FL: CRC/Lewis Publishers, 1997.

Tucker, R. K., "Problems dealing with petroleum contaminated soils: A New Jersey perspective," in *Petroleum Contaminated Soils*, Vol. I, (edited by Kostecki, P. T., and Calabrese, E. J.), Boca Raton, FL: CRC/Lewis Publishers, 1989.

World Resources Institute, *World Resources 1992-93*, New York: Oxford University Press, 1992.

Suggested Readings

Abdul, S. A., "Migration of petroleum product through sandy hydrologic systems." *Ground Water Monitoring Review*, 8, No...4:73-81, 1988.

Birkeland, P. W., *Soils and Geomorphology*, New York: Oxford University Press, 1984.

Blackman, W. C., Jr., *Basic Hazardous Waste Management*, Boca Raton, FL: Lewis Publishers, 1993.

Fetter, C. W., *Contaminant Hydrogeology*, New York: Macmillan, 1993.

Hillel, D., *Fundamentals of Soil Physics*, New York: Academic Press, 1980.

Holmes, G., Singh, B. R., and Theodore, L., *Handbook of Environmental Management and Technology*, New York: John Wiley and Sons, 1993.

Kehew, A. E., *Geology for Engineers and Environmental Scientists*, 2nd ed., Englewood Cliffs, NJ: Prentice-Hall, 1995.

Palmer, C. M., *Principles of Contaminant Hydrogeology*, Boca Raton, FL: Lewis Publishers, 1992.

Testa, S. M. and Winegarden, D. L., *Restoration of Petroleum-Contaminated Aquifer*, Boca Raton, FL: CRC/Lewis Publishers, 1991.

U.S. Environmental Protection Agency, *Ground Water, Vol. I: Ground Water and Contamination*, Washington, DC: U.S. Environmental Protection Agency, 1990.

CHAPTER 21

SOIL POLLUTION CONTROL TECHNOLOGY

The response to subsurface contamination in the United States, as in many other countries, has been a massive effort to define the extent of contamination and to remediate the subsurface—starting with the most contaminated sites. This environmental campaign has been driven by governmental regulations dealing with waste handling and disposal, as well as with many other potentially contaminating activities. One side effect of this endeavor has been a boom in the employment market for engineers and environmental scientists that specialize in detection, monitoring, and remediation of subsurface contamination. A new industry has been created and the performance of this work seems sure to continue well into the twenty-first century. —Kehew

CHAPTER OUTLINE

- Introduction
- USTs: The Problem
- Risk Assessment
- Exposure Pathways
- Remediation of UST Contaminated Soils

INTRODUCTION

Soil or *subsurface remediation* is a still-developing branch of environmental science and engineering. Because of the regulatory programs of CERCLA (Comprehensive Environmental Response, Compensation, and Liabilities Act of 1980—better known as Superfund) and RCRA (Resource Conservation and Recovery Act of 1976—better known as the "cradle to grave act"), remediation has not only been added to the environmental vocabulary, but also has become common and widespread. Just how common and widespread? To best answer this question, follow the response of venture capitalists in their attempts to gain a foothold in this new technological field. Regarding this issue, MacDonald (1997) points out:

> *In the early 1990s, venture capitalists began to flock to the market for groundwater and soil cleanup technologies, seeing it as offering significant new profit potential. The market appeared large; not only was $9 billion per year being spent on contaminated site cleanup, but existing technologies were incapable of remediating many serious contamination problems.*

NOTE: From MacDonald's comments, soil remediation technology appears a booming enterprise, with unlimited potential, virtually a "can't miss" proposition; however, inherent problems limit its potential. Remediation technology is a double-edged sword, as we will discuss later.

Since CERCLA and RCRA, numerous remediation technologies (also commonly known as Innovative Cleanup Technologies) have been developed and have become commercially available. In this chapter, we discuss these technologies, especially those designed and intended to clean up sources of subsurface contamination caused by underground storage tanks (USTs). We focus on technology used in contamination from failed USTs primarily because these units have been the cause of the majority of contamination events and remediation efforts to date. As a result, enormous volumes of information have been recorded on this remediation practice, both from the regulators and the private industries involved with their cleanup.

Keep in mind that (though it likely is the goal of the regulatory agency monitoring a particular UST cleanup effort) no matter what the contaminant, removing every molecule of contamination and restoring the landscape to its natural condition is highly unlikely.

USTs: THE PROBLEM

In Chapter 18 we discussed USTs and the subsurface contamination problems associated with them. In this section we take a closer, more in-depth look at USTs and the remediation technologies used to clean up contamination produced by them.

No one knows for sure the exact number of underground storage tank (UST) systems installed in the U.S. However, all present-day estimates range in the millions. Several thousand of these tanks (including ancillaries such as piping) are currently leaking.

USTs leak for several reasons: (1) corrosion, (2) faulty tank construction, (3) faulty installation, (4) piping failure, (5) overfills and spills, or (6) incompatibility of UST contents.

USTs: Corrosion Problems

The most common cause of tank failure is corrosion. Many older tanks were constructed of a single shell made of unprotected bare steel, and have leaked in the past (and have hopefully been removed), are leaking at present, or (if not removed or rehabilitated) will leak in the future. If undetected or ignored, such a leak (even a small one) can cause large amounts of petroleum product to be lost to the subsurface.

USTs: Faulty Construction

As with any material item, USTs are only as good as their construction, workmanship, and materials. If a new washing machine is improperly assembled, it will likely fail, sooner than later. If a ladder is made of a material not suited to handle load-bearing, it may fail, and the consequences may result in injury. There are many different lists, some longer than others, of possible failures resulting from poor or substandard construction or workmanship.

USTs are no different than any other manufactured item. If they are not constructed properly, if workmanship is poor, they will fail. It's that simple, although the results of such failure are not so simple.

To understand how USTs can fail because of poor construction and workmanship—and the end result—see Case Study 21.1, which relates an actual occurrence (only names and locations have been changed) of tank failure.

Case Study 21.1
Roundaway Tanks

AJ Roundaway Tank Company, located in Southport, Virginia, had been in the tank building business for 60 years. Roundaway built all sorts of tanks—large square tanks, rectangular tanks, round tanks, even uniquely designed special order tanks.

In its earliest days, Roundaway exclusively built bare, unprotected steel tanks, until government agencies moved in with new regulations concerning tank leakage problems. When this occurred, Roundaway saw the writing on the wall. Building tanks that wouldn't rust or corrode would be best. So, Abe Parsons, owner of Roundaway, did a little research and came up with the idea of retrofitting his plant, changing from making bare steel tanks (the old standbys) and bringing on line a relatively new process for making a tank out of new materials—materials that could not corrode, could not rust, and could not possibly leak.

The new tank building material Abe incorporated into his process is known as fiberglass reinforced plastic (FRP). Abe knew that other tank building companies had already started to manufacture and sell FRP tanks, so he had some catching up to do.

He also had to modify his plant somewhat and train his fourteen employees on the new technique. And that's what he did. It took about six months to retrofit the plant, and during this process his employees were thoroughly trained on all aspects of constructing FRP tanks.

Abe sold FRP tanks for several years with no complaints; his customers were pleased with his products. Abe was pleased.

Abe's pleasure was soon shattered, however. A week before the Christmas holidays in 1991, with most of Abe's full-time employees on vacation, only a skeleton crew of four workers was manning the plant. During this week, Roundaway received an order for eighteen 6,000 gallon round fuel tanks for a petroleum storage facility. At first, Abe thought he could wait to start this project after the first of the year. However, the customer made it clear that it needed delivery ASAP, and the sequence of events leading to failure began.

With only four workers on hand, Abe had to make a decision. Should he call his other employees in from their vacations so that they could fabricate their normal three tanks at one time—a fabrication process that would take them all day—or should he use the people he had on hand to fabricate one tank at a time? Unfortunately, Abe made the latter choice.

Of the four employees Abe had available, two were material formers who had about the same length of experience (less than two years); the other two employees were apprentices with only limited experience in FRP fabrication. Abe wasn't worried, though; he would be right there to supervise the operation, and the workers would simply follow his directions.

This scenario might have worked, except that while Abe was definitely the owner of the company, he was not an expert in FRP tank fabrication—even though he thought he was.

He gathered his four workers together and laid out his plan and instructions. "We'll be able to get a good head start on this order, and by the time the regulars get back, the ball will be rolling," he told them.

The ball got rolling in good fashion. Two of the workers worked the press forming devices used to form the plastic half-shells; another worker trimmed and measured the seams between the halves to ensure proper tolerance; the fourth worker, the newest member of the workforce with exactly three weeks experience, mainly watched the other three at work.

After forming the pieces, the four workers worked together to fit each piece in place, preliminary to the heat fusion process that binds the halves into a whole, and the fiberglass wrap and resin treatment that came next. After wrapping the tank with fiberglass cloth over a bed of resin and letting it set up, the three more experienced workers went back to starting the tank fabrication process on another tank while the newest worker was given a pot of resin and tape pieces and told to place fiberglass and resin at the critical joint sections, smooth them out, and let them set before grinding off the rough edges.

The new employee was enthusiastic about being given the responsibility for doing something, especially in this case, because making sure the joints are properly sealed is the most important part of the job. Abe, of course, kept his eye on his new employee just to make sure he didn't mess things up. It would have helped, however, if Abe had had some notion of just how the joints should be properly sealed.

The trouble began when they got to the fourth tank. It was the day of Christmas Eve. The workers (including Abe) were anxious to finish the day's work and get home to their families.

Abe let the first worker leave early—about ten o'clock that morning. He reasoned that the fourth tank's half-shells were already formed (they looked good to him). It just made good sense to let one of the tank formers go. He kept the assistant materials former a while longer, but let him go an hour later.

Abe and the other two workers went ahead and put the pieces together and taped them in place. The three of them worked to seal the top joint section first (a mistake—Abe's experienced workers would have worked both sides at once, ensuring a tight fit) until about 2 p.m. Then Abe decided that they would turn the tank over to the other side, and let the new employee do a couple hours more work on the second and final set of joints while he and the other worker left for the holiday.

The new employee worked alone and quite diligently. But then a problem developed. When he tried to force the other seam of the two half shells together (to decrease the gap between them), they wouldn't budge but half an inch or so. What was he going to do? At first he didn't know. Then, while sitting on a stool in front of the tank, drinking eggnog his girlfriend had fixed him, the solution dawned on him.

"I'll form a narrow plastic wedge and put it in the gap and use a bit more fiberglass and a bunch more resin. Then I'll wait until it sets up and take off the rough spots."

He filled in the three-quarter inch gap with a thin strip of plastic held in place with a liberal amount of resin, then placed fiberglass cloth strips the entire length of the gap and heaped on several coats of resin. A couple of hours later, he had a tank that, to the eye, was completely sealed. He looked at his watch, noticed it was about two hours later than his regular quitting time, and moved quickly to put what he considered to be the last touches to the tank. He would come back after Christmas, grind down the rough spots, and coat it again.

The day after Christmas, the four workers who were not on extended vacation reported back to work. The new employee made sure that Abe and the workers got a look at his handiwork on the mis-fitted tank. They gave it only a cursory look, because at a glance it looked okay—and because they had other tanks to start on before the full shift reported to work the day after New Year's.

Abe told the new employee to go ahead, grind off the rough spots and add some more resin where needed, then they would move the tank out of the way and let it cure. Abe also told the him to stamp the tank with the stencil set. Part of the tank fabricating process included giving each tank a stenciled number and born-on date.

He did as instructed and after several more coats of resin, he hooked the tank to the overhead lift and moved it to the temporary curing and storage trunnions. Once on the trunnions he applied the stenciled information as directed—Tank # 91-606 (meaning: 606th tank of 1991)—12/24/91

Tank 91-606 cured for about a week, then the regular tank inspector, back from vacation, inspected it the day after New Year's. He noticed that the quality of workmanship didn't seem to be up to normal standard, but through bloodshot eyes it looked okay to him. His next move was to set the tank up for a hydrostatic (hydro) test to ensure that it didn't leak and that it could withstand a standard amount of pressure.

During the hydro, the inspector noticed that the air compressor had some type of problem—after the accumulator (where air is stored for ready use) emptied too soon, the compressor came on for a while and shut down before the accumulator filled and had to be re-started each time. He did restart it several times, but lunch time was five minutes away, he had a heck of a hangover, so he told himself that the tank was okay—and besides, he would have to call an electrician to check out the compressor. He went to lunch.

After lunch, the inspector had seen just about all he wanted to see of tank 91-606, so he stamped it "inspected" and signed the paperwork.

Later that day, tank 91-606 was moved to the warehouse and put alongside the other tanks that would later be shipped to the petroleum storage facility when all eighteen tanks were finished.

Two weeks passed and the eighteen tanks were completed, shipped, and delivered to the petroleum storage facility.

Tank 91-606 was the fifth tank put into the ground. Unfortunately, the contractor installing the tanks was too interested in getting the job done quickly so that he could move on to bigger and better things.

When 91-606 was craned from its storage pallet, it was jerked about, banged against the ground and two other tanks (slightly breaking the bond and forming a hairline crack in the patch that sealed the three-quarter inch gap), and then dumped into the ground. The ground had not been properly prepared either. Once the tank was semi-level and all pipes were connected, the backhoe moved in and buried it, then moved on to another tank.

Within two hours of having been filled with diesel fuel, tank 91-606 began to leak. At first the contents just seeped out of a slight crack in the section with the uneven three-quarter inch gap, the one that had been filled with fiberglass cloth and resin. About an hour later the slight crack enlarged to a gaping crack running the entire length of the tank along the uneven seam.

The petroleum storage facility noticed the leak within two days. Their regulation-required leak detection equipment had done its job; it had sounded an alarm indicating that the tank was leaking. The problem was that the tank had not just leaked; it had emptied.

The petroleum storage facility had an environmental nightmare on their hands—6,000 gallons of diesel fuel was now in the ground. Complicating the situation, beneath the tank, the subsurface was composed of unconsolidated material that abruptly ended at a clay interface (aquitard). The clay stopped the flow of oil vertically, but did not stop it horizontally. The oil, following the path of least resistance, flowed horizontally (actually downgradient) until it came out at an outcropping a few feet above Cedar Creek, where it emptied, contaminating the creek.

Today, AJ Roundaway Tank Fabricating Company is no longer in business. The lawsuit, with litigation costs, the resulting penalty costs, and fines from local, state, and federal officials, bankrupted the company.

Faulty Installation

In Case Study 21.1, during installation of tank 91-606, we referred to the mishandling of the tank during installation, and improper site preparation. Along with careful handling of the tank itself and any appurtenances, the tank bed must be specially prepared to receive the tank for burial.

Probably the most important step in tank installation is to ensure that adequate backfilling is provided to ensure that no possible movement of the tank can occur after it is placed in the ground. Any such movement might not only damage the tank (especially FRP tanks), but could also jar loose any pipe connections or separate pipe joints. Experience has shown that failure to use special care in this installation process results in leaks.

Care must also be taken to ensure that underground leak detection devices are carefully and correctly installed. Obviously, if a tank is leaking, knowing it as soon as possible is best so that remediation can be initiated quickly, before a minor spill turns into a nasty environmental contamination incident.

Piping Failures

We have mentioned tank piping failure resulting from improper installation, but piping can fail in other ways as well. Before we discuss them, note that the EPA and other investigators clearly indicate that piping failure is one of the most common causes of "larger" UST spills.

If metal piping is used for connecting tanks together or to delivery pumps or fill drops, or for any other reason, the danger of corrosion from rust or from electrolytic action is always present. Electrolytic action occurs because threaded pipes (or other metal parts made electrically active by threading) have a strong tendency to corrode if not properly coated or otherwise protected. To prevent electrolytic action, usually cathodic protection is installed to negate the electrolytic action.

Piping failures are often caused by poor workmanship, which usually appears around improperly fitted piping joints (both threaded and PVC types), incomplete tightening of joints, construction accidents, and improper installation of cover pad. Figure 21.1 shows a typical service station tank and piping layout.

Figure 21.1 Typical service station tank and piping layout (source: EPA, Leak Lookout: Musts for USTs, *Report No. 530/UST-88/006 and 530/UST-88/008, Washington, DC: Office of Underground Storage Tanks, 1988)*

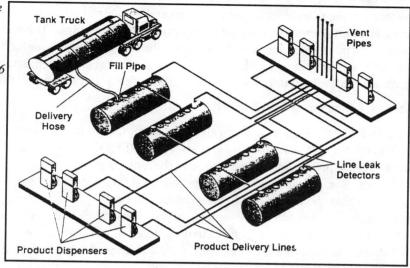

Spills and Overfills

All UST facilities are subject to environmental pollution occurring as the result of spills and overfills—usually the result of human error. Though the EPA has promulgated tank filling procedures in its 40 CFR 280 regulations, and the National Fire Protection Association (NFPA) has issued its NFPA-385 Tank Filling Guidelines, spills from overfilling still frequently occur. Overfilling a UST is bad enough in itself, but the environmental contamination problem is further compounded when such actions occur repeatedly. Petroleum products or hazardous wastes can literally saturate the spill area and can intensify the corrosiveness of soils (Blackman, 1993).

We pointed out earlier that the most common cause of UST spills from overfilling is human failure. See Case Study 21.2 for an example (an actual occurrence, but names and locations have been changed).

Case Study 21.2
Human Error

The tanker truck arrived at the plant site early Monday morning. The driver stopped her truck, got out and walked over to the plant office. At the office, she was greeted by the plant clerk who asked how he could help her. The driver stated that she had a load of #2 fuel oil to deliver, and asked for directions to the receiving tank. The clerk said that it would be easier for the truck driver to follow him in her truck as he walked to the tank and showed her exactly where it was located.

At the tank, the driver parked the truck, got out and ran the filling hose from the truck to the filling port. She uncapped the filling port and inserted the measuring stick to determine the amount of fuel that was in the tank. After determining that the 5,000 gallon tank was almost empty, she inserted the hose nozzle into the fill port and then walked back to the truck where she activated the truck filling pump, which in turn charged the hose to the tank. When she verified that the hose was secure (not leaking and in proper position), she activated the hose nozzle trigger device and the tank began to fill.

She stood at the nozzle for a minute or two and then returned to the truck, got into the cab, lit up a cigarette and peered off into space, her mind in deep thought. If her eyes had been focused on objects in front of her, she would have seen the warning sign mounted to a five foot pole just in front of her (see Figure 21.2). However, she did not see the sign, and after a few puffs on the cigarette, she decided to rest her eyes. Within a few minutes, she was fast asleep.

Meanwhile, the tank steadily filled.

The sign shown in Figure 21.2 is a standard warning sign used at some facilities to alert delivery truck drivers that the facility is well aware of NFPA-385, which provides a list of guidelines directed at the proper procedures to use in the tank filling process. The driver was aware of these guidelines. She'd seen them (and even followed them properly) many times before—the ones that stated that the driver must stay alert and in control of the filling process at all times, and must stand within reach of the emergency shut-off device in case the tank overfills. She was aware of these requirements and usually paid attention to what she was doing, but familiarity with them and confidence in her own abilities lessened their power. She had never had a tank overfilling problem in her six years on-the-job, so guidelines were really not all that important to her anymore. And besides, her mind was elsewhere.

About 30 minutes later, while the driver snored away, the tank filled to capacity, and then overflowed—to the only place it could—the grass-covered ground surrounding the tank area. Not until approximately 16,000 gallons of #2 fuel oil had overflowed and created a small lake around the filling area, did a plant operator who just happened to walk by notice that a lake of fuel oil was now covering what used to be the grassy area.

At first the operator just stood there, shocked. Surprised and somewhat hesitant about what to do, she quickly realized she needed to find out where the truck driver was—and she did. The operator found the driver lying in the front seat of the truck cab, sound asleep.

The operator yelled at the driver, shook her vigorously to wake her, and stated the problem. The driver, a bit groggy, jumped from the truck and tripped the emergency switch to shut off the pump. It was too late, of course, but better late than wait until the entire truckload emptied.

The aftermath of this spill was routine. The plant manager, somewhat upset, informed the pertinent state regulatory agencies about the spill while the plant superintendent called out the plant's HazMat team to respond to the spill.

The HazMat team (well-trained under NFPA 472-473 guidelines and OSHA's HAZWOPER standard) did the best they could to first contain the leak by using the plant's oil boom (too short to surround the entire leak area; subsequently they augmented it by shoveling dirt to build an earthen dike system surrounding the spill) and by using hand pumps and 55-gallon drums to pump the liquid pool dry—but most of the spill seeped into the ground.

The plant manager, after having notified all key parties about the spill, called a local HazMat contractor in to help with the cleanup. For the next three days, the contractor excavated the soil, which was deposited into dump truck after dump truck and hauled away for disposal, and when excavation had removed what appeared to be all of the contaminated soil, the 20 foot deep excavation was left open to "air out."

While the excavation was open, the plant manager hired a tank inspection company to come in and do a complete inspection of the tank. The tank integrity was still fine, so about a week later,

NOTICE

THIS FACILITY FULLY COMPLIES WITH AND ENFORCES REQUIREMENTS OF NFPA-385. MOREOVER, ALL OUTSIDE VENDORS ARE EXPECTED TO COMPLY.

Figure 21.2 NFPA-385 notice alerting fuel truck drivers that the facility is aware of proper fuel tank filling requirements as specified by the National Fire Protection Guidelines

the contractor came back with several loads of fill material and packed it in the excavated area. The plant manager ordered the plant superintendent to plant more grass and add a couple of young trees to the new soil.

The total monetary cost of this particular incident came to $19,855.00. The plant paid the bill and was later reimbursed by the fuel truck company. The truck driver was fired.

———

That the incident described in Case Study 21.2 could actually occur may seem silly to some, ridiculous to others, criminal to regulators, and embarrassing to those involved. Isn't this often the case whenever human error comes into play? However you describe this incident, a few things are certain: it happened, it continues to happen, and it will happen again.

Compatibility of Contents and UST

Obviously, materials must be stored in containers that will contain and hold them. Although it seems obvious, we must not forget that placing highly corrosive materials into containers not rated to contain them is asking for trouble. New chemicals (including fuels) are being developed all the time. Usually the motive for developing such fuels is to achieve improved air quality, but improving air quality does little good if it is at the expense of the water and/or soil—if they are threatened by a new fuel that is incompatible with a particular storage tank.

Many of the USTs presently in use are the FRP tanks that were put in place to replace the old, unprotected, bare steel tanks. FRPs are rated (or can be modified using a different liner) to safely store the fuel products now in common use. The problem occurs when a new, exotic blend of fuel is developed and then placed in an incompatible FRP-type tank.

Common incompatibility problems that have been observed include blistering, internal stress, cracking, or corrosion of the underfilm. To help prevent FRP-constructed or lined tank problems, the American Petroleum Institute has put together a standard that should be referred to whenever existing tanks are to be used for different fuel products.

RISK ASSESSMENT

The problems associated with hydrocarbon spillage or disposal are complex. This complexity is somewhat eased by using the *risk assessment process*, which enables scientists, regulatory officials, and industrial managers to evaluate the public health risks associated with the hydrocarbon releases (or any other toxic chemical release) to soil and groundwater. The risk assessment process consists of the following four steps.

1) *Toxicological evaluation* (hazard identification): should answer the question of whether the chemical has an adverse effect. The factors that should be considered during the toxicological evaluation for each contaminant include routes of exposure (ingestion, absorption, and inhalation), types of effects, reliability of data, dose, mixture effects, and the strength of evidence supporting the conclusions of the toxicological evaluation.

2) *Dose-Response Evaluation*: once a chemical is toxicologically evaluated and the result indicates it is likely to cause a particular adverse effect, the next step is to determine the potency of the chemical. The *dose-response* curve (see Chapter 5) is used to describe the relationship that exists between degree of exposure to a chemical (dose) and the magnitude of the effect (response) in the exposed organism.

3) *Exposure Assessment*: conducted to estimate the magnitude of actual and/or potential human exposures, the frequency and duration of these exposures, and the pathways by which humans are potentially exposed.

4) *Risk Characterization:* the final step in risk assessment. It is the process of estimating the incidence of an adverse health effect under the conditions of exposure found and described in the exposure assessment (Ehrhardt et al., 1986; ICAIR, 1985; Blackman, 1993).

EXPOSURE PATHWAYS

Along with performing and evaluating the findings from the four steps involved in risk assessment, determining exposure pathways resulting from the performance of the remediation option chosen to mitigate a particular UST leak or spill is also important. Exposure pathways may be encountered during site excavation, installation, operations, maintenance, and monitoring. They consist of two categories: (1) direct human exposure pathways, and (2) environmental exposure pathways. These two categories are subdivided into primary and secondary exposure pathways.

Primary exposure pathways directly affect site operations and personnel (e.g., skin contact during soil sampling) or directly affect cleanup levels, which must be achieved by the remedial technology (e.g., when soil impact is the principal issue at a site, soil impact sets the cleanup level and corresponding time frame when cleanup ceases).

Secondary exposure pathways occur as a minor component during site operations (e.g., wind blown dust), and exhibit significant decreases with time as treatment progresses (EPRI *and* EEI, 1988).

REMEDIATION OF UST CONTAMINATED SOILS

Before petroleum-contaminated soil from a leaking UST can be remediated, preliminary steps must be taken. *Soil sampling* is important, not only to confirm that a tank is actually leaking, but also to determine the extent of contamination. Any petroleum product remaining within a UST should be pumped out into above ground holding tanks or containers before the tank area is excavated and the tank removed. Removing any residual fuel before excavation is undertaken because of the potential damage to the tank during removal (See Figure 21.3).

When site sampling is completed, the range and extent of contamination determined, and the UST removed, a determination must be made as to what type of remediation technology to employ in the actual cleanup effort.

Figure 21.3 Why it is important to drain any residual chemical or fuel from a UST prior to its removal

Technical investigations and evaluations of the various aspects of remediation methods for petroleum hydrocarbons in soil, fate and behavior of petroleum hydrocarbons in soil, and economic analyses have been performed by various organizations, environmental industries, and regulatory agencies. Certainly one of those industries in the forefront in conducting such studies is the electric utility industry. This particular industry owns and operates many USTs, as well as facilities for using, storing, or transferring petroleum products, primarily motor and heating fuels.

The EPA has developed Federal regulations for reducing and controlling environmental damage from UST leakage, and many state and localities have developed and implemented strict regulations governing USTs and remedial actions for product releases to soil and groundwater. As a result, the Electric Power Research Institute (EPRI), the Edison Electric Institute (EEI), and the Utility Solid Waste Activities Group (USWAG), in a cooperative effort, conducted a technical investigation. From their findings, they developed a report entitled *Remedial Technologies for Leaking Underground Storage Tanks*. This 1988 report focuses on one of the major components of the technical investigation, which, as its title suggests, described and evaluated available technologies for remediating soil and groundwater that contain petroleum products released from an underground storage tank leak.

The EPRI-EEI/USWAG report provides a general introduction to state-of-the-art cleanup technology, and serves as a reference in determining the feasible methods, a description of their basic elements, and discussion of the factors to be considered in their selection and implementation for a remedial program.

The available technologies for remediating soil and groundwater containing petroleum products listed by EPRI-EEI/USWAG are divided into two categories: in situ treatment and non-in situ treatment. In situ treatment refers to treatment of soil in place. The remedial technologies are further subdivided within these categories as follows:

In Situ Technologies
- Volatilization
- Biodegradation
- Leaching and Chemical Reaction
- Passive Remediation
- Isolation/Containment

Non-In Situ Technologies
- Land Treatment
- Thermal Treatment
- Asphalt Treatment
- Solidification/Stabilization
- Chemical Extraction
- Excavation

Each of these remedial technologies is briefly described in the following sections using information adapted from the 1988 EPRI-EEI/USWAG study (which has become the standard reference, as current today as it was in 1988).

In Situ Technologies

The remedial technologies discussed in the following sections address only those technologies that can be performed in place at the site. Because no excavation is required, exposure pathways are minimized to those that result from the actual streams produced by the in situ technologies, and not those associated with the handling and transport involved in the non-in situ technologies, discussed later in the section entitled "Non-In Situ Technologies."

In Situ Volatilization (ISV)

In situ volatilization (ISV), or in situ air stripping, utilizes forced or drawn air currents through in-place soil to remove volatile compounds (see Figure 21.4). ISV has a successful track record for both effectiveness and cost efficiency.

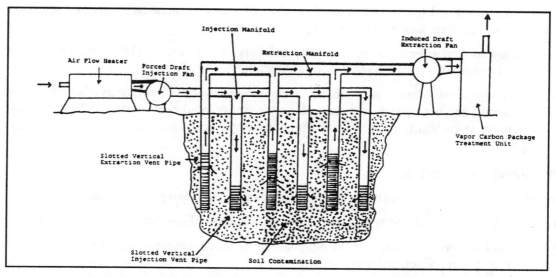

Figure 21.4 Schematic of the type of system used to enhance subsurface ventilation and volatilization of volatile organic compounds (adapted from EPRI-EEI, Remedial Technologies for Leaking Underground Storage Tanks, 1988, p. 19)

A common ISV system used to enhance subsurface ventilation and volatilization of volatile organic compounds consists of the following operations:

1) A pre-injection air heater warms the influent air to raise subsurface temperatures and increase the volatilization rate.

2) Injection and/or induced draft forces establishes airflow through the unsaturated zone.

3) Slotted or screened pipe is used to allow airflow through the system and to restrict entrainment of soil particles.

4) A treatment unit (usually activated carbon) is used to recover volatilized hydrocarbon, minimizing air emissions.

5) Miscellaneous airflow meters, bypass and flow control valves, and sampling ports are generally incorporated into the design to facilitate airflow balancing and assess system efficiency.

Certain factors influence volatilization of hydrocarbon compounds from soils. These factors fall into four categories: soil, environment, chemical, and management (Jury, 1986).

Soil factors

Soil factors include water content, porosity/permeability, clay content, and adsorption site density.

1) *Water content* influences the rate of volatilization by affecting the rates at which chemicals can diffuse through the vadose zone. An increase in soil water content will decrease the rate at which volatile compounds are transported to the surface via vapor diffusion.

2) *Soil porosity* and *permeability* factors relate to the rate at which hydrocarbon compounds volatize and are transported to the surface. A function of the travel distance and cross-sectional area available for flow, diffusion distance increases and cross-sectional flow area decreases with decreasing porosity.

3) *Clay content* affects soil permeability and volatility. Increased clay content decreases soil permeability, which inhibits volatilization.

4) *Adsorption site density* refers to the concentration of sorptive surface available from the mineral and organic contents of soils. An increase in adsorption sites indicates an increase in the ability of the soils to immobilize hydrocarbon compounds in the soil matrix.

Environmental Factors

Environmental factors include temperature, wind, evaporation, and precipitation.

1) *Temperature* increase will increase the volatilization of hydrocarbon compounds.

2) *Wind* increase will decrease the boundary layer of relatively stagnant air at the ground/air interface, which can assist volatilization.

3) *Evaporation* of water at the soil surface is a factor controlling the upward flow of water through the unsaturated zone, which can assist volatilization.

4) *Precipitation* provides water for infiltration into the vadose zone.

Chemical Factors

Not surprisingly, chemical factors are critical players in affecting the way in which various hydrocarbon compounds interact with the soil matrix. Solubility, concentration, octanol-water participating coefficient, and vapor pressure are the primary chemical properties that affect the susceptibility of chemicals to the in situ volatilization process.

Management Factors

Management factors are related to soil management techniques (fertilization, irrigation, etc.) that decrease leaching, increase soil surface contaminant concentrations, assist volatilization, or maximize soil aeration.

Environmental Effectiveness of ISV

Site-specific conditions (soil porosity, clay content, temperature, etc.) drive the effectiveness of in situ volatilization techniques. Pilot studies and actual experience confirm the following effects:

- In situ volatilization has been successful for remediation in unsaturated zones containing highly permeable sandy soils with little or no clay.
- Recovery periods are typically on the order of 6 to 12 months.
- Gasoline (which is light and volatile) has the greatest recovery rate.
- In situ volatilization can be used in conjunction with product recovery systems.
- Because ultimate cleanup levels are site-dependent and cannot be predicted, they are usually set by regulatory agencies.

In Situ Biodegradation

In situ biodegradation utilizes the naturally occurring microorganims in soil to degrade contaminants to another form. Most petroleum hydrocarbons can be degraded to carbon dioxide and water by microbial processes (Grady, 1985). For hydrocarbon removal, this process is enhanced by stimulating their growth and activities, primarily through the addition of oxygen and nutrients. Factors such as temperature and pH influence their rate of growth.

Based upon documentation and significant background information related to successful land treatment of refinery waste, biodegradation has proven its worth as an efficient and cost-effective method for the reduction of hydrocarbons in soil.

Heyse et al. (1986) describe the biodegradation process (shown in Figures 21.5 and 21.6) as follows:

1) A submersible pump transports groundwater from a recovery well to a mixing pump.

2) Nutrients such as nitrogen, phosphorous, and trace metals are added to the water in a mixing tank. These nutrients are then transported by the water to the soil to support microbial activity.

3) Hydrogen peroxide is added to the conditioned groundwater from the mixing tank just prior to reintroduction to the soil. As hydrogen peroxide decomposes, it provides the needed oxygen for microbial activity.

4) Groundwater is pumped to an *infiltration gallery* and/or injection well, which reintroduces the conditioned water to the aquifer or soils.

5) Groundwater flows from the infiltration galleries or injection wells through the affected area, then back to the recovery wells. The flow of the water should contact all soils containing degradable petroleum hydrocarbons.

6) The water is drawn to the recovery well and pumped to the mixing tank to complete the treatment loop.

7) Groundwater in which hydrocarbon concentrations have been reduced to very low levels is often sent through a carbon adsorption process for removal of the residual hydrocarbons.

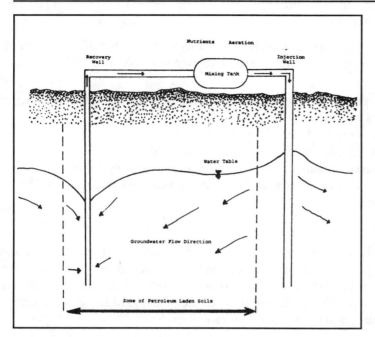

Figure 21.5 Schematic for in situ biodegredation—injection well (adapted from Weston, "Report for Petroleum Marketers Association of America, 1986" in EPRI-EEI Remedial Technologies for Leaking Underground Storage Tanks, 1988, p. 39)

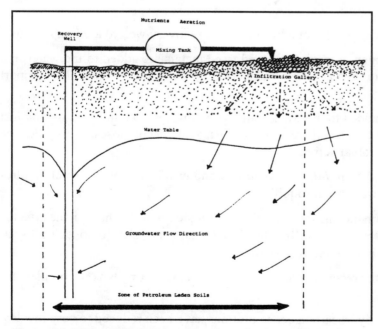

Figure 21.6 Schematic for in situ biodegredation—infiltration gallery (adapted from Weston, "Report for Petroleum Marketers Association of America, 1986" in EPRI-EEI Remedial Technologies for Leaking Underground Storage Tanks, 1988, p. 40)

Environmental Factors

The environmental factors that influence biodegradation in soils are temperature and microbial community.

1) *Temperature* is important in biodegradation of contaminants in soils. In general, biodegradation of petroleum fractions increase as temperatures increase (up to 104°F) from increased biological activity (Bossert and Bartha, 1984).

2) A *microbial community* capable of degrading the target compound is important in the biodegradation process. Most in situ biodegradation schemes make use of existing microbial populations; however, attempts have been made to supplement these populations with additional organisms or engineered organisms.

Chemical Factors

While biodegradation is impossible if substrate concentrations are too high, biodegradation relies on a substantial substrate (target compound) presence to ensure microbes metabolize the target compound. Biodegradation is also limited by the solubility of a compound in water because most microbes need moisture to acquire nutrients and avoid desiccation.

Soil Factors

For the degradation of hydrocarbons in soil to occur, proper aerobic conditions are required. Moisture is also essential for microbial life; however, too much moisture (saturation) limits oxygen levels and can hinder biological activity. Bossert and Bartha (1984) report that moisture content between 50 and 80% of the water-holding capacity is considered optimal for aerobic activities. In the in situ biodegradation processes, oxygen transfer is a key factor; soils must be fairly permeable to allow this transfer to occur.

Another important soil factor is soil pH, which directly affects the microbial population supported by the soil. Biodegradation is usually greater in a soil environment with a pH of 7.8.

For optimal biodegradation of petroleum hydrocarbons to occur, nutrients (nitrogen and phosphorus) in the proper amounts are required.

Environmental Effectiveness

The effectiveness of in situ biodegradation is dependent upon the same site-specific factors as other in situ technologies, but the historical record for this technology is limited. However, several case studies suggest the following:

- In situ biodegradation is most effective for situations involving large volumes of subsurface soils.
- Significant degradation of petroleum hydrocarbons normally occurs in the range of 6 to 18 months (Brown et al., 1986).
- In situ biodegradation has most often been used for the remediation of groundwater impacted by gasoline.
- Research suggests limited biodegradation of benzene or toluene may occur under anaerobic conditions (Wilson et al., 1986).
- In soils, the remedial target level for in situ biodegradation could be in the low mg/l (ppm) level for total hydrocarbons (Brown et al., 1986).

In Situ Leaching and Chemical Reaction

The *in situ leaching and chemical reaction* process uses water mixed with a surfactant (a surface-active substance—soap) to increase the effectiveness of flushing contaminated soils in the effort to leach the contaminants in the groundwater. The groundwater is then collected downstream of the leaching site, through a collection system for treatment and/or disposal (see Figure 21.7).

Figure 21.7 Schematic of a leachate recycling system (source: EPA, "Review of In-Place Treatment Techniques for Contaminated Surface Soils," Volume 1: Technical Evaluation, Washington, DC, EPA/540/2-84-003, 1984)

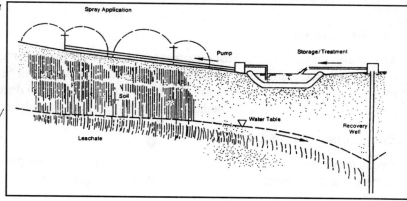

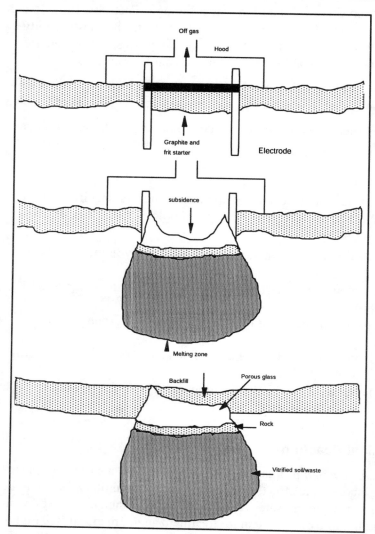

Figure 21.8 In situ vitrification process sequence (adapted from Pacific Northwest Laboratories, Application of in situ vitrification to PCB-contaminated soils. *Prepared for Electric Power Research Institute EPRCS-4834, 1986)*

Environmental Effectiveness

The in situ leaching and chemical reaction process is not commonly practiced. Little performance data on its environmental effectiveness exists.

In Situ Vitrification

The *in situ vitrification* process employs electrical current passed through electrodes (driven into the soil in a square configuration), which produces extreme heat and converts soil into a durable glassy material (see Figure 21.8). The organic constituents are pyrolized in the melt and migrate to the surface where they combust in the presence of oxygen. Inorganics in the soil are effectively bound in the solidified glass (Johnson and Cosmos, 1989).

Environmental Effectiveness

Organic materials are combusted and/or destroyed by the high temperatures encountered during the vitrification process. The in situ vitrification process is a developing technology. The jury is still out in determining its environmental effectiveness.

In Situ Passive Remediation

The *in situ passive remediation process* is the easiest to implement and the least expensive, mainly because it involves no action at the site; however, it is generally unacceptable to the regulatory agencies. It relies upon several natural processes to destroy the contaminant. These natural processes include biodegradation, volatilization, photolysis, leaching, and adsorption.

Environmental Effectiveness

Because passive remediation depends upon a variety of site-specific and constituent-specific factors, the environmental effectiveness of passive remediation must be decided on a case-by-case basis.

In Situ Isolation/Containment

As the name implies, isolation/containment methods are directed toward preventing migration of liquid contaminant or leachates containing contaminants. Accomplished by separating the contamination area from the environment and by installation of impermeable barriers to retain liquid containments within the site, successful application of these methods is usually contingent on the presence of an impervious layer beneath the contaminant to be contained, and the attainment of a good seal at the vertical and horizontal surfaces.

Experience has shown that the containment devices discussed in this section adequately isolate the contamination. However, destruction of the contaminant is not accomplished.

Containment Methods

Slurry walls are fixed underground physical barriers formed in an excavated trench by pumping slurry, usually a bentonite or cement and water mixture (see Figure 21.9).

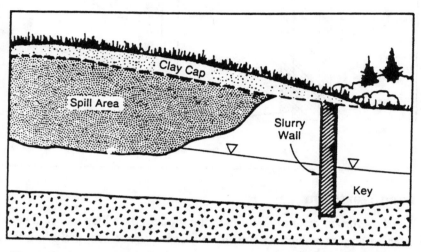

Grout curtains (similar to slurry walls) are suspension grouts composed of Portland cement or grout are injected under pressure to form a barrier (see Figure 21.10).

Figure 21.9 Keyed-in slurry wall (source: EPA, Slurry Trench Construction for Pollution Control, EPA-540/2-84-001, 1984)

Sheet piling construction involves physically driving rigid sheets, pilings of wood, steel, or concrete into the ground to form a barrier.

Environmental Effectiveness

Isolation/containment systems are effective in physically preventing or impeding migration, but the contaminant is not removed or destroyed.

Non-In Situ Technologies

Unlike in-situ techniques, *non-in situ techniques* require the removal (usually by excavation) of contaminated soils. These soils can either be treated on-site or hauled off-site and treated. Another difference that must be taken into consideration when employing non-in situ techniques is the exposure pathways associated with the handling and/or transport of contaminated soil.

The non-in situ technologies for soils discussed in this section include land treatment, thermal treatment, asphalt incorporation, solidification/stabilization, chemical extraction, and excavation.

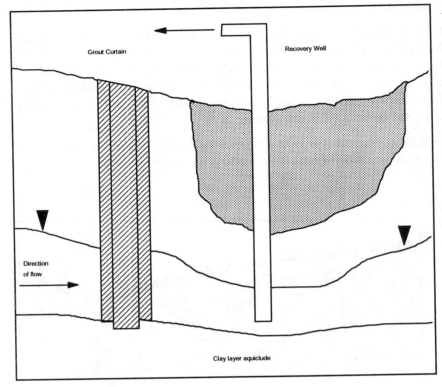

Figure 21.10 Upgradient grout curtain (adapted from Sommerer and Kitchens, "Task 1 Literature Review on Groundwater and Diversion Barriers," Atlantic Research Corporation, 1980. Prepared for U.S. Army Hazardous Materials Agency, Aberdeen Proving Grounds, Maryland)

Land Treatment

Land treatment or *land farming* is the process by which affected soils are removed and spread over an area to enhance naturally-occurring processes, including volatilization, aeration, biodegradation, and photolysis.

The land treatment process involves tilling and cultivating soils to enhance biological degradation of hydrocarbon compounds. The basic land treatment operation is shown in Figure 21.11 and is described as follows:

- The area to be used for land treatment is prepared by removing surface debris, large rocks, and brush.
- The area is graded to provide positive drainage, and surrounded by a soil berm to contain run-off within the land treatment area.
- The pH is adjusted with lime (if necessary) to provide a neutral pH.
- If the site is deficient in nutrients, fertilizer is added.
- The petroleum contaminated soil is spread uniformly over the surface of the prepared area.
- The contaminated material is incorporated into the top 6 to 8 in. of soil (to increase contact with microbes) with a tiller, disc harrow, or other plowing device.
- Reapplication of soils that contain petroleum products is done at proper intervals to replenish hydrocarbon supply.
- Hydrocarbon and nutrient levels and soil pH are monitored to assure that the hydrocarbons are properly contained and treated in the land treatment area.

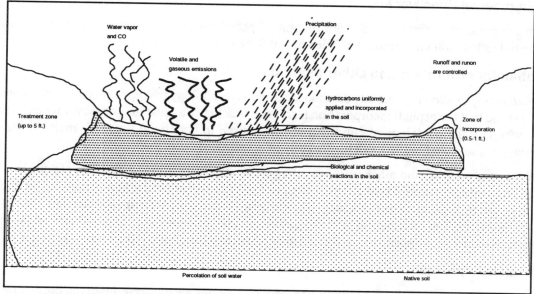

Figure 21.11 Mechanics that occur during land treatment (adapted from American Petroleum Institute, "Land Treatability" of Appendix VII of Constituents Present in Petroleum Industry Wastes, 1984)

Environmental Effectiveness

The effectiveness of land treatment or land farming is highly dependent on site-specific conditions. Several years of experience with treating petroleum compounds using this technology confirm the following:

- Land treatment is an effective means of degrading hydrocarbon compounds.
- Continuous treatment of petroleum-laden soils can result in accumulation of metals in the soil matrix.
- Ultimate degradation rates are site-dependent and cannot be predicted.

Thermal Treatment

Thermal treatment of contaminated soils requires special equipment, but is capable of providing complete destruction of the petroleum-laden contaminant. Affected soils are removed from

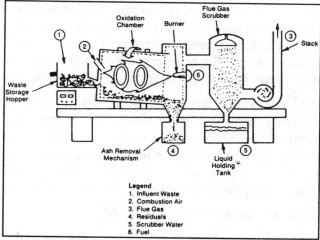

Figure 21.12 Schematic of rotary kiln incinerator (source: EPA, Handbook: Remedial Action of Waste Disposal Sites (Revised), EPA/625/6-85-006, 1985)

the ground and exposed to excessive heat in one of various types of incinerators currently available. These include rotating kilns (see Figure 21.12), fluidized bed incinerators, fixed kilns or hearths, rotating lime or cement kilns, and asphalt plants.

Environmental Effectiveness

High temperature incineration for destruction of petroleum product-laden soil is well documented. Destruction and removal efficiencies of 99% can be expected.

Asphalt Incorporation and Other Methods

Asphalt incorporation is a recently developed remedial technology that goes beyond remediation, in the sense that the asphalt incorporation technique is actually a reuse and/or recycling technology, whereby the containment entrained in soil is used in beneficial reuse (to make asphalt, cement products, and bricks), not just destroyed or disposed.

Asphalt incorporation and other reuse/recycling technologies involve assimilation of petroleum-laden soils into hot or cold asphalt processes, wet or dry cement production processes, or brick manufacturing. During these processes, the petroleum-laden soils are mixed with other constituents to make the final product. In turn, the petroleum contaminants are either volatilized during some treatments, or trapped within the substance, thereby limiting contaminant migration.

Asphalt Incorporation

The conversion of asphalt into asphalt concrete or bituminous concrete involves producing a material that is plastic when being worked and that sets up to a specified hardness sufficient for its end use. The incorporation of contaminated soil into bituminous end products is accomplished by two conventional processes, *Cold-Mix Asphalt Processes (CMA)* and *Hot-Mix Asphalt processes (HMA)* (Testa, 1997).

Cold-Mix Asphalt Processes (CMA)

The cold-mix asphalt process (commonly referred to as environmentally processed asphalt) is a mobile or in-place process. It uses soils contaminated with a variety of contaminants (including petroleum hydrocarbons) to serve as the fine-grained component in the mix, along with asphalt emulsion and specific aggregates to produce a wide range of cold-mix asphaltic products. The mix is usually augmented with lime, Portland cement, or fly ash to enhance stability of the end product. The mixing or incorporation method is accomplished physically by either mixed-in-place methods for large quantities, or windrowing for smaller quantities.

The CMA process has several advantages: (1) a variety of contaminants can be processed, (2) large volumes of contaminated soil can be incorporated; (3) it possesses flexible mix design and specifications, (4) it is a mobile process, (5) it has minimal weather restrictions, (6) it is cost effective, (7) the product can be stockpiled and used when needed, and (8) processing can occur on site. The limitations of CMA include: (1) any volatiles present must be controlled, and (2) small volumes of contaminated soil may not be economically viable for mobile plants.

Hot-Mix Asphalt Processes (HMA)

The hot-mix asphalt process involves the incorporation of petroleum-laden soils into hot asphalt mixes as a partial substitute for aggregate. This mixture is most often applied to pavement. HMA is conventionally produced using either the batch or drum mixing processes. In either of these processes, both mixing and heating are used to produce pavement material.

During the incorporation process, the mixture, including the contaminated soils (usually limited to 5% of the total aggregate feed at any one time), is heated. This causes volatilization of the more-volatile hydrocarbon compounds at various temperatures. Compound migration is limited by incorporating the remainder of the compounds into the asphalt matrix during cooling.

The advantages associated with using the HMA process are: (1) the time required to dispose of hydrocarbon-laden material is limited only by the size of the batching plant (material may be excavated and stored until it can be used), and (2) it can process small volumes of affected soil easily. The disadvantages include: (1) the compound must be applied immediately after processing, (2) it has potential for elevated emissions, (3) it has emission restrictions, and (4) incomplete burning of light-end hydrocarbons can affect quality of end product.

Cement Production Process

Raw materials such as limestone, clay, and sand are incorporated into the cement production process. Once incorporated, these materials are usually fed into a rotary kiln. Contaminated soil may be introduced along with the raw materials, or dropped directly into the hot part of the kiln. The mix is then heated to up to 2700°F. Petroleum-laden soil chemically breaks apart during this process, whereas the inorganic compounds recombine with the raw materials and are incorporated into a clinker. The clinker results in dark, hard, golf-ball sized nodules of rapidly formed Portland cement, which is mixed with gypsum and ground to a fine powder (Testa, 1997).

The advantages of the cement production process include: (1) the technology is in place and has been tested, (2) raw materials are readily available, (3) the product has relatively low water solubility and low water permeability, and (4) it can accommodate a wide variety of contaminants and material. The disadvantages include: (1) odorous material limitations, (2) a wide range of volume increase, and (3) material restrictions, both technically and aesthetically.

Brick Manufacturing Processes

Petroleum-laden soil has been used as an ingredient in the production of bricks. The contaminated soil replaces either the shale and/or firing clay that is normally used in the brick manufacturing process. Normally, clay and shale are incorporated into a plasticized mixture, then extruded and molded into brick. When dried, the brick is fired in a kiln with temperatures of up to 2000°F during a 3-day residence period. When contaminated soil is added to the process, it is mixed with clay and shale, molded into brick, dried, and preheated. Then the brick is fired at 1700° to 2000°F for approximately 12 hours in the kiln. While in the kiln, high temperature and residence time destroys organics and incorporates inorganics into the vitrified end product.

The advantages associated with using the brick manufacturing process to reuse/recycle contaminated soils are: (1) fine-grained, low permeability soils can be accommodated, (2) the technology is in place and has been tested, and (3) processing can occur on-site. The disadvantage is that this process is restricted primarily to petroleum hydrocarbons and fly ash.

Solidification/Stabilization

Solidification/stabilization of petroleum-laden soils is used to immobilize contaminants by either encapsulating or converting them, but does not change the physical nature of the contaminant. This is not a commonly used practice for soils because ultimate destruction of the contaminants does not occur.

Solidification/stabilization processes can be performed either on- or off-site. Various stabilizers and additives are mixed with the material to be disposed. Figure 21.13 shows a generalized process for the manufacture of pozzolanic material (burnt shale or clay resembling volcanic dust

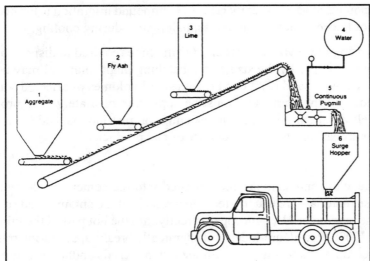

Figure 21.13 Typical lime-fly ash pozzolanic cement process (source: Transportation Research Board, 1976)

that will chemically react with calcium hydroxide at ordinary temperature to form compounds possessing cementitious properties) utilizing fly ash (Mehta, 1983; Transportation Research Board, 1976).

More commonly used to stabilize oily wastes and sludges contained in surface impoundments, solidification/stabilization processes accomplish this in two ways. In in situ surface impoundments, the stabilizing agent is added directly to the impoundment and thoroughly mixed. Treated in sections, as each solidifies, it is used as a base, which allows the equipment to reach further out into the impoundment.

The second method involves excavation of the sludges contained in the impoundment. In this procedure:

- Earth moving machines level piles of kiln dust into 6 to 12-inch deep layers
- A machine lifts the sludge from the impoundment and places it on top of the kiln dust
- Machines then mix the two materials together, and a pulverizing mixer is driven over the mixture until homogeneity is achieved
- The mixture is allowed to dry for about 24 hours, then compacted and field tested (Musser and Smith, 1984).

Usually, the layers are then stacked to build an in-place landfill, or the semi-solidified sludge can be trucked to another landfill location.

Chemical Extraction

Chemical extraction is the process in which excavated contaminated soils are washed to remove the contaminants of concern. This washing process typically is accomplished in a washing plant, which uses a water/surfactant or a water/solvent mixture to remove the contaminants. This method is very similar to the in situ leaching process described earlier. The primary difference is that by removing the soil from the ground, wash mixtures can be used that do not expose the environment to further contamination. This process increases product recovery and is a proven method for the removal of hydrocarbon contaminants from the soil.

Excavation

Excavation involves the physical removal of the contaminated soil for disposal at a hazardous waste or other disposal landfill site (see Figures 21.14 - 21.16). This process has been the mainstay of site remediation for several decades, but recently has been discouraged by newer regulations that favor alternative waste treatment technologies at the contaminated site. Today, excavation is generally considered a storage process instead of a treatment process and raises issues of future liability for the responsible parties regarding the ultimate disposal of the soils. One of the factors contributing to the regulators pushing for on-site treatment methodologies versus land-filling is that landfills are quickly reaching their fill limits, with fewer and fewer new landfilling sites authorized for construction and operation.

The EPA (1985) points out both the positive and negative aspects of excavation. On the positive side, excavation takes little time to complete and allows for complete cleanup of the site. The negative aspects are the necessary worker/operator safety considerations, the production of dust and odor, and the relatively high costs associated with the excavation, transportation, and ultimate disposal of the soil.

Figure 21.14 Excavation in progress to remove damaged chemical line and contaminated soil for proper non-in situ treatment

Figure 21.15 Ten sections of damaged (by corrosion) chemical piping that were removed from one excavation site.

Figure 21.16 Contaminated soil being piled for loading into trucks for transportation to non-in situ treatment facility

SUMMARY

At the beginning of this chapter, Jacqueline MacDonald made the point that "in the early 1990s venture capitalists began to flock to the market for groundwater and soil cleanup technologies, seeing it as offering significant new profit potential."

We also pointed out that even though MacDonald's comments appear to paint a glossy picture of new innovative environmental cleanup enterprises, remediation technology is a double-edged sword. On one side was the market's apparent promise; however, on the other side was actual experience, which showed that investments in remediation did not necessarily pay off.

Many of the early venture capitalists who invested in remediation technology lost money. MacDonald goes on to say that "today, capital investors are wary...because the market is too volatile and sluggish."

Despite a strong economy and increased spending on hazardous waste-site cleanup, private investment in innovative technologies for cleanup of groundwater and soil has declined. A 1997 National Research Council (NRC) report states that total venture capital investments in all industries increased by 87% between 1992 and 1995. Over the same time period, venture capital funding for remediation and other environmental technologies decreased by nearly 70%.

With the recent trend toward cleaning up the environment (especially hazardous waste sites in the United States), why is investment in innovative technology for cleanup of groundwater and soils waning?

According to the NRC, two key factors have constrained the U.S. remediation technologies market: segmentation, and no linkage between rapid cleanup of contaminated sites and the financial self-interest of those responsible. The segmentation problem is created by the many regulatory programs that oversee contaminated sites. These various programs overlap to a degree, but have different requirements for remediation technology approval. Because remediation technology is expensive and many of these overlapping programs require duplicated demonstrations to prove their technology works to the satisfaction of regulators in each of these programs, the expense is magnified many times. Many small companies lack the cash flow to wait out the delays.

The NRC points out that the second problem is the result of delays in site cleanup. Corporate financial managers have found that it is cheaper to delay cleanup through litigation than it is to clean up a contaminated site. "Most small companies cannot meet payroll and other costs without a steady stream of income generated on a predictable basis" (MacDonald, p. 563, 1997).

We have the developers and the innovative remediation technology we need to clean up contaminated soils. The only ingredient we lack is the will to do it (congressional pressure to require it) and the financial incentives for companies to clean up sites quickly. Both of these requirements are needed to ensure profitability for venture capitalists to invest further in innovative cleanup technology.

Cited References

American Petroleum Institute. *Landfarming: An Effective and Safe Way to Treat/Dispose of Oily Refinery Wastes*. Solid Waste Management Committee, 1980.

Blackman, W. C., Jr., *Basic Hazardous Waste Management*, Boca Raton, FL: Lewis Publishers, 1993.

Bossert, I. and Bartha, R., "The fate of petroleum in soil ecosystems," *Petroleum Microbiology*, R. M. Atlas, ed., New York: Macmillan Co., 1984.

Brown, R. S., Norris, R. D. and Estray, M. S., *In Situ Treatment of Groundwater*, Baltimore, Maryland: HazPro 86: Professional Certification Symposium and Exposition, 1986.

Ehrhardt, R. F., Stapleton, P. J., Fry, R. L., and Stocker, D. J., *How clean is clean?—Clean up Standards for Groundwater and Soil,* Washington, DC: Edison Electric Institute, 1986.

EPRI-EEI, *Remedial Technologies for Leaking Underground Storage Tanks*, Chelsea, Michigan: Lewis Publishers, 1988.

Grady, P. C., "Biodegradation: Its Measurement and Microbiological Basis," *Biotechnology and Bioengineering*, Vol. 27, pp. 660-674, 1985.

Heyse, E., James, S. C., and Wetzel, R., "In Situ Aerobic Biodegradation of Aquifer Contaminants at Kelly Air Force Base." *Environmental Progress*, pp. 207-211, 1986.

ICAIR, Life Systems, Inc. *Toxicology Handbook*. Washington, DC: U.S. Environmental Protection Agency, 1985.

Johnson N. P., and Cosmos, M. G., "Thermal Treatment Technologies for HazWaste Remediation," *Pollution Engineering*, October: 79, 1989.

Jury, W. A., "Volatilization from soil," *Guidebook for Field Testing Soil Fate and Transport Models-Final Report*, Washington, DC: U.S. Environmental Protection Agency, 1986.

Kehew, A. E., *Geology for Engineers and Environmental Scientists*, 2nd ed., Englewood Cliffs, NJ: Prentice Hall, 1995.

MacDonald, J. A., "Hard times for innovation cleanup technology," *Environmental Science and Technology*, Vol. 31, No. 12, pp. 560-563, December 1997.

Mehta, P. K., "Pozzolanic and Cementitious By-Products as Miner Admixtures for Concrete—A Critical Review," in *Fly Ash, Silica Fume, Slag, and Other Mineral By-Products in Concrete*. Vol. 1, V. M. Malhotra, ed., American Concrete Institute, 1983.

Musser, D. T., and Smith, R. L., "Case Study: In Situ Solidification/Fixation of Oil Field Production Fluids—A Novel Approach," In *Proceedings of the 39th Industrial Waste Conference*, Purdue University, 1984.

National Research Council, *Innovations in Groundwater and Soil Cleanup: From Concept to Commercialization*, Washington, DC: National Academy Press, 1997.

Pacific Northwest Laboratories, *Application of In Situ Vitrification to PCB-Contaminated Soils*, Prepared for Electric Power Research Institute, EPRCS-4834, RP1263-24, 1986.

Sommerer, S., and Kitchens, J. F., Engineering and Development Support of General Decon Technology for the DARCOM Installation Restoration Program. *Task 1 Literature Review on Groundwater Containment and Diversion Barriers (Draft)*, Aberdeen Proving Ground, Maryland: U.S. Army Hazardous Materials Agency.

Testa, S. M., *The Reuse and Recycling of Contaminated Soil*, Boca Raton: FL: CRC/Lewis Publishers, 1997.

Transportation Research Board. *Lime-Fly Ash: Stabilized Bases and Subbases*, TRB-NCHRP Synthesis Report 37, 1976.

Weston, R. F., Underground Storage Tank Leakage Prevention Detection and Correction. *Report for Petroleum Marketers Association of America*, 1986.

Wilson, J. T., Leach, L. E., Benson, M., *and* Jones, J. N., "In Situ Biorestoration as a Ground Water Remediation Technique," *Ground Water Monitoring Review*, pp. 56-64, 1986.

U.S. Environmental Protection Agency, *Review of In-Place Treatment Techniques for Contaminated Surface Soils, Volume 1: Technical Evaluation*. Washington, DC: EPA/540/2-84-003.

U.S. Environmental Protection Agency, *Remedial Action at Waste Disposal Sites* (revised). Washington, DC: United States Environmental Protection Agency, 1985.

Suggested Readings

Ahmed, J. *Use of Waste Materials in Highway Construction*, Park Ridge, NJ: Noyes Data Corporation, 1993.

Alexander, M., *Biodegradation and Bioremediation*. San Diego: Academic Press, 1994.

Fetter, C. W., *Contaminant Hydrogeology*, New York: Macmillan, 1993.

Kostecki, P. T., and Calabrese, E. J., *Petroleum Contaminated Soils*, Vol. I, Chelsea, Michigan: Lewis Publishers, 1989.

Lar, R., Blum, W. H., and Valentine, C., *Methods for Assessment of Soil Degradation*, Boca Raton, FL: CRC Press/Lewis Publishers, 1997.

Palmer, C. M., *Principles of Contaminant Hydrogeology*, Boca Raton, FL: Lewis Publishers, 1992.

Tan, K. H., *Environmental Soil Science*, New York: Marcel Dekker, Inc., 1994.

U.S. Environmental Protection Agency, *Seminar on Transport and Fate of Contaminants in the Subsurface*, Washington, DC: EPA/CERI-87-45, 1987.

U.S. Environmental Protection Agency, *Ground Water, Vol. I: Ground Water and Contamination*. Washington, DC: EPA/625/6-90-016a, 1990.

PART V
SOLID AND HAZARDOUS WASTE

CHAPTER 22

SOLID WASTES

Humans have always been known for their garbage. Ancient garbage dumps (called middens) provide us with a wealth of information on our ancestors and their lives. From the historical perspective, this human garbage trail has allowed archaeologists to study humans from their earliest days, discovering many fascinating facts about them. From the garbage record, for example, we've determined that at every stage, humans have lived with enormous amounts of garbage, underfoot and all around them. Still true in the poorest parts of developing countries, this habit poses a significant hazard to human health and to the environment.

By the 20th century, most industrialized countries (with their modern approaches to sanitation) removed garbage and other human waste from many living and working environments. But how about the other environment, the one we abuse or rarely think about, the one that sustains our lives, the air, soil, and water of which are all critical to our very existence?

We transferred or diverted our garbage from our immediate living and working area into waterways, we heaped and often burned it in garbage dumps (today called surface impoundments), or we dumped it into areas called landfills (often former wetlands were filled in for anticipated future use).

The result has been so massive as to overwhelm and sometimes directly kill life in local environments Each year, in the United States alone, about 10 billion metric tons of non-agricultural solid waste is generated. Municipal solid waste alone accounts for more than 150 metric tons each year. An average U.S. citizen discards about 4 pounds of waste each day—that's nearly 1500 pounds per person, each year.

But what is new and threatening about modern garbage is not entirely the amount (great as it is), but its toxicity and persistence. Most waste in earlier times was biodegradable—that is, it could and did break down in the environment as part of natural processes (i.e., via biogeochemical cycles). However, today humans routinely use products made from or producing toxic chemicals and many other hazardous substances. Many of these are poisonous to start with; others become poisonous under certain situations—for example, when they are burned, or when they come into contact with certain other chemicals to form a chemical brew of unknown toxicity. This chemical brew also enters the food chain, and is passed along in concentration in the bodies of larger organisms. Many of these waste products, pesticides and plastics especially, persist in the environment for years, decades, and beyond.

The Wall Street Journal, the Washington Post, and the Christian Science Monitor have all called the garbage disposal situation in America a "crisis" (Peterson, 1987; Tongue, 1987; Richards, 1988). Is this a crisis that can be solved or will we continue to expand the historical garbage trail, leaving future archaeologists a record to study that will paint our generation as one of folly, of total disregard, of deliberate misuse—capable only of poor judgement?

CHAPTER OUTLINE

- Introduction
- Solid Waste Regulatory History (United States)
- Solid Waste Characteristics
- Sources of Municipal Solid Waste

INTRODUCTION

In the final part (Part Five) of this text, we discuss a growing and significant problem facing not only all practitioners of environmental science, but everyone else, as well: *Anthropogenically-produced wastes*. Specifically, what are we going to do with all the wastes we generate? What are the alternatives? What are the technologies available to us at present to mitigate the waste problem—a problem that grows with each passing day?

Before beginning our discussion, we should focus on an important question: When we throw waste away, is it really gone? Although we are faced today and in the immediate future with growing mountains of wastes we produce, an even more pressing problem is approaching—one related to the waste's toxicity and persistence.

We will discuss waste and the toxicity problem later, but now, think about the persistence of the wastes that we dispose. For example, when we excavate a deep trench and place within it several 55-gallon drums of liquid waste, then bury the entire sordid mess, are we really disposing of the waste in an earth-friendly way? Are we disposing of it permanently at all? What happens years later when the 55-gallon drums corrode and leak? Where does the waste go?

What are the consequences of such practices? Are they insignificant to us today because they are tomorrow's problems?

We need to ask ourselves these questions and determine the answers now. If we are uncomfortable with the answers we come up with, shouldn't we feel the same about the answers someone else (our grandchildren) will have to come up with later?

Waste is not easily disposed. We can hide it. We can move it from place to place. We can take it to the remotest corners of the earth. But because of its persistence, waste is not always gone when we think it is. It has a way of coming back, a way of haunting us, a way of reminding us—a way of persisting.

Waste is very persistent, as thousands of documented cases make clear. Case Study 22.1 recounts a personal experience with waste and its persistence taken from *The Science of Water* (1998). Keep in mind this is a minor example. Many much more serious examples of waste persistence have been documented.

Case Study 22.1
The Great Circle Route

If you have ever transited from the East Coast of the United States to Europe via ship across the Atlantic Ocean, you probably traveled the 'Great Circle Route.' This Great Circle Route is the shortest distance between the two continents. This is the case because the route follows the circular geometry of Earth.

During this transit, the ship's captain uses Satellite Navigation Aids (SatNav), other electronic aids, and standard ship's compass. Obviously, this is practicing good seamanship (and a lot of common sense).

During a transit of this same route [in 1980], when I had the distinct honor and pleasure of performing office-of-the-deck duties on a U.S. Navy warship, one thing became quite apparent to me almost instantly. I found out that if the ship's SatNav, electronic navigation aids, and compass failed, and if the sky was overcast for the entire voyage (this would not allow me to navigate by using the stars or sunlines, etc.), I would have little difficulty in finding my way, in navigating the ship to its European destination.

I would have had little difficulty in navigating because all I would have had to do was to follow the trail—the markers—the signposts—the remnants of previous ships following the same path. Never had I seen such a clearly marked seaway. Garbage, refuse of all descriptions, and plastic-like debris floating along. This floating trail was narrow but distinct. At night this garbage trail could easily be seen in the light shone from the ship's dim lights.

When I first saw this garbage trail, I just shook my head in disgust. There was no doubt whatsoever this route had been heavily traveled by others. This was apparent because man had left his footprint—man has a bad habit of doing this. What it really comes down to is "out of sight out of mind" thinking. Or when we abuse our resources in such a manner, maybe we think to ourselves: 'Why worry about it. Let God sort it out.'

This is exactly what worries a lot of people (including me); in the end God may do a lot of sorting out.

In this chapter, we define and discuss solid wastes. In particular we focus on a significant portion of solid wastes, *Municipal Solid Wastes (MSW)*, because these wastes are generated by people living in urban areas where many of the problems associated with solid waste occur. We also discuss another significant waste problem: Hazardous Wastes (Chapter 23). Chapter 24 presents waste control technologies related to waste minimization, treatment, and disposal.

SOLID WASTE REGULATORY HISTORY (UNITED STATES)

For most of the nation's history, municipal ordinances (rather than federal regulatory control) were the only solid waste regulations in effect. These local urban governments controlled solid waste almost from the beginning of each settlement because of the inherent severe health consequences derived from street disposal. Along with prohibiting dumping of waste in the streets, municipal regulations usually stipulated requirements for proper disposal in designated waste dump sites and mandated that owners remove their waste piles from public property.

The federal government did not get involved with regulating solid waste dumping until the nation's harbors and rivers were becoming overwhelmed with raw wastes. The federal government used its constitutional powers under the *Interstate Commerce Clause* of the constitution to enact the *Rivers and Harbors Act* in 1899. The U.S. Army Corps of Engineers was empowered to regulate and in some cases prohibit private and municipal dumping practices.

Not until 1965 did Congress finally get into the picture (as a result of strong public opinion) by adopting the *Solid Waste Disposal Act* of 1965, which became the responsibility of the U.S. Public Health Service to enforce. The intent of this act was to:

1)   Promote the demonstration, construction, and application of solid waste management and resource recovery systems that preserve and enhance the quality of air, water, and land resources.

2)   Provide technical and financial assistance to state and local governments and interstate agencies in the planning and development of resource recovery and solid waste disposal programs.

3)   Promote a national research and development program for improved management techniques; more effective organizational arrangements; new and improved methods of collection, separation, recovery, and recycling of solid wastes; and the environmentally safe disposal of non-recoverable residues.

4)   Provide for the promulgation of guidelines for solid waste collection, transport, separation, recovery, and disposal systems.

5)   Provide for training grants in occupations involving the design, operation, and maintenance of solid waste disposal systems (Tchobanoglous et al., 1993).

After Earth Day 1970, Congress became more sensitive to waste issues. In 1976, Congress passed solid waste controls as part of the *Resource Conservation and Recovery Act* (RCRA). "Solid waste" was defined as any garbage, refuse, sludge from a waste treatment plant, water supply treatment plant, or air-pollution control facility and other discarded material.

In 1980, Public Law 96-510, 42 U. S. C. Article 9601, the *Comprehensive Environmental Response, Compensation and Liability Act* (CERCLA) was enacted to provide a means of directly responding, and funding the activities of response, to problems at uncontrolled hazardous waste disposal sites. Uncontrolled MSW landfills are facilities that have not operated or are not operating under RCRA (EPA, 1989).

Many other laws that apply to the control of solid waste management problems are now in effect. Federal legislation and associated regulations have encouraged solid waste management programs to be implemented at the state level of government. It seems apparent that legislation will continue to be an important part of future solid waste management.

SOLID WASTE CHARACTERISTICS

Solid waste [also called refuse, litter, rubbish, waste, trash, and (as an over-simplification) garbage] refers to any of a variety of materials that are rejected or discarded as being spent, useless, worthless, or in excess. A useful waste classification system is shown in Table 22.1.

Table 22.1: Classification of Solid Waste

Type	Principal Components
Trash	Highly combustible waste paper, wood, cardboard cartons, including up to 10% treated papers, plastic or rubber scraps—commercial and industrial sources.
Rubbish	Combustible waste, paper, cartons, rags, wood scraps, combustible floor sweepings—domestic, commercial, and industrial sources.
Refuse	Rubbish and garbage—residential sources
Garbage	Animal and vegetable wastes, wastes from restaurants, hotels, markets—institutional, commercial, and club sources

Source: Adapted from Davis and Cornwell (1991), *Introduction to Environmental Engineering*, 2nd ed.

Solid waste is probably more correctly defined as "any material thing that is no longer wanted." O'Reilly (1992) points out that defining solid waste is tricky, because solid waste is a series of paradoxes:

- personal in the kitchen trashcan, but impersonal in a landfill
- what one individual may deem worthless (an outgrown or out-of-fashion coat, for example) and fit only for the trashcan, another may find valuable
- of little cost concern to many Americans, yet very costly to our society in the long term
- an issue of serious federal concern, yet a very localized problem from municipality to municipality

The popular adage is accurate—everyone wants waste to be picked up, but no one wants it to be put down. It goes almost without saying that the other adage, "Not In My Back Yard" (NIMBY) is also accurate. The important point, though, is that whenever a material object is thrown away, regardless of its actual or potential value, it becomes a solid waste.

Garbage (with its tendency to decompose rapidly and create offensive odors) is often used as a synonym for solid waste, but actually refers strictly to animal or vegetable wastes resulting from handling, storage, preparation, or consumption of food.

The collective and continual production of all refuse (i.e., the sum of all solid wastes from all sources) is referred to as the *solid waste stream*. As stated previously, an estimated 6 billion metric tons of solid waste are produced in the United States each year (Figure 22.1 A). The two largest sources of solid wastes are agriculture (animal manure, crop residues, and other agricultural by-products) and mining (dirt, waste rock, sand, and slag, the material separated from metals during the smelting process). About 10% of the total waste stream is generated by industrial activities (plastics, paper, fly ash, slag, scrap metal, and sludge or biosolids from treatment plants).

From Figure 22.1 A we can see that the other 3% of the solid waste stream is made up of municipal solid waste (MSW), which is the focus of this chapter and consists of refuse generated by households, businesses, and institutions. From Figure 22.1 B we can see that paper and pa-

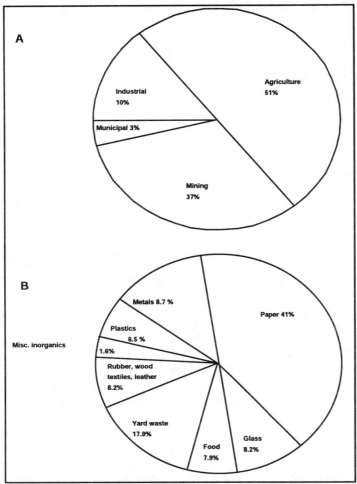

Figure 22.1 A: Sources of solid waste in the U.S.; B: Composition of solid waste discarded in a typical day by each American (source: EPA, Meeting the Environmental Challenge, *1990;* Characterization of Municipal Solid Waste in U.S., 1992 Update, *EPA/530-5-92-019)*

perboard account for the largest percentage (about 41%) of refuse materials by volume of MSW. Yard wastes are the next most abundant material, accounting for almost 18%. Glass and metals make up almost 17% of MSW, food wastes just under 8%, and plastics about 6.5%.

The EPA (1990) says that approximately 178 million metric tons of MSW were generated in the U.S. in 1990, equivalent to a bit more than 4 pounds per person per day. The EPA estimates that by the year 2000, waste generation in the U.S. will rise to more than 197 million metric tons annually, almost 4.5 pounds per person per day.

SOURCES OF MUNICIPAL SOLID WASTES

Sources of municipal solid wastes in a community are generally related to land use and zoning. MSW sources include residential, commercial, institutional, construction and demolition, municipal services, and treatment plants.

Residential Sources of MSW

Residential sources of MSW are generated by single and multifamily detached dwellings and apartment buildings. The types of solid wastes generated include food wastes, textiles, paper, cardboard, glass, wood, ashes, tin cans, aluminum, street leaves, and special bulky items including yard wastes collected separately, white goods (refrigerators, washers, dryers, etc.), batteries, oil, tires and household hazardous wastes.

Commercial Sources of MSW

Commercial sources of MSW are generated in restaurants, hotels, stores, motels, service stations, repair shops, markets, office buildings, and print shops. The types of solid wastes generated include paper, cardboard, wood, plastics, glass, special wastes such as white goods and other bulky items, and hazardous wastes.

Institutional Sources of MSW

Institutional sources of MSW are generated in hospitals, schools, jails and prisons, and government centers. The types of solid wastes generated by institutional sources are the same as those generated by commercial sources.

Construction and Demolition Sources of MSW

Construction and demolition sources of MSW are generated by new construction sites, the razing of old buildings, road repair/renovation sites, and broken pavement. The types of solid wastes generated by construction and demolition sources include standard construction materials such as wood, steel, plaster, concrete, and soil.

Municipal Services Sources of MSW

Municipal services (excluding treatment plants) sources of MSW are generated in street cleaning, landscaping, parks and beaches, recreational areas, and catch-basin maintenance and cleaning activities. The types of solid wastes generated by municipal services include rubbish; street sweepings; general wastes from parks, beaches, and recreational areas; and catch-basin debris.

Treatment Plant Site Sources of MSW

Treatment plant site sources of MSW are generated in water, wastewater, and other industrial treatment processes (e.g., incineration). The principal types of solid wastes generated by treatment plant sites are sludges or biosolids, fly ash, and general plant wastes.

SUMMARY

As we examine the problems associated with solid waste disposal, whether municipal, industrial, or hazardous, the answer to a question posed in the introduction of this chapter becomes more and more apparent: When we "throw away" waste, it is not gone. Dealing with the waste permanently has only been postponed. Sometimes this postponement means that when we go back, the wastes are rendered helpful and harmless (as with some biodegradable wastes), but more often, it means that the problems we must face will be worse—increased by chemistry and entropy. That 55-gallon drum was easier to handle before it rusted out.

We pay, somehow, for what we get or use, whether we see the charges or not. The price for our solid waste habits will soon be charged to us. In some places (big cities, for example), the awareness of the size of the bill is sinking in.

If we as a society are going to consume, build, and grow as we do, we have to pay the price for these actions. And that is sometimes going to mean that our waste is going to be "in our backyard." We will have to increase the amount of solid waste we reuse and recycle, we will have to spend tax dollars to solve the problems with landfills and trash incineration, and we will have to seriously look at how we live, how the goods we buy are packaged, and how our industries deal with their wastes; if we don't, the bill will be more than we can afford to pay.

Cited References

O'Reilly, J. T., *State & Local Government Solid Waste Management*, Deerfield, Illinois: Clark, Boardman, Callahan.

Peterson, C., "Mounting Garbage Problem," *The Washington Post*, April 5, 1987.

Richards, B., "Burning Issue," *The Wall Street Journal*, June 16, 1988.

Spellman, F. R., *The Science of Water: Concepts and Applications*, Lancaster, PA: Technomic Publishing Company, 1998.

Tchobanoglous, G., Theisen, H., and Vigil, S., *Integrated Solid Waste Management: Engineering Principles and Management Issues*, New York: McGraw-Hill, 1993.

Tonge, P., "All That Trash," *Christian Science Monitor*, July 6, 1987.

U.S. Environmental Protection Agency, *Decision-Maker's Guide to Solid Waste Management*, Washington, DC: EPA/530-SW89-072, 1989.

U.S. Environmental Protection Agency, *Characterization of Municipal Solid Waste in U.S.: 1992 Update*, Washington, DC: EPA/530-5-92-019.

Suggested Reading

Blumberg, L., and Gottlieg, R., *War on Waste: Can America Win Its Battle with Garbage?* Washington, DC: Island Press, 1989.

"Environmental Justice," *Christian Science Monitor*, March 15, 1994.

MacKay, D., Shiu, W. Y., and Ma, K-C., *Illustrated Handbook of Physical-Chemical Properties and Environmental Fate for Organic Chemicals*, Boca Raton, FL: CRC Press/Lewis Publishers, 1997.

Morris, D., "As if Materials Mattered," *The Amicus Journal* 13, No. 4, 17-21 (1991).

U.S. Environmental Protection Agency, *The Solid Waste Dilemma: An Agenda For Action— Background Document*, Washington, DC: EPA/530-SW-88-054A, 1988.

Wolf, N. and Feldman, E., *Plastics: America's Packaging Dilemma*, Washington, DC: Island Press, 1990.

CHAPTER 23

HAZARDOUS WASTES

The most alarming of all man's assaults upon the environment is the contamination of air, earth, rivers, and sea with dangerous and even lethal materials. This pollution is for the most part irrecoverable; the chain of evil it initiates not only in the world that must support life but in living tissues is for the most part irreversible. In this now universal contamination of the environment, chemicals are the sinister, and little-recognized partners of radiation in changing the very nature of the world—the very nature of life. —Rachel Carson

CHAPTER OUTLINE

- Introduction
- America: A Throwaway Society
- Hazardous Substances and Hazardous Wastes
- What Is a Hazardous Substance?
- What Is a Hazardous Waste?
- Where Do Hazardous Wastes Come from?
- Why Are We Concerned About Hazardous Wastes?
- Hazardous Waste Legislation

INTRODUCTION

Rachel Carson was able to combine the insight and sensitivity of a poet with the realism and observations of science more adeptly than anyone before her. Famous for her classic and highly influential book *Silent Spring*, it seems strange to us today that such a visionary was (after the publication of her magnum opus) ostracized, vilified, laughed at, lambasted, and disregarded. To those guilty of the sins that she revealed, Rachel Carson was an enemy to be derided—and silenced. She was not, however, disregarded by those who understood. To these concerned folks with conscience, her message was clear: waste, if not properly treated and handled, not only threatens human life in the short term, but the environment as a whole in the long term. Her plea was also clear: stop poisoning the earth.

Examined with the clear vision of retrospect, the environmental missionary Rachel Carson was well ahead of her time. The fears she expressed in 1962 were based on limited data, but have since been confirmed. Rachel Carson was right.

In this chapter, we discuss the hazards of hazardous materials and hazardous wastes (some of which drew Rachel Carson to pen and paper). We illustrate the nature of the substance, the problem, and the possible consequences.

AMERICA: A THROWAWAY SOCIETY

The main point discussed and illustrated in Chapter 22 is actually a small portrait of American society. The portrait displays and underscores a characteristic that might be described as habit, trend, custom, or practice—the tendency we have to discard those objects we no longer want. We simply throw them away—so much so and so often that we even call ourselves a "throwaway society."

When something is no longer of value because it is broken, worn out, out of style, or no longer needed for whatever reason, we feel discarding it should not be a big issue. But it is, particularly when the item we throw away is a hazardous substance—persistent, non-biodegradable, and poisonous.

What is the magnitude of the problem with hazardous substance/waste disposal? Let's take a look at a few facts.

- Hazardous substances—including industrial chemicals, toxic waste, pesticides, and nuclear waste—are entering the marketplace, the workplace, and the environment in unprecedented quantities.
- The United States produces almost 300 million metric tons of hazardous waste each year—with a present population of 260,000,000+, this amounts to more than one ton for every person in the country.
- Through pollution of air, soil and water supplies, hazardous wastes pose both short- and long-term threats to human health and environmental quality.

HAZARDOUS SUBSTANCES AND HAZARDOUS WASTES

Hazardous wastes can be informally defined as a subset of all solid and liquid wastes, which are disposed of on land rather than being shunted directly into the air or water, and which have the potential to adversely affect human health and the environment. We have the tendency to think of hazardous wastes as resulting mainly from industrial activities, but households also play a role in the generation and improper disposal of substances that might be considered hazardous wastes. Hazardous wastes, via Bhopal and other disastrous episodes, have been given much attention, but surprisingly little is known of their nature or the actual scope of the problem. In this section we examine definitions of hazardous materials, substances, wastes, etc., and attempt to bring hazardous wastes into perspective as a major environmental concern.

Unfortunately, defining a *hazardous substance* is largely a matter of "pick and choose," with various regulatory agencies and pieces of environmental legislation defining that term and related terms somewhat differently. Many of the terms are used interchangeably. Even experienced professionals in environmental health and safety fields, like Certified Hazardous Materials Managers, sometimes interchange these terms, though they are generated by different Federal agencies, by different pieces of legislation, and have somewhat different meanings dependent upon the nature of the problem addressed. To understand the scope of the dilemma we face in defining hazardous substance, let's take a look at the terms commonly used today, used interchangeably, and often thought to mean the same thing.

Hazardous Material

A hazardous material is a substance (gas, liquid, or solid) capable of causing harm to people, property, and the environment. The United States Department of Transportation (DOT) uses the term hazardous materials to cover nine categories identified by the *United Nations Hazard Class Number System*, including:

- Explosives
- Gases (compressed, liquefied, dissolved)
- Flammable Liquids
- Flammable Solids
- Oxidizers
- Poisonous Materials
- Radioactive Materials
- Corrosive Materials
- Miscellaneous Materials

Hazardous Substances

The term "hazardous substance" is used by the EPA for a chemical that, if released into the environment above a certain amount, must be reported, and depending on the threat to the environment, for which federal involvement in handling the incident can be authorized. The EPA lists hazardous substances in 40 CFR Part 302, Table 302.4.

The Occupational Safety and Health Administration (OSHA) uses the term hazardous substance in 29 CFR 1910.120 (which resulted from Title I of SARA and covers emergency response) differently than does the EPA. Hazardous substances (as defined by OSHA) cover every chemical regulated by both DOT and the EPA.

Extremely Hazardous Substances

Extremely hazardous substance is a term used by the EPA for chemicals that must be reported to the appropriate authorities if released above the *threshold reporting quantity (RQ)*. The list of extremely hazardous substances is identified in Title III of the *Superfund Amendments and Reauthorization Act* (SARA) of 1986 (40 CFR Part 355). Each substance has a threshold reporting quantity.

Toxic Chemicals

EPA uses the term *toxic chemical* for chemicals whose total emissions or releases must be reported annually by owners and operators of certain facilities that manufacture, process, or otherwise use listed toxic chemicals. The list of toxic chemicals is identified in Title III of SARA.

Hazardous Wastes

EPA uses the term hazardous wastes for chemicals regulated under the *Resource, Conservation and Recovery Act* (RCRA-40 CFR Part 261.33). Hazardous wastes in transportation are regulated by DOT (49 CFR Parts 170-179).

For our purposes, in this text we define a hazardous waste as any hazardous substance that has been spilled or released into the environment. For example, chlorine gas is a hazardous material. When chlorine is released to the environment, it becomes a hazardous waste. Similarly, when asbestos is in place and undisturbed, it is a hazardous material. When it is broken, breached, or thrown away, it becomes a hazardous waste.

Hazardous Chemicals

OSHA uses the term *hazardous chemical* to denote any chemical that poses a risk to employees if they are exposed to it in the workplace. Hazardous chemicals cover a broader group of chemicals than the other chemical lists.

WHAT IS A HAZARDOUS SUBSTANCE?

To form the strongest foundation for understanding the main topic of this chapter (hazardous waste), and because RCRA's definition for a hazardous substance can also be used to describe a hazardous waste, we use RCRA's definition. RCRA defines something as a hazardous substance if it possesses any of the following four characteristics: reactivity, ignitability, corrosiveness, or toxicity. Briefly,

- *Ignitability* refers to the characteristic of being able to sustain combustion and includes the category of flammability (ability to start fires when heated to temperatures less than 140°F, or 60°C).
- *Corrosive* substances (or wastes) may destroy containers, contaminate soils and groundwater, or react with other materials to cause toxic gas emissions. Corrosive materials provide a specific hazard to human tissue and aquatic life where the pH levels are extreme.
- *Reactive* substances may be unstable or have a tendency to react, explode, or generate pressure during handling. Pressure-sensitive or water-reactive materials are included in this category.
- *Toxicity* is a function of the effect of hazardous materials (or wastes) that may come into contact with water or air and be leached into the groundwater or dispersed in the environment.

Toxic effects that may occur to humans, fish, or wildlife are the principal concerns here. Toxicity, until 1990, was tested using a standardized laboratory test, called the *extraction procedure* (EP Toxicity Test). The EP Toxicity test was replaced in 1990 by the *Toxicity Characteristics Leaching Procedure (TCLP)* because the EP test failed to adequately simulate the flow of toxic contaminants to drinking water. The TCLP test is designed to identify wastes likely to leach hazardous concentrations of particular toxic constituents into the surrounding soils or groundwater as a result of improper management.

TCLP extracts constituents from the tested waste in a manner designed to simulate leaching actions that occur in landfills. The extract is then analyzed to determine if it possesses any of the toxic constituents listed in Table 23.1. If the concentrations of the toxic constituents exceed the levels listed in the table, the waste is classified as hazardous.

Table 23.1: Maximum Concentration of Contaminants for TCLP Toxicity Test

Contaminant	Regulatory Level (mg/l)
Arsenic	5.0
Barium	100.0
Benzene	0.5
Cadmium	1.0
Carbon tetrachloride	0.5
Chlordane	0.03
Chlorobenzene	100.0
Chloroform	6.0
Chromium	5.0
Cresol	200.0
2,4-D	10.0
1,4-Dichlorobenzene	7.5
1,5-Dichloroethane	0.5
2.4-Dinitrololuene	0.13
Endrin	0.02
Heptachlor	0.008
Hexachlorobnezene	0.13
Hexachloroethane	3.0
Lead	5.0
Lindane	0.4
Mercury	0.2
Methoxychlor	10.0
Methyl ethyl ketone	200.0
Nitrobenzene	2.0
Pentachlorophenol	100.0
Pyridine	5.0
Selenium	1.0
Silver	5.0
Tetrachloroethylene	0.7
Toxaphene	0.5
Trichloroethylene	0.5
2,4,5-Trchlorophenol	400.0
2,4,6-Trchlorophenol	2.0
2.4,5-TP (Silvex)	1.0
Vinyl chloride	0.2

Source: EPA (1990), 40 CFR 261.24.

WHAT IS A HAZARDOUS WASTE?

Recall our general rule of thumb that states that any hazardous substance spilled or released to the environment is no longer classified as a hazardous substance, but as a hazardous waste. We also said that the EPA uses the same definition for hazardous waste as it does for hazardous substance. The four characteristics described in the previous section (reactivity, ignitability, corrosivity, or toxicity) can be used to identify hazardous substances as well as hazardous wastes.

Note that the EPA lists substances that it considers hazardous wastes. These lists take precedence over any other method used to identify and classify substances as hazardous (i.e., if a substance is listed in one of the EPA's lists described below, it is legally a hazardous substance, no matter what).

EPA Lists of Hazardous Wastes

EPA-listed hazardous wastes are organized into three categories: *Nonspecific source wastes*, *specific source wastes*, and *commercial chemical products*; all listed wastes are presumed to be hazardous regardless of their concentrations. The EPA developed these lists by examining different types of wastes and chemical products to determine whether they met any of the following criteria:

- Exhibit one or more of the four characterizations of a hazardous waste.
- Meet the statutory definition of hazardous waste.
- Are acutely toxic or acutely hazardous.
- Are otherwise toxic.

These lists are described briefly, as follows

- *Nonspecific source wastes* are generic wastes, commonly produced by manufacturing and industrial processes. Examples from this list include spent halogenated solvents used in degreasing, and wastewater treatment sludge from electroplating processes, as well as dioxin wastes, most of which are "acutely hazardous" wastes because of the danger they present to human health and the environment.
- *Specific source wastes* are from specially identified industries such as wood preserving, petroleum refining, and organic chemical manufacturing. These wastes typically include sludges, still bottoms, wastewaters, spent catalysts, and residues, e.g., wastewater treatment sludge from pigment production.
- *Commercial chemical products* (also called "P" or "U" list wastes because their code numbers begin with these letters) include specific commercial chemical products or manufacturing chemical intermediates. This list includes chemicals such as chloroform and creosote, acids such as sulfuric and hydrochloric, and pesticides such as DDT and kepone (40 CFR 261.31, 32 and 33).

Note that the EPA ruled that any waste mixture containing a listed hazardous waste is also considered a hazardous waste and must be managed accordingly. This applies regardless of what percentage of the waste mixture is composed of listed hazardous wastes. Wastes derived from hazardous wastes (residues from the treatment, storage, and disposal of a listed hazardous waste) are considered hazardous waste as well (EPA, 1990).

WHERE DO HAZARDOUS WASTES COME FROM?

Hazardous wastes are derived from several waste generators. Most of these waste generators are in the manufacturing and industrial sectors and include chemical manufacturers, the printing industry, vehicle maintenance shops, leather products manufacturers, the construction industry, metal manufacturing, etc. These industrial waste generators produce a wide variety of wastes, including strong acids and bases, spent solvents, heavy metal solutions, ignitable wastes, cyanide wastes, etc.

WHY ARE WE CONCERNED ABOUT HAZARDOUS WASTES?

From the environmental scientist's perspective, any hazardous waste release that could alter the environment in any way is of major concern. The specifics of their concern lie in acute and chronic toxicity to organisms, bioconcentration, biomagnification, genetic change potential, etiology, pathways, change in climate and/or habitat, extinction, persistence, and esthetics (visual impact).

We have stated consistently that when a hazardous substance or hazardous material is spilled or released into the environment, it becomes a hazardous waste. This is important because specific regulatory legislation has been put in place regarding hazardous wastes, responding to hazardous waste leak/spill contingencies, and for proper handling, storage, transportation, and treatment of hazardous wastes—the goal being protection of the environment, and, ultimately, ourselves.

Why so much concern about hazardous substances and hazardous wastes? This question is relatively easy to answer because of hard lessons we have learned in the past. Our answers are based on experience—actual hazardous materials incidents that we know of and have witnessed have resulted in tragic consequences, not only to the environment, but also to human life. Consider as an example Case Study 23.1[excerpted from *Surviving An OSHA Audit* (1998)]. Maybe it will provide a better explanation of why the control of hazardous substances and wastes is critically important to us all.

Case Study 23.1
Market Day

Day rose heavy and hot, but the wind whispered in the field beyond the sod house as if murmuring delightful secrets to itself. A light breeze entered open windows and gently touched those asleep inside. A finger of warmth, laden with the rich, sweet odor of earth, lightly touched Juju's cheek—rousing her this morning as it had often in her nine years of life. On most days, Juju would lay on her straw mat and daydream, languishing in the glory of waking to another day on Mother Earth. But nothing was normal on this morning. This day was different—full of surprises and excitement. Juju and her mother Lanruh were setting out on adventure today—and Juju couldn't wait.

As she stood at the foot of her make-shift bed, Juju swiftly tucked the folds of thin fabric around her slender waist and let the fall of cloth hang to her feet. She pulled her straight black hair tight in a knot at the back of her neck before she draped the end of the sari over her head.

While Juju dressed, Lanruh performed the same ritual in her small room, next to Juju's. Lanruh was excited about the day's events, too—She knew Juju was thrilled and she was delighted in her daughter's pleasure and excitement. Lanruh chuckled to herself as she remembered the many times over the last few years that Juju had begged to be included, to be taken to the Grand Market in town. Lanruh understood Juju's excitement. Going into the town, taking it all in—Market thrilled Lanruh, too.

As they stepped out of the sod house and onto the dirt road, the scented breeze that had touched Juju's cheek earlier greeted them. They walked together, hand-in-hand toward town, three kilometers to the south.

Juju bubbled with anticipation, but she held it in, presenting the calm, serene face expected of her. Even so, every nerve in her young body reverberated with excitement.

As they walked along the road, Juju, fascinated by everything she saw, took in everything they passed in this extension of her small world. People and cattle everywhere—she had never seen so many of either! Her world had grown, suddenly—and it felt good to be alive.

As they neared town, Juju could see tall buildings. How big and imposing they were—and so many of them! In town, in places they passed, some of the streets were actually paved. Juju had never seen paved streets. This trip to town was her first city experience, and she was enthralled by all the strange and wonderful sights. As they walked along the street leading to the marketplace, Juju was over-awed by the tall buildings and warehouses. "What could they all be used for?" she wondered. Some of them had sign boards above their doors, but little good that did for Juju—she couldn't read.

The light, following breeze had escorted Juju and Lanruh since they left home, and it was still with them as they turned toward the market. Juju could see the entrance, and the throngs of bustling people ahead, and her eyes snapped with excitement.

Suddenly, with one breath of that sweet air (was it the same sweet air that had touched her into waking only two hours earlier?) Juju began coughing. She clutched her throat with both hands, falling to her knees in sudden agony. Her mother was also fallen, gasping for air. The breeze that had begun her day now ended it—delivering an agent of death. But Juju didn't have time to realize what was happening. She couldn't breath. She couldn't do anything—except die—and she did.

Juju, Lanruh, and over two thousand others died within a very few minutes.

Those who died that December 3, 1984 day never knew what killed them. The several hundred others who died soon after did not know what killed them, either. The several thousand inhabitants who lived near the marketplace, near the industrial complex, near the pesticide factory, near the chemical spill, near the release point of that deadly toxin—knew little, if any of this. They knew only death and killing sickness that sorry day.

Those who survived that day were later told that a deadly chemical had killed their families, their friends, their neighbors, their acquaintances. They were killed by a chemical spill that today is infamous in the journals of hazardous materials incidents. Today, this incident is studied by everyone who has anything to do with chemical production and handling operations. We know it as Bhopal.

The dead knew nothing of the disaster—and their deaths were the result.

HAZARDOUS WASTE LEGISLATION

A few people (Rachel Carson for one) could have predicted that a disaster on the scale of Bhopal was ripe to occur. But humans are strange in many ways. We may know that a disaster is possible, even likely. We may predict it, but often enough we do not act. We don't think about the human element.

So what do we do about it? We legislate, of course.

Because of Bhopal and other similar (but less catastrophic) chemical spill events, the United States Congress (pushed by public concern) developed and passed certain environmental laws and regulations to regulate hazardous substances/wastes in the U.S. This section focuses on the two regulatory acts most crucial to the current management programs for hazardous wastes. The first (mentioned several times throughout the text) is the Resource Conservation and Recovery Act (RCRA). Specifically, RCRA provides guidelines for prudent management of new and future hazardous substances/wastes. The second act (more briefly mentioned) is the Comprehensive Environmental Response, Compensation, and Liability Act (CERCLA), otherwise known as Superfund, which deals primarily with mistakes of the past: inactive and abandoned hazardous waste sites.

Resource Conservation and Recovery Act

The Resource Conservation and Recovery Act (RCRA) is the U.S.'s single most important law dealing with the management of hazardous waste. RCRA and its amendment Hazardous and Solid Waste Act (HSWA-1984) deal with the ongoing management of solid wastes throughout the country—with emphasis on hazardous waste. Keyed to the waste side of hazardous materials, rather than broader issues dealt with in other acts, RCRA is primarily concerned with land disposal of hazardous wastes. The goal is to protect groundwater supplies by creating a "cradle-to-grave" management system with three key elements: a tracking system, a permitting system, and control of disposal.

1) A *tracking system*—a manifest document accompanies any waste that is transported from one location to another.

2) A *permitting* system—helps assure safe operation of facilities that treat, store, or dispose of hazardous wastes.

3) A *disposal control* system—restrictions and controls governing the disposal of hazardous wastes onto or into the land (Masters, 1991).

RCRA regulates five specific areas for the management of hazardous waste (with the focus on treatment, storage, and disposal). These are:

1) Identifying what constitutes a hazardous waste and providing classification of each.

2) Publishing requirements for generators to identify themselves, which includes notification of hazardous waste activities and standards of operation for generators.

3) Adopting standards for transporters of hazardous wastes.

4) Adopting standards for treatment, storage, and disposal facilities.

5) Providing for enforcement of standards through a permitting program and legal penalties for noncompliance (Griffin, 1989).

Arguably, RCRA is our single most important law dealing with the management of hazardous waste—it certainly is the most comprehensive piece of legislation that the EPA has promulgated to date.

CERCLA

The mission of the Comprehensive Environmental Response, Compensation, and Liabilities Act of 1980 (Superfund or SARA) is to clean up hazardous waste disposal mistakes of the past, and to cope with emergencies of the present. More often referred to as the Superfund Law, as a result of its key provisions a large trust fund (about $1.6 billion) was created. Later, in 1986, when the law was revised, this fund was increased to almost $9 billion. The revised law is designated as the Superfund Amendments and Reauthorization Act of 1986 (SARA). The key requirements under CERCLA are listed in the following:

1) CERCLA authorizes the EPA to deal with both short-term (emergency situations triggered by a spill or release of hazardous substances) and long-term problems involving abandoned or uncontrolled hazardous waste sites for which more permanent solutions are required.

2) CERCLA has set up a remedial scheme for analyzing the impact of contamination on sites under a hazard ranking system. From this hazard ranking system, a list of prioritized disposal and contaminated sites is compiled. This list becomes the *National Priorities List (NPL)* when promulgated. The NPL identifies the worst sites in the nation, based on such factors as the quantities and toxicity of wastes involved, the exposure pathways, the number of people potentially exposed, and the importance and vulnerability of the underlying groundwater.

3) CERCLA forces those parties who are responsible for hazardous waste problems to pay the entire cost of cleanup.

4) Title III of SARA requires federal, state, and local governments and industry to work together in developing emergency response plans and in reporting on hazardous chemicals. This requirement is commonly known as the *Community Right-To-Know Act*, which allows the public to obtain information about the presence of hazardous chemicals in their communities and releases of these chemicals into the environment.

SUMMARY

Advancements in technology have made our lives more comfortable, safer, healthier, and in many cases more enjoyable. Some would say that progress is not without cost. Can we afford the consequences if these costs include more incidents like Bhopal, Times Beach, Love Canal, or the Exxon Valdez? If such disasters are to be included as a cost of progress, then we must say that the cost outweighs the gain.

What we need to do to ensure a balance between technological progress and its environmental results is to use technological advances to ensure that progress is not too costly or life-threatening, to both our environment and to us.

In Chapter 24 we discuss the technologies currently available for handling and disposing of hazardous wastes in ways that ensure we "progress" safely.

Cited References

Carson, R., *Silent Spring*. Boston: Houghton Mifflin Company, 1962.

DOT 49 CFR—170-179, U.S. Department of Transportation

40 CFR 261.24, U.S. Environmental Protection Agency, (1990)

40 CFR 261.31, .32, .33

40 CFR 264.1

40 CFR 264.52b

40 CFR 302.4

Griffin, R. D., *Principles of Hazardous Materials Management*, Chelsea, Michigan: Lewis Publishers, 1989.

Masters, G. M., *Introduction to Environmental Engineering and Science*, New York: Prentice-Hall, 1991.

RCRA, Public Law 98-616, *Hazardous and Solid Wastes Act*, amendments PL 94-580 (42 USC 6901), 1984.

SARA (CERCLA), Public Law 99-499, *Superfund Amendments & Reauthorization Act* (1986) (Amended 142 USC 9601, 1980).

Spellman, F. R., *Surviving an OSHA Audit: A Manager's Guide*, Lancaster, PA: Technomic Publishing Company, 1998.

U.S. Environmental Protection Agency, *RCRA Orientation Manual*, Washington, DC: United States Environmental Protection Agency, 1990.

Suggested Readings

Blackman, W. C., *Basic Hazardous Waste Management*, Boca Raton, FL: Lewis Publishers, 1993.

Coleman, R. J. and Williams, K. H., *Hazardous Materials Dictionary*, Lancaster, PA: Technomic Publishing Company, 1988.

Kharbanda, O. P., and Stallworthy, E. A., *Waste Management*, UK: Grower, 1990.

Knowles, P-C., (ed.), *Fundamentals of Environmental Science and Technology*, Rockville, Maryland: Government Institutes, Inc., 1992.

Lindgren, G. F., *Managing Industrial Hazardous Waste: A Practical Handbook*, Chelsea, Michigan: Lewis Publishers, 1989.

Portney, P. R., (ed.), *Public Policies for Environmental Protection*, Washington, DC: Resources for the Future, 1993.

Wentz, C.A., *Hazardous Waste Management*, New York: McGraw-Hill, Inc., 1989.

CHAPTER 24

WASTE CONTROL TECHNOLOGY

How to handle society's toxic chemical waste now ranks among the top environmental issues in most industrial countries. Without concerted efforts to reduce, recycle, and reuse more industrial waste, the quantities produced will overwhelm even the best treatment and disposal systems. —Sandra Postel

CHAPTER OUTLINE

- Introduction
- Waste Minimization
- Recycling
- Treatment Technologies
- Ultimate Disposal

INTRODUCTION

One of the most challenging and pressing current environmental concerns confronting environmental scientists and others is what we should do with all of the solid and hazardous wastes our throwaway society produces. In simple (and simplistic) terms, we could say that we should shift from a throwaway society to a recycling one, which would help restore a gain in our living standards. We could also say that since disposing hazardous waste is so expensive and risky, to make the situation better we should follow RCRA's Waste Management Hierarchy (in descending order of desirability) to (1) stop producing waste in the first place; (2) if we cannot avoid producing it, then produce only minimum quantities; (3) recycle it; (4) if it must be produced, but cannot be recycled, then treat it: (5) if it cannot be rendered non-hazardous, dispose of it in a safe manner; and (6) once it is disposed, continuously monitor it to ensure that there are no adverse effects to the environment.

All of these statements have merit. The question is, are they realistic? To a point, yes. We have developed several different strategies to curb the spread of hazardous substances/wastes. One approach is treatment of hazardous wastes to neutralize them or make them less toxic. However, a better strategy would be to reduce or eliminate the use of toxic substances and the generation of hazardous waste. To a degree, we can accomplish this, but to think that we can simply do away with all our hazardous materials, processes that use hazardous materials, and processes that produce hazardous materials is, at the present time, wishful thinking.

What we need to do is refine our waste reduction programs as much as possible and develop technologies that will better treat waste products that we are not able to replace, do away with, or reduce. We have such technologies and/or practices available to us today. Environmental science and technology can be put to work to develop and use measures and practices by which hazardous chemical wastes can be minimized, recycled, treated, and disposed. We review these measures, practices, and technologies in this chapter.

WASTE MINIMIZATION

Waste minimization (or *source reduction measures*) is accomplished in a variety of ways, and includes feedstock or input substitution, process modifications, and good operating practices. Note that before any source reduction measure can be put into place, considerable amounts of information must be gathered.

One of the first steps to be taken in the information gathering process is determining the exact nature of the waste produced. The waste must initially be characterized and categorized by type, composition, and quantity, a task accomplished by performing a *chemical process audit* or *survey* of the chemical process. Keep in mind that during this information-gathering survey, looking closely for any off-specification input materials that might produce defective outputs, any inadvertent contamination of inputs, process chemicals, and outputs, and any obsolete chemicals (which should be properly disposed of) is important.

During the survey, particular attention should be given to problem areas—excessive waste amounts per unit of production, excessive process upsets or bad batches, or frequent off-specification inputs. The effect of process variables on the waste stream created, and the relationship of waste stream composition to the input chemicals and process methods used should be looked at. For example, determine exactly how much process water is used. Can the amount of water used be reduced? Can process water be reused? Questions like these should be addressed during the chemical process survey (Lindgren, 1989).

To determine the feasibility of reuse, recycling, materials recovery, waste transfer, or proper methods of waste disposal, the exact nature of the waste must also be determined. Usually accomplished through sampling the wastestream and then analyzing the sample in the laboratory, the nature of the waste can yield valuable information about the industrial process and the condition of process equipment.

Substitution of Inputs

After completing the chemical process survey, the information gathered may suggest or justify the substitution of certain chemicals, process materials, or feedstock to enable the process's hazardous wastes to be reduced in volume or no longer be produced. Note that input substitutions are often inseparable from process modifications.

A few specific examples of possible input substitutions include:

- use of synthetic coolants in place of emulsified oil coolants
- use of water-based paints instead of solvent-based paints
- use of noncyanide-based electroplating solutions
- use of cartridge filters in lieu of earth filters

Process Modifications

One of the key benefits derived by performing the chemical process audit or survey is that audits often point to or suggest modifications to production systems that work to minimize hazardous waste stream production. Any time a chemical process can be made more efficient, a reduction in the volume and toxicity of the residuals usually results.

Good Operating Practices

Reducing wastage, preventing inadvertent releases of chemicals, and increasing the useful lifetime of process chemicals are all directly related to *good operating practices*. Ensuring good operating practices by workers can only be accomplished through effective worker training. This training should not only include proper process operations, but also effective spill response training.

RECYCLING

Various strategies have been developed to recycle (and thus minimize) the volume of hazardous wastes that must be disposed of. These strategies recover or recycle resources, either materials or energy, from the waste stream. The key point to note in chemical process recycling is that the product must receive some processing before reuse. Wastes generally recognized as having components of potential value include:

- flammable and combustible liquids
- oils
- slags and sludges
- precious metal wastes
- catalysts
- acids
- solvents

From the list above, we can see that one such recycling or recovery effort involves the reclamation of organic solvents. Usually accomplished by using highly effective distillation techniques, solvents contaminated with metals and organics are heated to produce a liquid phase and a vapor phase. Lighter components with high volatiles rise to the top of the liquid phase and begin to vaporize. By carefully controlling the waste mixture's temperature, the desired substance can be vaporized and recovered by condensation, leaving the heavier contaminants behind. What remains is a concentrated, highly toxic mixture (far reduced in volume) referred to as *still bottoms*. Bottoms may contain usable metals and other solvents. As distillation technology improves, more of these bottom materials will be recovered and possibly reused.

TREATMENT TECHNOLOGIES

Because of the 1984 and 1991 amendments to RCRA, hazardous wastes must be treated prior to ultimate disposal in a landfill. Even with process modifications, material substitution, and recycling, some portions of some waste streams may still be hazardous and must be properly contained. These hazardous waste components require additional treatment. Such treatment takes place in vessels (tanks), reactors, incinerators, kilns, boilers, or impoundments.

At the present time, several technologies are available for the treatment of hazardous waste streams. In this section, we discuss a few examples, including biological treatment, thermal treatment, activated carbon sorption, electrolytic recovery techniques, air stripping, stabilization and solidification, and filtration and separation treatment systems.

NOTE: Some of these treatment techniques were covered in greater detail earlier—for example, in situ and non-in situ soil contamination treatment; some technologies combine two of more of these basic technologies.

Biological Treatment

Several *biological treatment processes* are available for treating liquid hazardous waste streams (contaminated soils and solids are more difficult to treat), including activated sludge, aerobic lagoons, anaerobic lagoons, spray irrigation, trickling filters, and waste stabilization ponds. These processes are normally associated with biological treatment of municipal and industrial waste-water, and are generally used for removal of organic pollutants from wastewater. Generally effective with wastewater having low-to-moderate concentrations of simple organic compounds and lower concentrations of complex organics, these are generally ineffective in attacking mineral components, and are useless against heavy metals.

Biological treatment of toxic organic components requires considerably more sophisticated operational control (including pretreatment) than is necessary with nontoxic wastewater. Microorganisms used in biological treatment processes can easily be destroyed by rapid increases in rate of feed. Acclimation and development of a functional population of biota may require considerable time, and the system is continuously subject to upset (Blackman, 1993).

The two biological processes used for treatment of toxic waste are the *aerobic processes* (treatment is in the presence of oxygen—e.g., conventional aeration) and *anaerobic processes* (treatment occurs in the absence of oxygen—e.g., in a simple septic tank).

In aerobic treatment, organisms require both an energy and a carbon source for growth, and both affect what type of organisms will grow in a particular environment. Many hazardous-waste streams satisfy both basic requirements and, if appropriate nutrients are present, a thriving organic population for waste treatment can exist. Under these conditions, if pH and temperature are controlled, substances that are toxic to the active organisms can be eliminated.

The most important aspect limiting the applicability of aerobic biological treatment of hazardous waste is the biodegradability of the waste—its conversion by biological processes to simple inorganic molecules and to biological materials. Biodegradability of a particular waste is very system-specific, and the correct conditions for successful treatment (*detoxification* or biological conversion of a toxic substance to one less toxic) must be maintained to encourage the correct microbe mixture.

Anaerobic treatment of toxic waste streams has been effectively practiced on many different types of toxic waste streams. This form of treatment is basically a fermentation process in which organic waste is both oxidized and reduced.

Thermal Processes

Thermal treatment processes (incineration) are commonly used to treat both liquids and solids to either destroy the hazardous components or allow disposal of the process residue or treated waste in an EPA-approved hazardous waste landfill.

During incineration, carbon-based (organic) materials are burned at high temperatures—typically ranging from 1500°F to 3000°F—to break them down chiefly into hydrogen, carbon, sulfur, nitrogen, and chlorine. These constituent elements then combine with oxygen to form inorganic gases like water vapor, carbon dioxide, and nitrogen oxides. After combustion, the gases pass through a pollution control system to remove acidic gases and particulate matter prior to being released to the atmosphere.

The advantages of hazardous wastes incineration are twofold: (1) it permanently reduces or eliminates the hazardous character of the waste, and (2) it substantially reduces the volume of the waste being disposed.

Waste characteristics and treatment requirements determine the incinerator design to accommodate liquid or solid wastes. Temperature, turbulence, and retention time (commonly known as the 3 T's of incineration) are the prime factors determining incineration treatment design for both solid and liquid wastes.

Hazardous waste incinerators are regulated by the EPA and require a permit for operation. To receive an operating permit, an incineration facility must demonstrate a 99.99% destruction and removal efficiency (DRE) for each principal organic hazardous constituent in the feed material.

Non-in situ thermal processes (primarily incinerators) include designs such as liquid injection and boilers, rotary kilns, fluidized beds, and catalytics. More sophisticated and less common types of thermal treatment systems include wet oxidation, pyrolytic, and plasma processes. Some of these processes can be conducted in situ—with steam injection, radio-frequency heating, and vitrification (molten glass treatment) processes.

Activated Carbon Sorption

Organic substances may be removed from aqueous hazardous waste streams with *activated carbon* by *sorption*. Sorption is the transfer of a substance from a solution to a solid phase. In *adsorption* (a final chemical reaction that forms a cementitious precipitated sludge, not to be confused with *absorption*, which is defined as the physical process that does not chemically stabilize a waste material), chemical substances are removed from the waste stream onto a carbon matrix. The carbon may be used in either granular of powdered form, depending on the application.

The effectiveness of activated carbon in removing hazardous constituents from aqueous streams is directly proportional to the amount of surface area of the activated carbon; in some cases it is adequate for complete treatment. It can also be applied to pretreatment of industrial hazardous waste streams prior to follow-up treatment. Activated carbon sorption is most effective for removing from water those hazardous waste materials that are not water soluble.

Electrolytic Recovery Techniques

The *electrolytic recovery technique* (used primarily for recovery of metals from process streams, to clean process waters, or to treat wastewater prior to discharge) is based on the *oxidation-reduction* reaction, where electrode surfaces are used to collect the metals from the waste stream.

Typically, an electrolytic recovery system consists of a treatment vessel (tank, etc.) with electrodes, an electrical power supply, and a gas handling and treatment system. Recovered metal must be removed from the electrodes periodically when the design thickness is achieved for the recovered metal.

Air Stripping

The *air stripping* technique for removing hazardous constituents from waste streams, although not particularly effective, has been used for many years. Stripping is a means of separating volatile components from less volatile ones in a liquid mixture by partitioning the more volatile materials to a gas phase of air or steam. In air stripping, the moving gas is usually ambient air, which is used to remove volatile dissolved organic compounds from liquids, including ground and waste waters. Additional treatment must be applied to the exhaust vapors to destroy and/or capture the separated volatiles. The process is driven by the concentration gradient between air and liquid phase equilibrium for particular molecules according to *Henry's Law*, which states that at constant temperature, the weight of gas absorbed by a given volume of a liquid is proportional to the pressure at which the gas is supplied.

Stabilization and Solidification

Stabilization and *Solidification* are techniques used to convert hazardous waste from the original form to a physically and chemically more stable material. Accomplished by reducing the mobility of hazardous compounds in the waste prior to its land disposal, stabilization and solidification are particularly useful when recovery, removal, or converting hazardous components (as required by RCRA) from a waste prior to disposal in a landfill is not possible.

A wide variety of stabilization and solidification treatment processes use Portland cement as a binding agent. Waste/concrete composites can be formed that have exceptional strength and excellent durability, and that retain wastes very effectively (Blackman, 1993). Stabilization and solidification treatment processes improve handling and physical characteristics, and result in a reduction of solubility or limit the leachability of hazardous components with a waste.

Filtration and Separation

Filtration and *separation* hazardous waste treatment processes are physical processes. Filtration (the separation of solid particles from a liquid stream through use of semi-porous media) is driven by a pressure difference across the media. This pressure difference is caused by gravity, centrifugal force, vacuum, or elevated pressure.

Filtration applied to hazardous waste treatment falls into two categories—clarification and dewatering. *Clarification* takes place when liquids of less than 200 ppm are placed within a clarifier, and the solids are allowed to settle out, producing a cleaner effluent. *Dewatering* is performed on slurries and sludges. The goal of dewatering is to concentrate the solids into a semi-solid form for further treatment or land disposal.

ULTIMATE DISPOSAL

Most of us are familiar with open dumps. However, you might not be familiar with some of the dumping practices that occurred because of environmental legislation of the 1970s, which placed increasingly stringent controls on releases to the atmosphere and to the nation's waterways. To protect our atmosphere and our waterways, the 1970s legislative mindset drove us to dump hazardous materials into open dumps, the thinking being that land disposal is safer and proper. We didn't realize that this wasn't true until much later, when the tragic consequences of these practices became apparent to us.

We are now well aware that the land is not a bottomless sink that can be used to absorb all of our discards. We've learned that we must pretreat our wastes to detoxify them, to degrade them, to make them less harmful, to make them more earth-friendly—before we deposit them on or in the ground—our earth.

Regardless of the treatment, destruction, and immobilization techniques used, some residue(s) that must to be contained somewhere will always remain from hazardous wastes. This "somewhere" is burial in land, deep-well injection, surface impoundments, waste piles, and landfills. In this section we discuss each of these ultimate disposal methods.

Deep-Well Injection

The practice of *deep-well injection* is not new; it was used in the 1880s by the petroleum industry to dispose of saltwater produced when drilling for oil. However, disposing of hazardous materials by deep-well injection is a relatively recent development. The EPA estimates that about 9 billion gallons of the hazardous waste produced in the United States (about 22% of the total) is injected deep into the ground. Most of the deep-well injection sites are located in the Great Lakes region and along the Gulf Coast.

Deep-well injection involves the injection of liquid waste under pressure to underground strata isolated by impermeable rock strata where geologists believe they will be contained permanently and isolated from aquifers, typically at a depth of more than 700 m below the surface. A high-pressure pump forces the hazardous liquids into pores in the underground rock, where they displace the water, oil and gases originally present. Sandstone and other sedimentary rock formations are used because they are porous and allow the movement of liquids.

In theory, when properly constructed, operated, and monitored, deep-well injection systems may be the most environmentally sound disposal method for toxic and hazardous wastes currently available. However, as with anything else that in theory is "perfect" or affords us the "best available technology," deep-well injection has its problems. For example, though constructed at a depth below the groundwater table, fractures in the underground geology could allow waste to go where it is not wanted—namely, into the groundwater. The biggest problem with deep-well injection concerns the unknown. We are not certain of the exact fate of hazardous substances after injection—another example of the "we don't know what we do not know" syndrome.

Because of our uncertainty about the results of our hazardous waste disposal practices, the 1984 amendments to RCRA ban unsafe, untreated wastes from land disposal. For those land disposal facilities allowed to accept hazardous substances, the EPA (1986) implemented restrictions requiring

- The banning of liquids from landfills;
- The banning of underground injection of hazardous waste within 1/4-mile of a drinking water well;
- More stringent structural and design conditions for landfills and surface impoundments, including two or more liners, leachate collection systems above and between the liners, and groundwater monitoring;
- Cleanup or corrective action if hazardous waste leaks from a facility;
- Information from disposal facilities on pathways of potential human exposure to hazardous substances;
- Location standards that are protective of human health and the environment.

Surface Impoundments

Surface impoundments are diked or excavated areas used to store liquid hazardous wastes (see Figure 24.1). Because most surface impoundments are temporary, relatively cheap to construct, and allow easy access for treatment, they have been popular for many years.

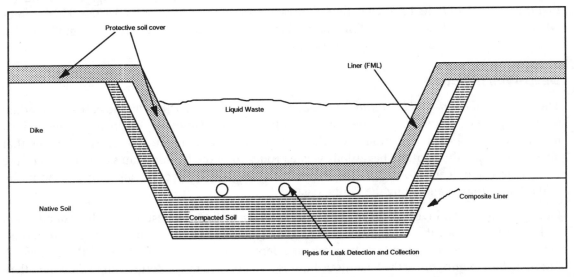

Figure 24.1 Cross-section of a surface liquid waste impoundment

Unfortunately, in the past, surface impoundments were poorly constructed (literally quickly dug out or diked and put into operation), poorly sited (built on a thin layer of permeable soil that allowed leachate to infiltrate to groundwater), located too close to sources of high-quality drinking water (wells or running water sources), and either poorly monitored or not monitored at all. In 1984, the EPA estimated that of the more than 180,000 surface impoundments surveyed, that prior to 1980 only about 25% were lined, and fewer than 10% had monitoring systems.

Because of the problems associated with poor siting, construction, and management of the early surface impoundments, EPA regulations have toughened the requirements for construction of new surface impoundments. Under the *Hazardous and Solid Waste Amendments (HSWA)* of 1984, for example, the EPA now requires new surface impoundments to include

- The installation of two or more liners;
- A leachate collection system between liners; and
- Groundwater monitoring.

Provisions must also be taken to ensure prevention of liquid escaping from overfilling or run-on, and prevention of erosion of dams and dikes. During construction and installation, liners must be inspected for uniformity, damage, and imperfections. These liners must also meet permit specifications for materials and thickness.

Waste Piles

Waste piles are normally associated with industrial sites, where common practice for years was to literally pile up industrial waste, and later, when the pile became too large, dispose it into a landfill. Industrial practice has been to list such piles as "treatment" piles, and even 40 CFR 264/265 subpart L refers to such piles as treatment or storage units.

The environmental problem with such piles is similar to the problems we discussed related to mining waste. Like mining waste, industrial waste piles are subject to weather exposure, including evaporation of volatile components to the atmosphere, and wind and water erosion. The most significant problem related to industrial waste piles is related to precipitation—leaching of contaminants (producing leachate), which may percolate into the subsurface.

RCRA specifications for waste piles are similar to those for landfills (to be discussed in the next section) and are listed in 40 CFR 264/265 subpart L. Under the RCRA guidelines, the owner or operator of a waste pile used for storage or treatment of non-containerized solid hazardous wastes, is given a choice between compliance either with the waste pile or landfill requirements. If the waste pile is used for disposal, it must comply with landfill requirements. The waste pile must be placed on an impermeable surface, and if leachate is produced, a control and monitor system must be in place. Waste piles must also be protected from wind dispersion.

Landfilling

Landfilling wastes has a history of causing environmental problems—including fires, explosions, production of toxic fumes, and storage problems when incompatible wastes are commingled. Landfills also have a history of contaminating surface water and groundwater (EPA, 1990).

Sanitary landfills are designed and constructed to dispose municipal solid wastes only. Not designed, constructed, or allowed to be operated for disposal of bulk liquids and/or hazardous wastes, the type of landfills that can legally receive hazardous wastes is known as *secure landfills* for hazardous wastes.

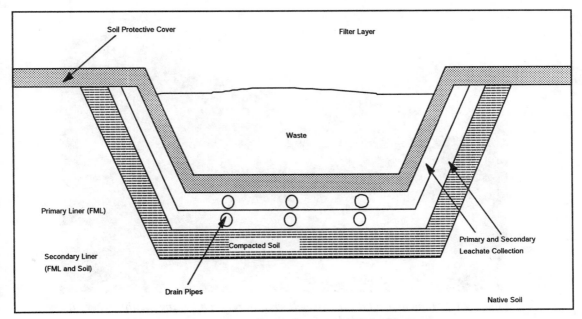

Figure 24.2 Cross-section of a secure landfill double liner system

Under RCRA, the design and operation of hazardous waste landfills has become much more technically sophisticated. Instead of the past practice of gouging out a huge maw from the subsurface and then dumping countless truck loads of assorted waste materials (including hazardous materials) into it until full, a hazardous waste landfill is now designed as a modular series of three-dimensional control cells. Design and operating procedures have evolved to include elaborate safeguards against leakage and migration of leachates.

Secure landfills for hazardous waste disposal are equipped with double liners. Leakage detection, leachate collection and monitoring, and groundwater monitoring systems are required (see Figure 24.2). Liners used in secure landfills must meet regulatory specifications. For example, the upper liner must consist of a 10 to 100 ml *flexible-membrane liner* (FML), usually made of sheets of rubber or plastic. The lower liner is usually FML, but recompacted clay at 3 ft. thick is also acceptable.

Examples of the landfilling process for MSW are shown in Figures 24.3 - 24.8.

Secure landfills must be constructed to allow the collection of leachate (usually via perforated drainage pipes with an attached pumping system) that accumulates above each liner. Leachate control is critical. To aid in this control process (especially from leachate produced by precipitation), a low permeability cap must be placed over completed cells. When the landfill is finally closed, a cap that will prevent leachate formation via precipitation must be put in place. This cap should be sloped to allow drainage away from the wastes.

When a landfill is filled and capped, it cannot be completely abandoned, ignored, or forgotten. The site must be monitored to ensure that leachate is not contaminating the groundwater. This is accomplished by installing test wells downgradient to assure detection of any leakage from the site.

Figure 24.3 Municipal solid waste is delivered to refuse handling facility, where it is piled up for further processing.

Figure 24.4 After initial sorting, what is left over is loaded into trucks for transport to landfill.

Figure 24.5 MSW is sorted for recycling.

Figure 24.6 MSW is delivered to landfill for disposal.

Figure 24.7 MSW is landfilled.

Figure 24.8 A portion of the landfill is filled and covered.

SUMMARY

RCRA's Waste Management Hierarchy sums up what could/should/would happen with waste—any kind of waste—in the best of all possible worlds. But though it is idealistic and simplistic to say we "should" follow these standards, in practical terms, we benefit in the long term by striving to achieve them.

Regulating the problem wastes, developing safe and environmentally friendly ways to dispose of them, and using the technologies we develop to control the future of such wastes is in the best interests of us all.

Cited References

Blackman, W. C., *Basic Hazardous Waste Management*, Boca Raton, FL: Lewis Publishers, 1993.

Lindgren, G. F., *Managing Industrial Hazardous Waste*, Chelsea, Michigan, 1989.

Postel, S., *Defusing the Toxics Threat: Controlling Pesticides and Industrial Wastes, Worldwatch Paper 79*, Washington, DC: Worldwatch Institute, pp. 36-37, 1987.

U.S. Environmental Protection Agency, *Solving the Hazardous Waste Problem: EPA's RCRA Program*, Washington, DC: EPA Office Of Solid Waste, 1986.

U.S. Environmental Protection Agency, *RCRA Orientation Manual, 1990 Edition*. Washington, DC: Government Printing Office, 1990.

Suggested Readings

Freeman, H. M., (ed.), *Hazardous Waste Minimization*, New York: McGraw-Hill, 1990.

Horan, N. J., *Environmental Waste Management: A European Perspective*, New York: Wiley, 1996.

Qasim, S. R., and Chiang, W., *Sanitary Landfill Leachate*, Lancaster, PA: Technomic Publishing Company, 1994.

Suthersan, S., *Remediation Engineering*, Boca Raton, FL: CRC Press/Lewis Publishers, 1997.

Tchobanoglous, G., Theisen, H., and Vigil, S., *Integrated Solid Waste Management: Engineering Principles and Management Issues*, New York: McGraw-Hill, 1993.

Wentz, C. A., *Hazardous Waste Management*, New York: McGraw-Hill, 1989.

SCIENCE AND TECHNOLOGY CAN HELP ACHIEVE A SUSTAINABLE EARTH

Our throwaway society sees the earth as a place of unlimited resources, unlimited space, and unlimited potential to absorb our discards—our contaminants. In a mentality that pervades society, we profess that by increasing production, consumption, and technology, our road to a better life for everyone will be smooth, well-marked, and without detour.

In the past, our mobility allowed us to pollute one area and then merely move on to another. From the earliest times, this has been our methodology—our pattern—especially when dealing with an environment we had fouled (from cave to forest to inner city). Today we would call this a frontier mentality. The problem is, of course, that we are short on frontier (and we wish to be more careful with the frontier we have left) and heavy on over-crowding, pollution, and a myriad of other environmental problems stemming from too much growth and increasing populations.

Many leash science and technology to the whipping post of criticism and lay on the lash for all the evils befalling the earth related to environmental problems—a trendy and inaccurate mindset. Another mindset professes the view (via blind faith) that technological innovation will eventually come to our rescue—that we'll solve these problems because we have solved problems before. Indeed, from a historical perspective, technology has helped to eradicate disease, expand our resource base, and raise our standard of living.

Although past technological success may portend well for the future, our optimism must be tempered by the inherent limitations of technological research and development. We may indeed be able to solve these problems, but not by merely thinking we can.

Humankind is resilient. When blown down by hurricane force winds; when flooded out by raging waters; when forced to abandon our location by volcanic action or earthquake; when pestilence enters the land and kills off many—we bounce back. We have learned to adapt to the natural whims of nature. We recover; eternally optimistic, we go forward.

The question is, can we bounce back from our own mistakes—mistakes that cause land to subside, rivers to overflow their inhabited banks, droughts throughout the land, pestilence to reign unabated, air thick enough to see and choke with one breath, soil so poisoned that even the lowest plant forms cannot grow, foul water that harbors disease? Do we have the same resilience against our own folly that we do against the occurrence of natural disasters?

To formulate solutions, we must first understand the problems. We must ask questions and determine answers. Another question we have to ask is: Would it not be wiser to unleash science and technology from the whipping post, and harness them to industry and government—to help us solve environmental challenges facing us now and in the future?

Through the proper use of science and technology we can solve any problem; we can sustain the earth and life within it. It is not our way to think otherwise. If it were, then we certainly would have lost our resiliency—and any hope for a sustainable future.

GLOSSARY

abiotic: The nonliving part of the physical environment (e.g., light, temperature, and soil structure).

Absorption: (1) Movement of a chemical into a plant, animal, or soil. (2) Any process by which one substance penetrates the interior of another substance. In chemical spill cleanup, this process applies to the uptake of chemical by capillaries within certain sorbent materials.

absorption units: Devices or units designed to transfer the contaminant from a gas phase to a liquid phase.

accidental spills: The unintended release of chemicals and hazardous compounds/materials into the environment.

acid: A hydrogen-containing corrosive compound that reacts with water to produce hydrogen ions; a proton donor; a liquid compound with a pH less than or equal to 2.

acid mine drainage: The dissolving and transporting of sulfuric acid and toxic metal compounds from abandoned underground coal mines to nearby streams and rivers when surface water flows through the mines.

acid rain: Precipitation that has become more acidic from falling through air pollutants (primarily sulfur dioxide) and dissolving them.

acidic deposition: See *acid rain.*

adiabatic: (without loss or gain of heat) When air rises, air pressure decreases and expands adiabatically in the atmosphere; since the air can neither gain nor lose heat its temperature will fall as it expands to fill a larger volume.

adiabatic lapse rate: The temperature profile or lapse rate, used as a basis for comparison for actual temperature profiles (from ground level) and hence for predictions of stack gas dispersion characteristics.

adsorption: (1) The process by which one substance is attracted to and adheres to the surface of another substance without actually penetrating its internal structure. (2) Process by which a substance is held (bound) to the surface of a soil particle or mineral in such a way that the substance is only available slowly.

adsorption site density: The concentration of sorptive surface available from the mineral and organic contents of soils. An increase in adsorption sites indicates an increase in the ability of the soils to immobilize hydrocarbon compounds in the soil matrix.

advanced wastewater treatment: Any treatment that follows primary and secondary wastewater treatment is considered advanced treatment.

advective wind: The horizontal air movements resulting from temperature gradients that give rise to density gradients and subsequently pressure gradients.

aerobic: Living in the air. Opposite of *anaerobic.*

aerobic processes: Many bio-technology production and effluent treatment processes are dependent on microorganisms that require oxygen for their metabolism. For example, water in an aerobic stream contains dissolved oxygen. Therefore, organisms utilizing this can oxidize organic wastes to simple compounds.

afterburners: A device that includes an auxiliary fuel burner and combustion chamber to incinerate combustible gas contaminants.

aggregate: Clusters of soil particles.

agricultural sources: Both organic and inorganic contaminants usually produced by pesticide, fertilizers, and animals wastes, all of which enter water bodies via runoff and groundwater absorption in areas of agricultural activity.

air: The mixture of gases that constitutes the earth's atmosphere.

air currents: Created by air moving upward and downward.

air mass: A large body of air with particular characteristics of temperature and humidity. An air mass forms when air rests over an area long enough to pick up the conditions of that area.

air pollutants: Generally include sulfur dioxide, hydrogen sulfide, hydrocarbons, carbon monoxide, ozone, and atmospheric nitrogen, but can include any gaseous substance that contaminates air.

air pollution: Contamination of atmosphere with any material that can cause damage to life or property.

air stripping: A mass transfer process in which a substance in solution in water is transferred to solution in a gas.

airborne contaminants: Any contaminant capable of being dispersed in air and/or carried by air to other locations.

airborne particulate matter: Fine solids or liquid droplets suspended and carried in the air.

albedo: The fraction of received radiation that is reflected by a surface.

algae: A large and diverse assemblage of eucaryotic organisms that lack roots, stems, and leaves, but have chlorophyll and other pigments for carrying out oxygen-producing photosynthesis.

alphiatic hydrocarbon: Compound comprised of straight chain molecules as opposed to a ring structure.

alkalinity: (1) The concentration of hydroxide ions. (2) The capacity of water to neutralize acids because of the bicarbonate, carbonate, or hydroxide content. Usually expressed in milligrams per liter of calcium carbonate equivalent.

alkanes: A class of hydrocarbons (gas, solid, or liquids depending upon carbon content). Its solids (paraffins) are a major constituent of natural gas and petroleum. Alkanes are usually gases at room temperature (methane) when containing less than 5 carbon atoms per molecule.

alkenes: A class of hydrocarbons (also called olefins); sometimes gases at room temperature, but usually liquids; common in petroleum products. Generally more toxic than alkanes, less toxic than aromatics.

alkynes: A class of hydrocarbons (formerly known as acetylenes). They are unsaturated compounds, characterized by one or more triple bonds between adjacent carbon atoms. Lighter alkynes, such as ethyne, are gases; heavier ones are liquids or solids.

amoebae (pl.) Amoeba (sing.): One of the simplest living animals, consisting of a single cell and belonging to the protozoa group. The body consists of colorless protoplasm. Its activities are controlled by the nucleus, and it feeds by flowing around and engulfing organic debris. It reproduces by binary fission. Some species of amoeba are harmless parasites.

anabolism: The process of building up cell tissue, promoted by the influence of certain hormones; the constructive side of metabolism as opposed to catabolism.

anaerobic: Not requiring oxygen.

anaerobic process: Any process (usually chemical or biological) carried out without the presence of air or oxygen, e.g., in a watercourse that is heavily polluted with no dissolved oxygen present.

analysis: The separation of an intellectual or substantial whole into its constituent parts for individual study.

animal feedlots A confined area where hundreds or thousands of livestock animals are fattened for sale to slaughter-houses and meat producers.

animal wastes: Consists of dung (fecal matter) and urine of animals.

anthropogenic sources: Generated by human activity.

anticyclone: High atmosphere areas characterized by clear weather and the absence of rain and violent winds.

apoenzyme: The protein part of an enzyme.

aqueous solution: Solution in which the solvent is water.

aquifer: Any rock formation containing water. The rock of an aquifer must be porous and permeable to absorb water.

aromatic hydrocarbons: Class of hydrocarbons considered to be the most immediately toxic; found in oil and petroleum products; soluble in water. Antonym: aliphatic.

asphalt incorporation: Soil remediation/recycling process whereby contaminated soil is removed from site and fed into an asphalt-making process as part of the aggregated filler substance.

atmosphere: The layer of air surrounding the earth's surface.

atom: A basic unit of physical matter indivisible by chemical means; the fundamental building block of chemical elements; composed of a nucleus of protons and neutrons, surrounded by electrons.

atomic number: Number of protons in the nucleus of an atom. Each chemical element has been assigned a number in a complete series from 1 to 100+.

atomic orbitals/electron shells: The region around the nucleus of an atom in which an electron is most likely to be found.

atomic weight: The mass of an element relative to its atoms.

auger: A tool used to bore holes in soil to capture a sample.

automatic samplers: Devices that automatically take samples from a wastestream.

autotrophic: An organism that can synthesize organic molecules needed for growth from inorganic compounds using light or another source of energy.

autotrophs: See *autotrophic.*

Avogadro's number: The number of carbon atoms in 12 g of the carbon-12 isotope (6.022045 x 1023). The relative atomic mass of any element, expressed in grams, contains this number of atoms.

bacilli (pl.), *bacillus* (sing): Members of a group of rodlike bacteria that occur everywhere in soil and air. Some are responsible for diseases such as anthrax or for causing food spoilage.

bacteria: One-celled microorganisms.

bacteriophage: A virus that infects bacteria; often called a *phage.*

baghouse filter: A closely woven bag for removing dust from dust-laden gas streams. The fabric allows passage of the gas while retaining the dust.

bare rock succession: An ecological succession process whereby rock or parent material is slowly degraded to soil by a series of bio-ecological processes.

base: A substance which when dissolved in water generates hydroxide (OH-) ions or is capable of reacting with an acid to form a salt.

beneficial reuse: The practice of reusing a typical waste product in a beneficial manner, e.g., wastewater biosolids to compost.

benthic: (benthos) The term originates from the Greek word for bottom and broadly includes aquatic organisms living on the bottom or on submerged vegetation.

best available technology (BAT): Essentially a refinement of best practicable means whereby a greater degree of control over emissions to land, air, and water may be exercised using currently available technology.

binomial system of nomenclature: A system used to classify organisms; organisms are generally described by a two-word scientific name, the *genus* and *species.*

bioaccumulation: The biological concentration mechanism whereby filter feeders such as limpets, oysters, and other shellfish concentrate heavy metals or other stable compounds present in dilute concentrations in sea or fresh water.

biochemical oxygen demand (BOD): The amount of oxygen required by bacteria to stabilize decomposable organic matter under aerobic conditions.

biodegradable: A material that is capable of being broken down, usually by microorganisms, into basic elements.

biodegradation: The process whereby natural decay processes break down man-made and natural compounds to their constituent elements and compounds, for assimilation in and by the biological renewal cycles, e.g., wood is decomposed to carbon dioxide and water.

biogeochemical cycles: *Bio* refers to living organisms and *geo* to water, air, rocks or solids. *Chemical* is concerned with the chemical composition of the earth. Biogeochemical cycles are driven by energy, directly or indirectly, from the sun.

biological oxygen demand (BOD): The amount of dissolved oxygen taken up by microorganisms in a sample of water.

biological treatment: Process by which hazardous waste is rendered non-hazardous or reduced in volume by the actions of microorganisms.

biological treatment process: Includes such treatment processes as activated sludge, aerated lagoon, trickling filters, waste stabilization ponds, and anaerobic digestion.

biology: The science of life.

biosolids treatment: Biosolids refers to water or sewage sludge. Biosolids treatment processes normally include conditioning, thickening, dewatering, disposal by incineration, composting, land application, or land burial.

biosphere: The region of the earth and its atmosphere in which life exists; an envelope extending from up to 6000 meters above to 10,000 meters below sea level that embraces all life from alpine plant life to the ocean deeps.

biostimulent: A chemical that can stimulant growth, e.g., phosphates or nitrates in a water system.

biota: The animal and plant life of a particular region considered as a total ecological entity.

biotic: Pertaining to life or specific life conditions.

biotic index: The diversity of species in an ecosystem is often a good indicator of the presence of pollution. The greater the diversity, the lower the degree of pollution. The biotic index is a systematic survey of invertebrate aquatic organisms and is used to correlate with river quality. It is based on two principles: (1) pollution tends to restrict the variety of organisms present at a point, although large numbers of pollution-tolerant species may persist, and (2) in a polluted stream, as the degree of pollution increase, key organisms tend to disappear in the following order: stone fly, mayflies, caddis fly, freshwater shrimp, bloodworms, tubificid worms.

blastospore: (or bud) Fungi spores formed by budding.

blowby: In an internal combustion engine, blowby occurs as gases from the piston ring area pass into the crankcase.

boiling point: The temperature when a substance changes from a liquid to a gas.

brackish water: Water (non-potable) containing between 100 and 10,000 ppm of total dissolved solids.

brick manufacturing process: In this text, contaminated soil recycling/remediation process whereby contaminated soil is added to the mix used to make brick.

brine: Water containing more than 100,000 ppm of total dissolved solids (salt—NaCl), which can yield salt after evaporation.

btu: British Thermal Unit, a measuring unit of heat.

budding: Type of asexual reproduction in which an outgrowth develops from a cell to form a new individual. Most yeasts reproduce in this way.

calorie: The amount of heat required to raise the temperature of one gram of water one degree centigrade.

capsule: Bacterial capsules are organized accumulations of gelatinous material on cell walls.

carbon adsorption: Process whereby activated carbon, known as the sorbent, is used to remove certain wastes from water by preferentially holding them to the carbon surface.

carbon cycle: The atmosphere is a reservoir of gaseous carbon dioxide, but to be of use to life this must be converted into suitable organic compounds, i.e., fixed, as in the production of plant stems, by the process of photosynthesis. The productivity of an area of vegetation is measured by the rate of carbon fixation. The carbon fixed by photosynthesis is eventually returned to the atmosphere as plants and animals die and the dead organic matter is consumed by the decomposer organisms.

carbon dioxide: A colorless, odorless inert gas—a by-product of combustion.

carbon monoxide A highly toxic and flammable gas that is a by-product of incomplete combustion. Very dangerous even in very low concentrations.

carbonate hardness: Temporary hard water caused by the presence of bicarbonates; when water is boiled, they are converted to insoluble carbonates that precipitate as scale.

catabolism: In biology, the destructive part of metabolism where living tissue is changed into energy and waste products.

catalysis: The acceleration (or retardation) of chemical or biochemical reactions by a relatively small amount of a substance (the catalyst), which itself undergoes no permanent chemical change, and which may be recovered when the reaction has finished.

catalyst: A substance or compound that speeds up the rate of chemical or biochemical reactions.

catalytic combustion: Operates by passing a preheated contaminant-laden gas stream through a catalyst bed, which promotes the oxidization reaction at lower temperatures. The metal catalyst (usually platinum) is used to initiate and promote combustion at much lower temperatures than those required for thermal combustion.

catalytic converter: A device fitted to the exhaust system of a motor vehicle to reduce toxic emissions from the engine. It converts harmful exhaust products to relatively harmless ones by passing the exhaust gases over a mixture of catalysts coated on a metal or ceramic honeycomb (a structure that increases the surface area and therefore the amount of active catalyst with which the exhaust gases will come into contact).

catchment: The natural drainage area for precipitation, the collection area for water supplies, or a river system. The area is defined by the notional line, or watershed, on surrounding high land.

cell: The basic biological unit of plant and animal matter.

cell membrane (cytoplasmic membrane): The lipid- and protein-containing, selectively permeable membrane that surrounds the cytoplasm in procaryotic and eucaryotic cells; in most types of microbial cell, the cell membrane is bordered externally by the cell wall. In microbial cells, the precise composition of the cell membrane depends on the species and on growth conditions and the age of the cell.

cell nucleus: Contained within a eucaryotic cell; a membrane-lined body that contains chromosomes.

cell wall: The permeable, rigid outermost layer of a plant cell composed mainly of cellulose.

cement production process: A contaminated soil recycling/remediation technology whereby contaminated soil is added to the mix in cement production.

CERCLA (Comprehensive Environmental Response, Compensation and Liability Act of 1980— aka, Superfund): Provides for cleanup and compensation, and assigns liability for the release of hazardous substances into the air, land, or water.

chemical bond: A chemical linkage that holds atoms together to form molecules.

chemical change: A transfer that results from the making or breaking of chemical bonds.

chemical equation: A shorthand method for expressing a reaction in terms of written chemical formulas.

chemical extraction: Process in which excavated contaminated soils are washed to remove contaminants of concern.

chemical formula: In the case of substances that consist of molecules, the chemical formula indicates the kinds of atoms present in each molecule and their actual number.

chemical oxygen demand (COD): A means of measuring the pollution strength of domestic and industrial wastes based upon the fact that all organic compounds, with few exceptions, can be oxidized by the action of strong oxidizing agents under acid conditions to carbon dioxide and water.

chemical precipitation: Process by which inorganic contaminants (heavy metals from groundwater) are removed by addition of carbonate, hydroxide, or sulfide chemicals.

chemical process audit/survey: A procedure used to gather information on the type, composition, and quantity of waste produced.

chemical reactions: When a substance undergoes a chemical change and is no longer the same substance; in a chemical reaction it becomes one or more new substances.

chemical weathering: A form of weathering brought about by a chemical change in the rocks affected, involving the breakdown of the minerals within a rock, and usually producing a claylike residue.

chemosynthesis: A method of making protoplasm using energy from chemical reactions, in contrast to the use of light energy employed for the same purpose in photosynthesis.

chloroflurocarbons (CFCs): Synthetic chemicals that are odorless, nontoxic, nonflammable, and chemically inert. CFCs have been used as propellants in aerosol cans, as refrigerants in refrigeration and air conditioners, and in the manufacture of foam packaging. They are partly responsible for the destruction of the ozone layer.

chlorophyll: A combination of green and yellow pigments, present in all "green" plants, which captures light energy and enables the plants to form carbohydrate material from carbon dioxide and water in the process known as photosynthesis. Found in all algae, phytoplankton, and almost all higher plants.

chloroplasts: A structure (or organelle) found with a plant cell containing the green pigment chlorophyll.

cilia: Small threadlike organs on the surface of some cells, composed of contractile fibers that produce rhythmic waving movements. Some single-celled organisms move by means of cilia. In multicellular animals, they keep lubricated surfaces clear of debris. They also move food in the digestive tracts of some invertebrates.

clarification: The process of removing solids from water.

clay content: The amount of clay (fine-grained sedimentary rock) within a soil.

Clean Air Act: The name given to two Acts passed by the U.S. Government. The Act of 1963 dealt with the control of smoke from industrial and domestic sources. It was extended by the Act of 1968, particularly to control gas cleaning and heights of stacks of installations in which fuels are burned to deal with smoke from industrial open bonfires. The 1990 Clean Air Act brought wide-ranging reforms, dealing with all kinds of pollution, from large or small mobile or stationary sources, including routine and toxic emissions, ranging from power plants to consumer products.

Clean Water Act (CWA): A keystone environmental law credited with significantly cutting the amount of municipal and industrial pollution fed into the nation's waterways. More formally known as the Federal Water Pollution Control Act Amendments, passed in 1972, it stems originally from a much-amended 1948 law aiding communities in building sewage treatment plants and has itself been much amended, most notably in 1977 and 1987.

clean zone: That upstream point in a river or stream before a single point of pollution discharge occurs.

climate: The composite pattern of weather conditions that can be expected in a given region. Climate refers to yearly cycles of temperature, wind, rainfall, and so on, not to daily variations.

coal gasification process: The conversion of coal (via destructive distillation or heated out) to gaseous fuel.

cocci (sing. coccus): Member of group of globular bacteria, some of which are harmful to humans.

cofactor: Nonprotein activator that forms a functional part of an enzyme.

cold front: The leading portion of a cold atmospheric air mass moving against and eventually replacing a warm air mass.

cold-mix asphalt process: A mobile or in-place process whereby contaminated soils are recycled/remediated by serving as the fine-grained component in the asphalt-making process.

collector: See *cyclone.*

colloidal material: A constituent of total solids in wastewater; consists of particulate matter with an approximate diameter range of from 1 millimicron to 1 micron.

color: A physical characteristic of water often used to judge water quality; pure water is colorless.

combined wastewater: The combination of sanitary wastewater and storm water runoff.

combustion: In chemical terms, the rapid combination of a substance with oxygen, accompanied by the evolution of heat and usually light. In air pollution control, combustion or incineration is a beneficial pollution control process in which the objective is to convert certain contaminants to innocuous substances such as carbon dioxide and water.

commercial chemical products: An EPA category listing of hazardous wastes (also called *P* or *U* listed wastes because their code numbers begin with these letters); includes specific commercial chemical products or manufacturing chemical intermediates.

commercial sources of MSW (Municipal Solid Waste): Solids generated in restaurants, hotels, stores, motels, service stations, repair shops, markets, office buildings and print shops.

Community Right-To-Know-Act: A part of SARA Title III under CERCLA. Stipulates that a community located near a facility storing, producing, or using hazardous materials has a right-to-know about the potential consequences of a catastrophic chemical spill or release of chemicals from the site.

composite sample: A sample formed by mixing discrete samples taken at periodic points in time or a continuous proportion of the flow. The number of discrete samples that make up the composite depends upon the variability of pollutant concentration and flow.

composting: A beneficial reuse biological process whereby waste (e.g., yard trimmings, wastewater biosolids, etc) is transformed into a harmless humus-like substance used as a soil amendment.

compound: A substance composed of two or more elements chemically combined in a definite proportion.

concentrated solution: Solute in concentration present in large quantities.

condensation: Used in air pollution control technology to remove gaseous pollutants from waste stream; a process in which the volatile gases are removed from the contaminant stream and changed into a liquid.

condenser: Air pollution control device used in condensation method to condense vapors to a liquid phase by either increasing the system pressure without a change in temperature or by decreasing the system temperature to its saturation temperature without a pressure change.

conduction: Heat flow of heat energy through a material without the movement of any part of the material itself.

confined aquifer: Consists of a water-bearing layer sandwiched between two less permeable layers; water flow is restricted to vertical movement only.

conidia: The asexual spores borne on aerial mycelia (e.g., actinomycetes bacteria).

construction and demolition sources of MSW (municipal solid wastes): Generated at new construction sites, razing of old buildings, road repair/renovation sites and broken pavement.

consumers: Organisms that cannot produce their own food and eat by engulfing or predigesting the fluids, cells, tissues, or waste products of other organisms.

contact condenser: Similar to a simple spray scrubber; it cools vapor stream by spraying liquid directly on the vapor stream.

control of disposal: A system of controls and restrictions governing the disposal of hazardous wastes onto or into the land. A key element of RCRA's goal of protecting groundwater supplies.

convection: Method of heat transfer whereby the heated molecules circulate through the medium (gas or liquid).

cooling tower method: A treatment method used to treat thermally polluted water by spraying the heated water into the air and allowing it to cool by evaporation.

corrosive: A substance that attacks and eats away other materials by strong chemical action.

covalent bond: A chemical bond produced when two atoms share one or more pairs of electrons.

cradle-to-grave-act: See *RCRA*.

crustacean: One of a class of arthropods that includes crabs, lobsters, shrimps, woodlice, and barnacles.

cultural eutrophication: Over-nourishment of aquatic ecosystems with plant nutrients resulting from human activities such as agriculture, urbanization, and industrial discharge.

cyclone: In air pollution control, a cyclone collector is used to remove particles from a gas stream by centrifugal force.

cytochrome: A class of iron-containing proteins important in cell metabolism.

cytoplasm: The jelly-like matter within a cell.

decomposers: Organisms such as bacteria, mushrooms, and fungi that obtain nutrients by breaking down complex matter in the wastes and dead bodies of other organisms into simpler chemicals, most of which are returned to the soil and water for reuse by producers.

decomposition: Process whereby a chemical compound is reduced to its component substances. In biology, the destruction of dead organisms either by chemical reduction or by the action of decomposers.

deep-well injection: In waste control technology, the ultimate disposal of liquid hazardous waste under pressure to underground strata isolated by impermeable rock strata to a depth of about 700 m.

density: The ratio of the weight of a mass to the unit of volume.

depletion: In evaluating ambient air quality, pertains to the fact that pollutants emitted into the atmosphere do not remain there forever.

desertification: Creation of deserts by changes in climate, or by human-aided processes.

detoxification: Biological conversion of a toxic substance to one less toxic.

dewatering: The physical or chemical process of removing water from sludge or biosolids.

diatom: Microscopic single-celled algae found in all parts of the world.

diffusion: (1) Mixing of substances, usually gases and liquids, due to molecular motion. (2) The spreading out of a substance to fill a space.

dilute solutions: A solution weakened by the addition of water, oil, or other liquid or solid.

dinoflagellates: Unicellular, photosynthetic protistan algae.

direct flame combustion (flaring): Used in air pollution control technology to burn off process off-gases (e.g., methane).

disinfection: Effective killing by chemical or physical processes of all organisms capable of causing infectious disease (e.g., chlorination is commonly employed in wastewater treatment process).

dispersion: The dilution and reduction of concentration of pollutants in either air or water. Air pollution dispersion mechanisms are a function of the prevailing meteorological conditions.

dissolved oxygen (DO): The amount of oxygen dissolved in a stream, river or lake; is an indication of the degree of health of the body of water and its ability to support a balanced aquatic ecosystem.

DNAPLs: Dense nonaqueous-phase liquids including carbon tetrachloride, creosote, trichloroethane, dichlorobenzene, and others, which can contaminate groundwater supplies.

domestic wastewater: Consists mainly of human and animal wastes, household wastes, small amounts of groundwater infiltration, and perhaps small amounts of industrial wastes.

dose-response curve: A visual means of determining, based on collected data, the percent mortality to dose administered.

dose-response evaluation: The toxological evaluation of the potency of a chemical.

dose-response relationship: Used by toxicologists to base toxicological considerations on. A dose is administered to test animals; depending on the outcome, it is increased or decreased until a range is found where at the upper end all animals die, and at the lower end all animals survive.

drainage basin: The geographical region drained by a river or stream.

dry adiabatic lapse rate: When a dry parcel of air is lifted in the atmosphere, undergoes adiabatic expansion and cooling, and results in a lapse rate (cooling) of -1°C/100m or 1-10°C/km.

dry tower method: A thermal pollution treatment technique whereby heated water is pumped through tubes and the heat is released into the air (similar to the performance of an automobile radiator).

dumps: An open locations where refuse and other waste materials are disposed of in a manner that does not protect the environment, is susceptible to open burning, or is exposed to the elements, vermin and/or scavengers.

dystrophic: Defective nutrition.

ecological toxicology: The branch of toxicology that addresses the effect of toxic substances not only on the human population but also the environment in general, including air, soil, surface water, and groundwater.

ecology: The study of the interrelationship of an organism or a group of organisms with its environment.

ecosystem: A self-regulating natural community of plants and animals interacting with one another and with their nonliving environment.

ecotoxicology: See *ecological toxicity.*

electrolytic recovery technique: Used primarily for recovery of metals from process streams, to clean process waters, or to treat wastewater prior to discharge; based on the oxidation-reduction reaction where electrode surfaces are used to collect the metals from the waste stream.

Electron Transport System: In metabolic transfer, a series of electron carriers that operate together to transfer electrons from donors such as NADH and $FADH_2$ to acceptors such as oxygen.

electron: A component of an atom; travels in a distant orbit around a nucleus.

electrostatic precipitation: Process using a precipitator to remove dust or other particles from air and other gases by electrostatic means. An electric discharge is passed through the gas, giving the impurities a negative electric charge. Positively charged plates are then used to attract the charged particles and remove them from the fast flow.

elements: The most simple substance that cannot be separated into more simple parts by ordinary means. There are more than 100 known elements.

emergency response: Relates primarily to OSHA's requirement under 29 CFR 1910.120 for chemical, industrial, storage, and waste sites to have a written Emergency Response Plan for any covered chemical release or spill to the environment that could jeopardize the good health and well-being of any worker. The EPA also requires an Emergency Response Plan for facilities handling, producing, or using covered chemicals in its Risk Management Plan requirements. Contingencies for fire and medical emergencies should also be included in emergency response plans.

emergent vegetation: A subdivision of the littoral zone of a pond; encompasses shoreline soil area and the immediate shallow water area where emergent plant life can take root underwater, grow, and surface above the waterline.

emergents: See *emergent vegetation.*

endergonic: A reaction in which energy is absorbed.

endoplasmic reticulum: Within a eucaryotic cell: a system of membranes that ramifies through the cytoplasmic region and forms the limiting boundaries, compartments, and channels whose lumina are completely isolated from the cytoplasm; the endoplasmic reticulum is a protein-containing lipid bilayer.

energy: A system capable of producing a physical change of state.

entropy: A measure of the disorder of a system.

environment: All the surroundings of an organism, including other living things, climate and soil, etc. In other words, the conditions for development or growth.

environmental degradation: All the limiting factors that act together to regulate the maximum allowable size or carrying capacity of a population.

environmental factors: Factors that influence volatilization of hydrocarbon compounds from soils. Environmental factors include temperature, wind, evaporation, and precipitation.

environmental science: The study of the human impact on the physical and biological environment of an organism. In its broadest sense, it also encompasses the social and cultural aspects of the environment.

environmental toxicology: The branch of toxicology that addresses the effect of toxic substances, not only on the human population, but also the environment in general, including air, soil, surface water, and groundwater.

enzymes: Proteinaceous substances that catalyze microbiological reactions such as decay or fermentation. They are not used up in the process but speed it up greatly. They can promote a wide range of reactions, but a particular enzyme can usually only promote a reaction on a specific substrate.

epilimnion: Upper layer of a lake; heated by the sun and lighter and less dense than the underlying water.

eucaryotic: An organism characterized by a cellular organization that includes a well-defined nuclear membrane.

euphotic: The surface layer of an ocean, lake, or other body of water through which there is sufficient sunlight for photosynthesis.

eutrophic lake: Lake with a large or excessive supply of plant nutrients (mostly phosphates and nitrates).

eutrophication: Natural process in which lakes receive inputs of plant nutrients as a result of natural erosion and runoff from the surrounding land basin.

evaporative emissions: Emission of fuel from internal combustion systems caused by diurnal losses, hot soak, and running losses.

evapotranspiration: Combination of evaporation and transpiration of liquid water in plant tissue and in the soil to water vapor in the atmosphere.

excavation: The physical removal of soil to construct a burial site for contaminants (landfill) and/or contaminated soil by mechanical means.

excavation and disposal: The removal of contaminated soil for treatment or ultimate disposal.

exergonic: Releasing energy.

exposure assessment: Measurement to estimate the magnitude of actual and/or potential human exposures, the frequency and duration of these exposures, and the pathways by which humans are potentially exposed.

exposure pathways: Consist of two categories: (1) direct human exposure pathways; and (2) environmental exposure pathways. Both of these categories are further subdivided into primary and secondary exposure pathways. Primary pathways are those that directly affect site operations and personnel (e.g., skin contact during soil sampling). Secondary exposure pathways occur as a minor component during site operations and exhibit significant decreases with time as treatment progresses (e.g., wind-blown dust).

extraction procedure (EP): A standardized laboratory test used to test for toxicity; replaced in 1990 by Toxicity Characteristics Leaching Procedure (TCLP).

extraction well: Used to lower the water table, creating a hydraulic gradient that draws a plume of contamination to the well so that the contaminant can be extracted.

extremely hazardous substance: An EPA term for those chemicals that must be reported to the appropriate authorities if released above the threshold reporting quantity.

facultative: Bacteria capable of growth under aerobic and anaerobic conditions.

Federal Water Pollution Control Act (Otherwise known as the Clean Water Act): Concerned with controlling and regulating the amount of municipal and industrial pollution fed into the nation's water bodies.

fermentation: The decomposition of organic substances by microorganisms and/or enzymes. The process is usually accompanied by the evolution of heat and gas, and can be aerobic or anaerobic.

fertilizer: Substance that adds essential nutrients to the soil and makes the land or soil capable of producing more vegetation or crops.

filtration: Technique by which suspended solid particles in a fluid are removed by passing the mixture through a filter. The particles are retained by the filter to form a residue and the fluid passes through to make up the filtrate.

first law of thermodynamics: In any chemical or physical change, movement of matter from one place to another, or change in temperature, energy is neither created nor destroyed, but merely converted from one form to another.

flagella: A threadlike appendage that gives some bacteria motility, extending outward from the plasma membrane and cell wall.

flare: See *direct flame combustion.*

flexible-membrane liner (FML): A rubber or plastic liner used in sanitary landfills.

floating leaf vegetation Part of the littoral zone in a lake or pond where vegetation rooted under the surface allows stems to produce foliage that is able to reach and float on the water surface.

fluoride: Fluoride salt is added to public drinking water supplies for improving resistance to dental caries.

food chain: A sequence of transfers of energy in the form of food from organisms in one trophic level to organisms in another trophic level when one organism eats or decomposes another.

food web: A complex network of many interconnected food chains and feeding interactions.

formula weight: The sum of the atomic weight of all atoms that comprise one formula unit.

friable: Readily crumbled in hand.

front: In meteorology, the boundary between two air masses of different temperature or humidity.

frustules: The distinctive two-piece wall of silica in diatoms.

fumigation: Results when emissions from a smokestack that is under an inversion layer head downward, leading to greatly elevated downwind ground-level concentrations of contamination.

fungi: Saprophytic or parasitic organisms that may be unicellular or made up of tubular filaments and that lack chlorophyll.

garbage: The generic name for waste emanating from households, containing mostly vegetable matter and paper.

gas: In the widest sense, applied to all aeriform bodies, the most minute particles of which exhibit the tendency to fly apart from each other in all directions. Normally these gases are found in that state at ordinary temperature and pressure. They can only be liquefied or solidified by artificial means, either through high pressure or extremely low temperatures.

gas laws: The physical laws concerning the behavior of gases. They include Boyle's law and Charles's law, which are concerned with the relationships between the pressure, temperature, and volume of an ideal (hypothetical) gas.

general biological succession: The process whereby communities of plant and animal species are replaced in a particular area over time by a series of different and usually more complex communities (aka *ecological succession*).

genome: A complete haploid set of chromosomes.

genus: A group of species with many characteristics in common.

geology: The science of the earth, its origin, composition, structure, and history.

geophysical testing: Used to evaluate the subsurface layers, locate the water table, and map contaminate contours using resistivity and conductivity meters.

geosphere: Consists of the inorganic (nonliving) portions of the earth, which are home to all the globe's organic (living) matter.

geothermal energy: The use of the earth's natural heat for human purposes; a massive form of alternative energy that is hard to tap.

geothermal power: See *geothermal energy*.

global warming: The long-term rise in the average temperature of the earth.

glycolysis: One of three phases of the catabolism of glucose to carbon and water process.

grab sample: An individual discrete sample collected over a period of time not exceeding 15 minutes.

gram: The basic unit of weight in the metric system; equal to 1/1000th of a kilogram; approximately 28.5 grams equal one ounce.

gravity: The force of attraction that arises between objects by virtue of their masses. On Earth, gravity is the force of attraction between any object in the Earth's gravitational field and the Earth itself.

gravity settlers: Used for the removal of solid and liquid waste materials from gaseous streams. Consists of an enlarged chamber in which the horizontal gas velocity is slowed, allowing particles to settle out by gravity.

greenhouse effect: The trapping of heat in the atmosphere. Incoming short-wave-length solar radiation penetrates the atmosphere, but the longer-wavelength outgoing radiation is absorbed by water vapor, carbon dioxide, ozone, and several other gases in the atmosphere and is reradiated to earth, causing an increase in atmospheric temperature.

greenhouse gases: The gases present in the earth's atmosphere that cause the greenhouse effect.

groundwater: Water collected underground in porous rock strata and soils; it emerges at the surface as springs and streams.

grout curtain: Used in in situ isolation and containment; consists of Portland cement or grout injected under pressure to form a barrier against contaminant movement in soil.

growth: Exponential bacterial growth.

growth curve: Plotting of bacterial growth cycles. The curve is divided into four phases designated as lag, exponential, stationary, and death. The lag phase, characterized by little or no growth, corresponds to an initial period of time when bacteria are first inoculated into a fresh medium. After the bacteria have adjusted to their new environment, a period of rapid growth, the exponential phase, follows. During this time, conditions are optimal and the population doubles with great regularity. As the bacteria food supply begins to be depleted, or as toxic metabolic products accumulate, the population enters the no-growth or stationary phase. Finally, as the environment becomes more and more hostile, the death phase is reached and the population declines.

guano: A substance composed chiefly of the dung of sea birds or bats, accumulated along certain coastal areas or in caves, and used as fertilizer.

habitat: The place or type of place where an organism or community of organisms naturally or normally thrives.

hardness: A water quality parameter. Water that does not lather easily with soap, and produces scale in pots, pans, and kettles. Caused by the presence of certain salts of calcium and magnesium.

Hazardous & Solid Waste Act/Amendments (1984): Part of RCRA that emphasizes the development and use of alternative and innovative treatment technologies that result in permanent destruction of wastes or reduction in toxicity, mobility, and volume. Land disposal is greatly restricted under the 1984 RCRA amendments.

hazardous chemical: An explosive, flammable, poisonous, corrosive, reactive, or radioactive chemical requiring special care in handling because of hazards it poses to public health and the environment.

hazardous material: A substance in a quantity or form posing an unreasonable risk to health, safety, and/or property when transported in commerce; a substance that by its nature, containment, and reactivity has the capability for inflicting harm during an accident occurrence; characterized as being toxic, corrosive, flammable, reactive, an irritant, or a strong sensitizer, and thereby poisonous. A threat to health and the environment when improperly managed.

hazardous substance: An EPA term used for certain listed chemicals that, when released into the environment above a certain amount, must be reported.

hazardous waste: Waste materials or mixtures of waste that require special handling and disposal because of their potential to damage health and the environment.

hazardous wastestream: A gaseous or liquid wastestream that contains any type of hazardous substance.

heat: A condition of matter caused by the rapid movement of its molecules. Energy has to be applied to the material in sufficient amounts to create the motion, and may be applied by mechanical or chemical means.

heat balance: The constant trade-off that takes place when solar energy reaches the earth's surface and is absorbed, then must return to space to maintain the earth's normal heat balance.

heat islands: Large metropolitan areas where heat generated has an influence on the ambient temperature (adds heat) in and near the area.

heavy metals: A group of elements whose compounds are toxic to humans when found in the environment; examples are cadmium, mercury, copper, nickel, chromium, lead, zinc, and arsenic.

Henry's law: Governs the behavior of gases in contact with water.

heterotrophic: A category of organism that obtains its energy by consuming the tissue of other organisms.

heterotroph: See *heterotrophic*.

holoenzyme: A complete enzyme consisting of an apoenzyme and a coenzyme.

horizon: In soil, a layer of soil approximately parallel to the soil surface, differing in properties and characteristics from adjacent layers below or above it.

hot soak: Evaporative emissions from heat from an internal combustion engine after the engine is shut off.

hot-mix asphalt process: A remedial technology whereby a contaminant is entrained in soil in beneficial reuse to make asphalt. In the hot-mix process, the petroleum-laden soil is added as part of the aggregate to hot asphalt and then mixed to make the final product.

humidity: The amount of water vapor in a given volume of the atmosphere (absolute humidity), or the ration of the amount of water vapor in the atmosphere to the saturation value at the same temperature (relative humidity).

humus: That more or less stable fraction of the soil organic matter remaining after the major portions of added plant and animal residues are decomposed. Usually dark in color.

hydraulic gradient: The difference in hydraulic head divided by the distance along the fluid flow path. Groundwater moves through an aquifer in the direction of the hydraulic gradient.

hydrocarbon: A chemical containing only carbon and hydrogen atoms. Crude oil is a mixture largely of hydrocarbons.

hydrological cycle: The means by which water is circulated in the biosphere. Evapotranspiration from the landmass plus evaporation from the oceans is counterbalanced by cooling in the atmosphere and precipitation over both land and oceans.

hydrosphere: The portion of the earth's surface covered by the oceans, seas, and lakes.

hypha (pl. hyphae): In fungi, a tubular cell that grows from the tip and may form many branches.

hypolimnion: The cold, relatively dense bottom layer of water in a stratified lake.

ideal gas law: A hypothetical gas that obeys the gas laws exactly in regard to temperature, pressure, and volume relationships.

igneous: Rock formed by the cooling and solidification of hot, molten material.

ignitability: One of the characteristics used to classify a substance as hazardous.

impaction: In air pollution control technology, a particle collection process whereby the center of mass of a particle diverging from a fluid strikes a stationary object and is collected by the stationary object.

impoundment: A lake classification; an artificial, man-made lake made by trapping water from rivers and watersheds.

in situ biodegradation: Uses naturally occurring microorganisms in soil to degrade contaminants to another form.

in situ isolation/containment: In soil remediation, this method prevents the migration of liquid contaminant or leachates containing contaminants.

in situ leaching and chemical reaction: Soil remediation process whereby water mixed with a surfactant is used to leach contaminants from the soil into the groundwater. The groundwater is then collected downstream of the leaching site, through a collection system for treatment and/or disposal.

in situ passive remediation: The easiest to implement and least expensive remediation methodology because it involves no action at the site; it lets nature takes its course, but is not readily or normally accepted by regulators.

in situ technologies: Remedial technologies performed in place at the site.

in situ vitrification: Employs electrical current passed through electrodes driven into the soil, which produces extreme heat and converts soil into a durable glassy material. The organic constituents are pyrolized in the melt and migrate to the surface where they combust in the presence of oxygen. Inorganics in the soil are effectively bound in the solidified glass.

in situ volatilization: Commonly known as air stripping, this process uses forced air or drawn air currents through in place soil to remove volatile compounds.

incineration: The application of high temperatures (800—3000°F) to break down organic wastes into simpler forms and to reduce the volume of waste needing disposal. Energy can be recovered from incineration heat.

inclusion: Storage granules often seen within bacterial cells.

industrial practices: Practices that can lead to soil contamination including from USTs, oil field sites, chemical sites, geothermal sites, manufactured gas plants, mining sites, and environmental terrorism.

industrial wastewater: The liquid wastes produced by industry.

infiltration galleries: Technique used in in situ biodegradation process to reintroduce conditioned groundwater to the soil or aquifer.

infrared radiation: Invisible electromagnetic radiation of wavelength between about 0.75 micrometers and 1 mm—that is, between the limit of the red end of the visible spectrum and the shortest microwaves.

injection well: In groundwater remediation, used to raise the level of the water table and to push a contaminated plume away from a potable water system (well).

innovative cleanup technology: Any new or developing soil remediation technology.

inorganic substance: A substance that is mineral in origin and does not contain carbon compounds, except as carbonates, carbides, etc.

insolation: The amount of direct solar radiation incident per unit horizontal area at a given level.

institutional sources of MSW (municipal solid wastes): Wastes generated in hospitals, schools, jails and prisons, and government centers.

interception: In particle collection technology, interception occurs when the particle's center of mass closely misses the object, but, because of its finite size, the particle strikes the object, and is collected.

Interstate Commerce Clause: The clause in the U.S. Constitution which the federal government used to enact the Rivers and Harbors Act of 1988, enabling the U.S. Army Corp of Engineers to regulate and in some cases prohibit private and municipal dumping practices.

ionic bonds: A chemical bond in which electrons have been transferred from atoms of low ionization potential to atoms of high electron affinity.

irrigation: Artificial water supply for dry agricultural areas by means of dams and channels.

isobar: A line drawn on maps and weather charts linking all places with the same atmospheric pressure (usually measured in millibars).

jet stream: A narrow band of very fast wind found at altitudes of 6-10 miles in the upper troposphere or lower stratosphere.

Kelvin: Temperature scale used by scientists. It begins at absolute zero and increases by the same degree intervals as the Celsius scale; that is, 0°C is the same as 273K and 100°C is 373K.

Krebs Cycle or *citric acid cycle:* The final part of the chain of biochemical reactions by which organisms break down food using oxygen to release energy (respiration).

land farming: Another name for land treatment whereby various contaminants are spread on soil and worked into the surface and subsurface to allow biodegradation to take place.

land treatment: See *land farming.*

landfilling: An ultimate disposal technique whereby solid and hazardous wastes are disposed of in excavated sites.

landfill: land waste disposal site located without regard to possible pollution of groundwater and surface water resulting from runoff and leaching; waste is covered intermittently with a layer of earth to reduce scavenger, aesthetic, disease, and air pollution problems.

lapse rate: The rate of change of air temperature with increasing height.

latent heat of fusion: The amount of heat required to change one gram of a substance from the solid to the liquid phase at the same temperature.

latent heat of vaporization: The amount of heat required to change one gram of a substance from the liquid to the gas phase at the same temperature.

law of conservation of mass: In any ordinary physical or chemical change, matter is neither created nor destroyed but merely changed from one form to another.

laxative effect: The consumption of hard water combined in the presence of magnesium sulfates sometimes leads to development of laxative effect on new consumers.

leach liquors: Liquid leached from a substance via water circulation through or over it.

leachate: The liquid formed when rainwater percolates downward through landfilled wastes, picking up contaminants that might then enter the surrounding environment.

lead: A heavy metal, the accumulation of which in organic tissue can produce, in animals and humans, behavioral changes, blindness, and ultimately death.

lead-mine scale: Generally occurs in geothermal process equipment such as in piping where scale-buildup leads to process equipment failure.

Leaking Underground Storage Tanks (LUST): The 1986 U.S. UST cleanup fund.

lentic (calm waters): Lakes, ponds, and swamps.

limited: Limiting nutrients such as carbon, nitrogen, and phosphorous.

limiting factor: Factors such as temperature, light, water, or a chemical limit the existence, growth, abundance, or distribution of an organism.

limiting nutrient: See *limited.*

limnetic: The open water surface layer of a lake through which there is sufficient sunlight for photosynthesis.

limnology: The study of lakes and other bodies of open fresh water in terms of their plant and animal biology and their physical properties.

liquid: A state of matter between a solid and a gas.

liter: A metric unit of volume, equal to one cubic decimeter (1.76 pints).

lithosphere: The earth's crust—that is, the layers of soil and rock that comprise the earth's crust.

litter: The intact and partially decayed organic matter lying on top of the soil; discards thrown about without regard to the environment.

littoral: The shallow zone of waters near the shore of a body of water.

LNAPLs: Light Nonaqueous-Phase Liquids such as gasoline, heating oil, and kerosene.

loam: The textural-class name for soil having a moderate amount of sand, silt, and clay. Loam soils contain 7-27% clay, 28-50% silt, and 23-52% sand.

lotic: Running fresh water systems, e.g., rivers, streams, etc.

magma: The molten rock material within the earth's core.

management factors: The management techniques (fertilization, irrigation, etc.) employed in land and soil management that work to decrease leaching, increase soil surface contaminant concentrations, or maximize soil aeration against volatilization.

manifest: See *tracking system.*

mass: The quantity of matter and a measurement of the amount of inertia that a body possesses.

mass balance equations: Used to track pollutants from one place to another.

materials balance: The law of conservation of mass/matter, which says that matter has to go somewhere, but is neither created nor destroyed in the process.

mature pond: A pond that reaches maturity, characterized by being carpeted with rich sediment, with aquatic vegetation extending out into open water and a great diversity of plankton, invertebrates, and fishes.

maximum contaminant levels (MCLs): Primary drinking water standard and maximum contaminant levels allowed based on health related criteria.

maximum sustainable yield: The highest rate at which a renewable resource can be used without impairing or damaging its ability to be fully renewed.

melting point: The temperature at which a substance changes from solid to liquid.

meromictic: Chemically stratified lakes in which different dissolved chemicals are partly mixed.

mesosome: A common intracellular structure found in the bacterial cytoplasm; an invagination of the plasma membrane in the shape of tubules, vesicles, or lamellae.

mesosphere: An atmospheric layer that extends from the top of the stratosphere to about 56 miles above the earth.

mesotrophic lake: A term used to distinguish between an oligotrophic and eutrophic lake.

metabolic transformation: The assembly-line like activities that occur in microorganisms during the processing of raw materials into finished products.

metabolism: The chemical processes of living organisms; a constant alternation of building up and breaking down. For example, green plants build up complex organic substances from water, carbon dioxide, and mineral salts (photosynthesis); by digestion, animals partially break down complex organic substances, ingested as food, and subsequently resynthesize them in their own bodies.

metalloids: An element that exhibits the properties of both metals and nonmetals.

metals: Elements that tend to lose their valence electrons.

metamorphic: A type of rock that forms when rocks lying deep below the earth's surface are heated to such a degree that their original crystal structure is lost. As the rock cools, a new crystalline structure is formed.

meteorology: The scientific observation and study of the atmosphere to help accurately forecast weather.

meter: The standard of length in the metric system, equal to 39.37 inches or 3.28 feet.

methane: (CH4) The simplest hydrocarbon of the paraffin series. Colorless, odorless, and lighter than air, it burns with a bluish flame and explodes when mixed with air or oxygen. Methane is a greenhouse gas.

microbial community: The community of microbes available to biodegrade contaminants in the soil.

microbial degradation: The natural process whereby certain microbes in soil can degrade contaminants into harmless constituents.

microbiology: The study of organisms that can only be seen under the microscope.

middens: Primitive dunghills or refuse heaps.

midnight dumping: The illegal dumping of solid or hazardous wastes into the environment.

mining waste: The earth and rock (including minerals and/or chemicals within) from a mine, discarded because the mineral or fuel content is too low to warrant extraction. This waste may become an environmental problem if toxic substances are leached from it into a river, stream, groundwater, or the soil.

mitochondria: A microscopic body found in the cells of almost all living organisms and containing enzymes responsible for the conversion of food to usable energy.

mixture: In chemistry, a substance containing two or more compounds that still retain their separate physical and chemical properties.

mobile sources: Non-stationary sources of gaseous pollutants, such as locomotives, automobiles, ships, and airplanes.

mobilization: The mobilizing of metals in soil by the acidity of precipitation.

modeling: The use of mathematical representations of contaminant dispersion and transformation to estimate ambient pollutant concentrations.

molar concentration (molarity): In chemistry, a solution that contains one mole of a substance per liter of solvent.

mole: SI unit (symbol mol) of the amount of a substance. The amount of a substance that contains as many elementary entities as there are atoms in 12 g of the isotope carbon-12.

molecular weight: The weight of one molecule of a substance relative to 12C, expressed in grams.

molecule: The fundamental particle that characterizes a compound. It consists of a group of atoms held together by chemical bonds.

monitor wells: Installed wells specifically designed to provide a means to monitor a contaminant plume in soil/groundwater.

monitoring: Process whereby a contaminant is tracked.

Montreal Protocol: Required signatory countries to reduce their consumption of CFCs by 20% by 1993, and by 50% by 1998.

morphogenesis: Evolutionary development of the structure of an organism or part.

motility: An organism's mobility; ability to move.

municipal services sources of MSW (municipal solid wastes): Wastes generated in restaurants, hotels, stores, motels, service stations, repair shops, markets, office buildings, and print shops.

municipal solid wastes (MSW): Consists of municipally derived wastes including paper, yard wastes, glass and metals, and plastics.

mycelium: An interwoven mass of threadlike filaments or hyphae forming the main body of most fungi. The reproductive structures, or "fruiting bodies," grown from the mycelium.

mycology: The branch of botany that deals with fungi.

National Ambient Air-Quality Standards: (NAAQS) Established by the EPA at two levels: Primary and Secondary. Primary standards are required to be set at levels that will protect public health and include an "adequate margin of safety," regardless of whether the standards are economically or technologically achievable. Primary standards must protect even the most sensitive individuals, including the elderly and those with respiratory ailments. Secondary air quality standard are meant to be even more stringent than primary standards. Secondary standards are established to protect public welfare (e.g., structures, crops, animal, fabrics).

National Priorities List (NPL): The NPL identifies the worst waste sites in the nation based on such factors as the quantities and toxicity of wastes involved, the exposure pathways, the number of people potentially exposed, and the importance and vulnerability of the underlying groundwater.

nektons: In a water environment, the free-swimming organisms.

neustons: In a water environment, the organisms living on the surface.

neutrally stable atmosphere: An intermediate class between stable and unstable conditions. Will cause a smoke-stack plume to cone in appearance as the edges of the plume spread out in a V-shape.

neutron: Elementary particles that have approximately the same mass as protons but have no charge. They are one constituent of the atomic nucleus.

niche: The functional role of an organism within its community. It is the complete ecological description of an individual species (including habitat, feeding requirements, etc.).

nitrates: In fresh water pollution, a nutrient (usually from fertilizer) that enters the water system and can be toxic to animals and humans in high concentrations.

nitrification: The process that takes place in soil when bacteria oxidize ammonia, turning it into nitrates.

nitrogen cycle: The natural circulation of nitrogen through the environment.

nitrogen dioxide (NO_2): A reddish-brown, highly toxic gas with a pungent odor. One of the seven known nitrogen oxides that participates in photochemical smog and primarily affects the respiratory system.

nitrogen fixation: Nature accomplishes nitrogen fixation by means of nitrogen-fixing bacteria.

nitrogen oxide (NO): A colorless gas used as an anaesthetic; it is mainly formed by soil bacteria in decomposing nitrogenous material.

non-in situ technology: Remediation/recycling technology that takes place away from the contamination site.

noncarbonate hardness: A property of water where the hardness cannot be removed by boiling and is classified as permanent.

nonmetals: An element that tends to gain electrons to complete its outer shell.

nonpoint source: Source of pollution in which wastes are not released at one specific, identifiable point but from a number of points that are spread out and difficult to identify and control.

nonpoint source pollution: Pollution that cannot be traced to a specific source but rather comes from multiple generalized sources.

nonrenewable resources: Resources that exist in finite supply or are consumed at a rate faster than the rate at which they can be renewed.

nonspecific source wastes: Generic wastes commonly produced by manufacturing and industrial processes, e.g., spent solvents.

nonvolatile: A substance that does not evaporate at normal temperatures when exposed to the air.

normal lapse rate: The rate of temperature change with height is called the lapse rate. On average, temperature decreases - 65°C/100m or -6.5°C/km, which is the normal lapse rate.

nucleoid: The primitive nuclear region of the procaryotic cell.

nutrient cycles: See *biogeochemical cycles.*

nutrients: Elements or compounds needed for the survival, growth, and reproduction of a plant or animal.

nutrition: The process of nourishing or being nourished.

oligotrophic lake: A lake with a low supply of plant nutrients.

organelle: A specialized part of a cell that resembles and functions as an organ.

organic chemistry: The branch of chemistry concerned with compounds of carbon.

organic matter: Includes both natural and synthetic molecules containing carbon, and usually hydrogen. All living matter is made up of organic molecules.

organic substance: Any substance containing carbon.

overgrazing: Consumption of vegetation on rangeland by grazing animals to the point that the vegetation cannot be renewed or is renewed at a rate slower than it is consumed.

oxidation: The process in which electrons are lost.

oxidation-reduction: The (redox) process where electrons are lost and gained.

oxidize: To combine with oxygen.

oxygen: An element that readily unites with materials.

oxygen sag curve: The oxygen content in a stream or river system after organic pollution is introduced into the water body; organic pollution causes a profusion in growth of organisms that tends to decrease the amount (sag) of oxygen available.

ozone: The compound O_3. Found naturally in the atmosphere in the ozonosphere, a constituent of photochemical smog.

ozone holes: Holes created in the ozone layer because of chemicals, especially CFCs.

packed tower: A remediation method (scrubber) employed to clean a contaminated gaseous wastestream by exposing the wastestream to biological media or chemical scrubbing agents.

parasite: Primary or secondary (or higher) consumer that feeds over an extended period of time on a plant or animal known as a host.

parent material: The unconsolidated and more-or-less chemically weathered mineral or organic matter from which the solum of soils is developed by pedogonic processes.

particulate matter: Normally refers to dust and fumes; travels easily through air.

pascal (Pa): A unit of pressure equal to one newton per square meter.

pathogen: Any disease-producing organism.

pedologist: A person who studies soils.

peds: A unit of soil structure such as an aggregate, crumb, prism, block, or granule, formed by natural processes.

pellicle: A *Euglena* structure that allows for turning and flexing of the cell.

period: An interval of geologic time that is a subdivision of an era and made up of epochs; a horizontal row of the periodic table, elements that have approximately the same energy.

periodic law: The properties of elements are periodic functions of the atomic number.

periodic table: A list of all elements arranged in order of increasing atomic numbers and grouped by similar physical and chemical characteristics into periods; based on the chemical law that physical or chemical properties of the elements are periodic functions of their atomic weights.

permanent pond: Actually a misnomer because there are no "permanent" ponds. The designation is used anyway to describe a pond that is shallow enough to permit aquatic plants to penetrate the surface anywhere over its entire mass; its mass is not so great as to allow formation of large waves that could erode the shoreline; it has no temperature layering, rather a gradient of temperatures extending from the surface to bottom.

permitting system: A key element of RCRA is a permitting system that works to ensure safe operation of facilities that treat, store, or dispose of hazardous wastes.

perpetual resource: A resource such as solar energy that comes from an essentially inexhaustible source and thus will always be available on a human time scale regardless of whether or how it is used.

persistent substance: Refers to a chemical product that has a tendency to persist in the environment for quite some time, e.g., plastics.

pesticide: Any chemical designed to kill weeds, insects, fungi, rodents, and other organisms that humans consider to be undesirable.

pH: A numerical designation of relative acidity and alkalinity; a pH of 7.0 indicates precise neutrality, high values indicate increasing alkalinity and lower values indicate increasing acidity.

phosphates: A nutrient substance obtained from fertilizers.

phosphorous cycle: A biogeochemical cycle in which phosphorous is converted into various chemical forms and transported through the biosphere.

photochemical reaction: A reaction induced by the presence of light.

photochemical smog: A complex mixture of air pollutants produced in atmosphere by the reaction of hydrocarbons and nitrogen oxides under the influence of sunlight.

photosynthesis: A complex process that occurs in the cells of green plants whereby radiant energy from the sun is used to combine carbon dioxide (CO_2) and water (H_2O) to produce oxygen (O_2) and simple sugar or food molecules such as glucose.

physical change: The process that alters one or more physical properties of an element or compound without altering its chemical composition. Examples include changing the size and shape of a sample of matter and changing a sample of matter from one physical state to another.

physical weathering: The physical changes produced in rocks by atmospheric agents (wind, precipitation, heat, cold, etc.).

pioneer community: The first successfully integrated set of plants, animals, and decomposers found in an area undergoing primary ecological succession.

piping failure: A common equipment component failure in many different systems; the most common cause of UST spills.

plankton: Microscopic floating plant and animal organisms of lakes, rivers, and oceans.

planktonic: See *plankton.*

plasma membrane: See *cytoplasmic membrane.*

plate tower: In absorption scrubbing, plate towers contain perforated horizontal plates or trays designed to provide large liquid-gas interfacial area. The polluted air stream rises up through the perforations in each plate; the rising gas prevents liquid from draining through the openings rather than through a downpipe. During continuous operation, contact is maintained between air and liquid, allowing gaseous contaminants to be removed, with clean air emerging from the top of the tower.

plume: (1) The column of non-combustible products emitted from a fire or smokestack. (2) A vapor cloud formation having shape and buoyancy. (3) A contaminant formation dispersing through the subsurface.

point source: Discernable conduits such as pipes, ditches, channels, sewers, tunnels, or vessels from which pollutants are discharged.

point source pollution: Pollution that can be traced to an identifiable source.

pollute: To impair the quality of some portion of the environment by the addition of harmful impurities.

pond: A still body of water smaller than a lake, often of artificial construction.

pond succession: Pond transformation process whereby a young pond is formed, develops over time to a mature pond, then to a senescent (old) pond.

pool zone: In a body of moving water (river or stream), the quiet or still water portion.

positive crankcase ventilation (PCV): Technology used to control crankcase emissions.

preliminary treatment: (1) In wastewater, involves treatment prior to primary treatment. (2) In industrial applications, pretreatment of the wastestream before it becomes plant effluent and then influent into wastewater treatment plant for further treatment.

pressure: Force per unit area.

pressure gradient force: A variation of pressure with position.

primary consumers: In the food chain, organisms that consume producers (autotrophs).

primary exposure pathways: In site remediation, the exposure pathways that directly affect site operations and personnel or directly affect cleanup levels that must be achieved by the remedial technology.

primary pollutants: Pollutants emitted directly into the atmosphere, where they exert an adverse influence on human health or the environment. The six primary pollutants are carbon dioxide, carbon monoxide, sulfur oxides, nitrogen oxides, hydrocarbons, and particulates. All but carbon dioxide are regulated in the U.S.

primary standards: The Clean Air Act (NAAQS) air quality standards covering criteria pollutants.

primary treatment: Wastewater treatment process in which mechanical treatment is employed to screen out large solids and settle out suspended solids.

procaryotic: A type of primitive cell lacking a membrane-delimited nucleus.

producers: Organisms that use solar energy (green plant) or chemical energy (some bacteria) to manufacture their own organic substances (food) from inorganic nutrients.

profundal: Deep-water zone of a lake, not penetrated by sunlight.

proton: A component of a nucleus, 2000 times more massive than an electron; differs from a neutron by its positive (+1) electrical charge. The atomic number of an atom is equal to the number of protons in its nucleus.

protozoa: Single-celled microorganisms; includes the most primitive forms of animal life.

pumping well system: In control technology for leaking USTs, the preferred method used to recover free product from the water table when the spill is deep.

radiation: The emitting of energy from an atom in the form of particles of electromagnetic waves; energy waves that travel with the speed of light, and upon arrival at a surface are either absorbed, reflected, or transmitted. Not absorbed by air and traveling in straight lines, it is the most serious cause of fire spreading from one structure to another.

radiative inversions: In temperature inversions, a nocturnal phenomenon caused by cooling of the earth's surface. Inversions prompt the formation of fog and simultaneously trap gases and particulates, creating a concentration of pollutants.

radioactive material: Any material that spontaneously emits ionizing radiation.

rapids zone: The turbulent zone of a stream or river in which water is agitated by subsurface obstructions, causing turbulence and aeration of water.

reactive: The tendency of a material to react chemically with other substances.

recharge area: The area in which precipitation percolates through to recharge groundwater.

recovery zone: The zone in a stream or river where contamination is reduced by the self-purification process.

recycle: The process of recovery and reuse of materials from waste streams.

recycling: See *recycle*.

recycling technology: The technology available to recycle or reuse waste products; processes such as composting and hot- and cold-mix asphalt incorporation.

reduction: Removal of oxygen from a compound; lowering of oxidation number resulting from a gain of electrons.

refuse: Rubbish and garbage; residential sources.

relative humidity: The percentage of moisture in a given volume of air at a given temperature in relation to the amount of moisture the same volume of air would contain at the saturation point.

renewable resources: Resources that can be depleted in the short run if used or contaminated too rapidly, but that normally will be replaced through natural processes.

representative sample: A sample of a universe or whole, such as a waste pile, lagoon, or groundwater, which can be expected to exhibit the average properties of the whole.

reservoir: A large and deep human-created standing body of freshwater.

residential sources of MSW (municipal solid wastes): Municipal solid wastes from households consisting primarily of paper, glass, vegetable waste, and paperboard, ash, tin cans, etc.

Resource Conservation and Recovery Act (RCRA): 1976 act passed by Congress to control dumping of waste materials—cradle-to-grave.

resource: Something that serves a need, is useful, and is available at a particular cost.

reuse: To use a product again and again in the same form, as when returnable glass bottles are washed and refilled.

ribosomes: In bacterial cytoplasm, minute, rounded bodies made of RNA and loosely attached to the plasma membrane; they are the site of protein synthesis and are part of the translation process.

risk assessment: Evaluation of the threat to public health and the environment posed by a hazardous waste facility; considering the probability of an incident and its effects.

risk characterization: Final step in risk assessment process whereby an estimate of the incidence of an adverse health effect under the conditions of exposure found and described in the exposure assessment is determined.

Rivers and Harbors Act (1899): Initiated the first legislative authority given to a federal agency (U.S. Army Corps of Engineers) to prevent dumping wastes into rivers and harbors.

rotifer: A minute multicellular aquatic organism having at the anterior end a wheellike ring of cilia.

rubbish: Consists of combustible waste, paper, cartons, rags, wood scraps, combustible floor sweepings; domestic, commercial, and industrial sources.

running losses: Evaporative emissions from internal combustion engines as a result of driving; losses also occur when the fuel is heated by the road surface, and when fuel is forced from the fuel tank while the vehicle is being operated and the fuel tank becomes hot.

runoff: Surface water entering rivers, freshwater lakes, or reservoirs from land surfaces.

Safe Drinking Water Act (SDWA): Mandated the EPA to establish drinking-water standards for all public water systems serving 25 or more people, or having 15 or more connections.

saline water: Water with excessive salt content.

salt spreading: The practice of spreading salt on roadways during winter to help reduce ice and snow accumulation; road salts contaminate soil during runoff.

sanitary landfill: A method of solid waste disposal designed to minimize water pollution from runoff and leaching; waste is covered with a layer of soil within a day after being deposited at the landfill site.

sanitary wastewater: Separate sewer system designed to remove domestic wastes from residential areas.

saprophyte: An organism that uses enzymes to feed on waste products of living organisms or tissues of dead organisms.

SARA: Superfund Amendments and Reauthorization Act of 1986; see *CERCLA.*

saturated zone: Subsurface soil saturated with water; the water table.

scaling: When carbonate hard water is heated and calcium carbonate and magnesium hydroxide are precipitated out of solution, they form a rock-hard scale that clogs hot-water pipes and reduces the efficiency of boilers, water heaters, and heat exchangers.

science: The observation, identification, description, experimental investigation, and theoretical explanation of natural phenomena.

scientific method: A systematic form of inquiry that involves observation, speculation, and reasoning.

sea level rise: The natural rise of sea level that occurs in cyclical patterns throughout history; currently may be affected by or the result of human impact on global warming.

second law of thermodynamics: Natural law dictating that in any conversion of heat energy to useful work, some of the initial energy input is always degraded to a lower-quality, more dispersed, less useful form of energy, usually low-temperature heat that flows into the environment; you can't break even in terms of energy quality.

secondary drinking water standards: The unenforceable guidelines based on both aesthetics such as taste, odor, and color of drinking water, as well as nonaesthetic characteristics such as corrosivity and hardness.

secondary exposure pathways: In on-site remediation, occurs as a minor component during site operations and exhibits significant decreases with time as treatment progresses (e.g., wind-blown dust).

secondary standards: Refers to NAAQS requirement to protect public welfare.

secondary treatment (of sewage): The removal of impurities from water by the digestive action of various small organisms in the presence of air or oxygen.

secure landfill: A land site for the storage of hazardous solid and liquid wastes normally placed in containers and buried in a restricted-access area that is continually monitored. Such landfills are located above geologic strata that are supposed to prevent the leaching of wastes into groundwater.

sedimentary (rock form): A rock formed from materials deposited from suspension, or precipitated from solution and usually being more or less consolidated. The principal sedimentary rocks are sandstones, shales, limestones, and conglomerates.

sediments: Soil particles dislodged by rain drops that travel via runoff into streams, rivers, lakes, or oceans and are deposited there as sediments.

self-purification: The natural phenomenon occurring in running water systems (streams and rivers) whereby physical, chemical, and biological processes work to purify the water.

senescent pond: A pond that has reached old age.

separation: A hazardous waste treatment technology (filtration and separation) whereby filtration is used to separate solid particles from a liquid stream through use of semi-porous media; driven by a pressure difference across the media, and caused by gravity, centrifugal force, vacuum, or elevated pressure.

septic zone: In the self-purification process that takes place in running water bodies (streams or rivers), the zone characterized by heavy organic pollution and low DO levels.

sheet piling: In in situ isolation/containment technology, the physical driving of rigid sheets or pilings of wood, steel, or concrete into the ground to form a barrier for containment.

silage liquor: The liquid drained or leached from fodder prepared by storing and fermenting green forage plants in a silo.

slope: A soil property in which the steepness of the soil layer is directly related to the degree of erosion that may occur.

slope wind: Winds that move through a typical river valley; they flow downhill into the valley floor.

slurry walls: In in situ isolation/containment, fixed underground physical barriers formed in an excavated trench by pumping slurry, usually a bentonite or cement and water mixture.

smog: Visible air pollution; a dense, discolored haze containing large quantities of soot, ash, and gaseous pollutants such as sulfur dioxide and carbon dioxide.

soft water: Water with a hardness of less than 50 ppms.

soil: A dynamic natural body in which plants grow, composed of mineral and organic materials and living forms.

soil boring: Using a boring tool (such as an auger) to take soil samples for analysis.

soil factor: In in situ soil remediation, soil factors include water content, porosity/permeability, clay content, and adsorption site density.

soil fertility: The quality of a soil that enables it to provide essential chemical elements in quantities and proportions for the growth of specified plants.

soil forming process: The mode of origin of the soil, with special reference to the processes or soil-forming factors responsible for the development of the solum, or true soil, from the unconsolidated parent material.

soil horizon: A layer of soil, approximately parallel to the soil surface, differing in properties and characteristics from adjacent layers below or above it.

soil pollution: Contamination of the soil and subsurface by the addition of contaminants or pollutants.

soil profile: A vertical section of the soil from the surface through all its horizons, including C horizons.

soil remediation: The use of various techniques and/or technologies to decontaminate or dispose contaminated soil.

soil sampling: Conducted to determine through analysis the type, texture, and structure of a soil; used to collect samples of contaminated soil to determine degree and extent of contamination and for analysis.

soil structure: The combination or arrangement of primary soil particles into secondary particles, units, or peds. These secondary units may be, but usually are not, arranged in the profile in such a manner as to give a distinctive characteristic or pattern. The secondary units are characterized and classified on the basis of size, shape, and degree of distinctness into classes, types, and grades, respectively.

soil texture: The relative proportions of the various soil separates in a soil.

soil washing and extraction: In pollution control technology used for USTs, used to leach contaminants from the soil into a leaching medium, after which the extracted contaminants are removed by conventional methods.

Solid Waste Disposal Act (1965): This act was the first major step taken by U.S. legislators to promote (among other things) the demonstration, construction, and application of solid waste management and resource recovery systems to preserve and enhance the quality of air, water, and land resources.

solid: Matter that has a definite volume and a definite shape.

solid waste: Any normally solid material resulting from human or animal activities that is useless or unwanted.

solid waste stream: A stream of solid waste materials as a whole.

solidification: A stabilization technique used to convert hazardous waste from its original from to a physically and chemically more stable material. This is accomplished by reducing the mobility of hazardous compounds in the waste prior to its land disposal.

solidification/stabilization: See *solidification.*

solubility: The ability of a substance to mix with water.

solute: The dissolved substance in a solution.

solvent: The substance in excess in a solution.

sorption: Process of adsorption or absorption of a substance on or in another substance.

species: A group of individuals or populations potentially able to interbreed and unable to produce fertile offspring by breeding with other sorts of animals and plants.

specific gravity: Ratio of the weight of the volume of liquid or solid to the weight of an equal volume of water.

specific heat: The amount of heat energy in calories necessary to raise the temperature of one gram of the substance one degree Celsius.

specific source wastes: Wastes from specifically identified industries such as wood preserving, petroleum refining, and organic chemical manufacturing. Typically includes sludges, still bottoms, wastewaters, spent catalysts, and residues.

spirilla: Bacteria shape characterized as being nonflexible helical and curved.

spoil: Material removed from an excavation.

sporangiospore: Spores that form within a sac called a sporangium. The sporangia are attached to stalks called sporangiophores.

spore: Reproductive stage of fungi.

spring overturn: The lake phenomenon whereby the entire body of water within the lake overturns due to changes in water density.

stability: Atmospheric turbulence; a function of vertical distribution of atmospheric temperature.

stability class: Term used to classify the degree of turbulence in the atmosphere.

stabilization: See *Solidification/Stabilization.*

stable atmosphere: Marked by air that is cooler at the ground than aloft, by low wind speeds, and consequently by a low degree of turbulence.

standard temperature & pressure (STP): As the density of gases depends on temperature and pressure, it is customary to define the pressure and temperature against which the volume of gases are measured. The normal reference point is standard temperature and pressure 0°C at a standard atmosphere of 760 millimeters of mercury. All gas volumes are referred to at these standard conditions.

stationary sources: Source of air pollution emanating from any fixed or stationary point.

still bottoms: What remains after a spent solvent is distilled (for recycling); composed of a concentrated, highly toxic mixture, far reduced in volume.

stockpile: Certain chemical products (such as road salt) kept in quantity for possible use, runoff from which may contribute to soil pollution.

stormwater: Normal stormwater containing grit and street debris but no domestic or sanitary wastes.

stratification: Temperature-density relationship of water in temperate lakes (>25 ft in depth) that leads to stratification and subsequent turnover or overturn.

stratosphere: A region of the atmosphere based on temperature between approximately 10 and 35 miles in altitude.

stripping: A waste control technology whereby volatile compounds are separated from less volatile ones in a liquid mixture by the partitioning the more volatile materials to a gas phase of air or steam.

subadiabatic: The ambient lapse rate when it is less than the dry adiabatic lapse rate.

submerged vegetation: In a pond, the submerged plants that grow where light can penetrate the water surface and reach them.

subsidence inversion: A type of inversion usually associated with a high pressure system, known as anticyclones, which may significantly affect the dispersion of pollutants over large regions.

subsoil: That part of the soil below the plow layer.

substrate: The material or substance upon which an enzyme acts.

suggested levels: Nonenforceable guidelines for secondary drinking water standards, regarding public welfare.

sulfur cycle: The natural circulation of sulfur through the environment.

sulfur dioxide: A primary pollutant originating chiefly from the combustion of high-sulfur coals.

sulfurous smog: The haze that develops in the atmosphere when molecules of sulfuric acid accumulate, growing in size as droplets until they become sufficiently large to serve as light scatterers.

summer stagnation: In lake stratification, a state that occurs in some lakes when the top layer of water is warmer than the bottom layer. It results in layers of different density, the top light, the bottom heavy. With increased temperature, the top layer becomes even lighter, and a thermocline forms. From top to bottom, we now have the lightest and warmest on top, medium weight and relatively warm in the middle, and the heaviest and coldest below, with a sharp drop in temperature at the thermocline. There is no circulation of water in these three layers. If the thermocline is below the range of effective light penetration, the oxygen supply becomes depleted in the hypolimnion, since both photosynthesis and the surface source of oxygen are cut off.

superadiabatic: Describes the lapse rate when a parcel of air starting at 1000m at 20°C, for example, starts moving downward and becomes cooler and denser than its surroundings. Because the ambient air is unstable, it will continue to sink.

Superfund Law: See *CERCLA and SARA.*

Superfund: See *CERCLA and SARA.*

surface condenser: In air pollution control technology, a type of condensation equipment, normally a shell-and-tube heat exchanger. It uses a cooling medium of air or water where the vapor to be condensed is separated from the cooling medium by a metal wall. Coolant flows through the tubes, while the vapor is passed over and condenses on the outside of the tubes, and drains off to storage.

surface impoundment: (1) Another name for a garbage dump. (2) Diked or excavated areas used to store liquid hazardous wastes.

surface origins: The surface origins of soil contaminants that include gaseous and airborne particulates; infiltration of contaminated surface water; land disposal of solid and liquid waste materials; stockpile, tailings, and spoil; dumps; salt spreading on roads; animal feedlots; fertilizers and pesticides; accidental spills; and composting of leaves and other wastes.

surface water: Water that is on the earth's surface and exposed to the atmosphere and that is mostly the product of precipitation.

surfactant: A surface-active substance (soap).

symbiotic: A close relationship between two organisms of different species; one where both partners benefit from the association.

synthesis: The formation of a substance or compound from more elementary compounds.

tailings: The residual fine-grained waste rejected after mining and processing of ore, usually after washing.

taste and odor: A water quality parameter.

TCLP (Toxicity Characteristics Leaching Procedure): Replaced the EP Toxicity Test; it is designed to identify wastes likely to leach hazardous concentrations of particular toxic constituents into the surrounding soils or groundwater.

temperature: A measure of the average kinetic energy of the molecules.

temperature inversion: A condition characterized by an inverted lapse rate.

thermal circulation: The result of the relationship based on a law of physics whereby the pressure and volume of a gas are directly related to its temperature.

thermal incinerator (or afterburner): A device used in combustion whereby the contaminant airstream passes around or through a burner and into a refractory-lined residence chamber where oxidation occurs. Flue gas from a thermal incinerator is at high temperature and contains recoverable heat energy.

thermal inversion: A layer of cool air trapped under a layer of less dense warm air, thus preventing reversing the normal situation.

thermal Nox: Created when nitrogen and oxygen in the combustion air (e.g., within an internal combustion engine) are heated to a high temperature (above 1000K) to cause nitrogen (O_2) and oxygen (O_2) in the air to combine.

thermal pollution: Increase in water temperature with harmful ecological effects on aquatic ecosystems.

thermal radiation: Heat energy directly radiated into space from the earth's surface and atmosphere.

thermal treatment: In non-in situ soil pollution control technology, the complete destruction (by incineration) of petroleum-laden contaminants.

thermal treatment processes: In waste control technology, incineration of wastes.

thermocline: The fairly thin transition zone in a lake that separates an upper, warmer zone from a lower, colder zone.

thermosphere: A region of the atmosphere based on temperature between approximately sixty to several hundred miles in altitude.

threshold of effect: In the dose-response relationship, the level of "no effect."

threshold reporting quantity: Level set by the EPA for extremely hazardous substances that if exceeded during spill or release to environment must be reported to appropriate authorities.

tilth: The physical condition of soil as related to its ease of tillage, fitness as a seedbed, and its impedance to seedling emergence and root penetration.

topsoil: The layer of soil moved in cultivation.

Total Kjeldahl Nitrogen (TKN): The total concentration of organic and ammonia nitrogen in wastewater.

total dissolved solids: The solids residue after evaporating a sample of water or effluent; expressed in mg/liter.

toxic chemical: Term used by the EPA for chemicals whose total emissions or releases must be reported annually by owners and operators of certain facilities that manufacture, process, or otherwise use a listed toxic chemical. The list of toxic chemicals is identified in Title III of SARA.

toxic metals: Metals such as arsenic, cadmium, lead, and mercury that are all cumulative toxins and particularly hazardous to human health.

toxic or hazardous substance: Substances injurious to the health of individual organisms and sometimes fatal.

toxicity: The degree of poisonousness.

toxicological evaluation: Part of risk assessment that should answer the question: Does the chemical have an adverse effect?

toxin: A poison produced by a plant or animal.

tracking system: In hazardous waste management, a manifest document that accompanies any waste transported from one location to another.

transformation: Chemical transformations that takes place in the atmosphere, for example, the conversion of the original pollutant to a secondary pollutant such as ozone.

trash: Highly combustible waste paper, wood, cardboard cartons, including up to 10% treated papers, plastic, or rubber scraps.

treatment plant site sources of MSW (municipal solid wastes): Wastes generated in water, wastewater, and other industrial treatment processes (e.g., incineration ash, sludges or biosolids, and general plant wastes).

trench method: In pollution control technology for USTs, a method used to capture the entire leading edge of the contaminant plume.

trenching: See *excavation.*

trophic level: The feeding position occupied by a given organism in a food chain, measured by the number of steps removed from the producers.

troposphere: A region of the atmosphere based on temperature between the earth's surface and 10 miles in altitude.

turbidity: Reduced transparency of the atmosphere, caused by absorption and scattering of radiation by solid or liquid particles other than clouds, and held in suspension.

turbulence: (1) Uncoordinated movements and a state of continuous change in liquids and gases; (2) one of the 3 Ts of combustion.

turnover: The mixing of the upper and lower levels of a lake, which most often occurs during the spring and fall, caused by dramatic changes in surface water temperature.

unconfined aquifer: An aquifer not underlain by an impermeable layer.

underground storage tanks (USTs): Underground tanks designed to store chemicals, especially fuels.

United Nations Hazard Class Number System: A system used to designate and label hazardous materials using a dedicated number system.

unsaturated zone: Lies just beneath the soil surface and is characterized by crevices that contain both air and water; water contained therein is not available for use.

unstable atmosphere: Characterized by a high degree of turbulence.

vacuole: A small cavity in the protoplasm of a cell.

vadose water: Water in the unsaturated zone which is essentially unavailable for use.

valence: The net electric charge of an atom or the number of electrons an atom can give up (or acquire) to achieve a filled out shell.

valley winds: At valley floor level, slope winds transform into valley winds which flow down-valley, often with the flow of a river.

venting: In pollution control technology, a method of remediating hydrocarbon (gasoline) spills or leaks from USTs.

Venturi: A short tube with a constricted throat used to determine fluid pressures and velocities by measurement of differential pressures generated at the throat as a fluid traverses the tube.

vernal ponds: Spring ponds, usually of short duration.

virus: An infectious agent with a simple acellular organization, a protein coat, a single type of nucleic acid, and reproducing only within living host cells.

volatile organic compounds (VOCs): Organic compounds that evaporate and contribute to air pollution directly or through chemical or photochemical reactions to produce secondary pollutants, principally ozone.

volatile: When a substance (usually a liquid) evaporates at ordinary temperatures if exposed to the air.

volatilization: When a solid or liquid substance passes into the vapor state.

volume: Surface area multiplied by a third dimension.

warm front: Marks the advance of a warm air mass as it rises up over a cold one.

waste minimization: An umbrella term that refers to industrial practices that minimize the volume of products, minimize packaging, extend the useful life of products, and minimize the amount of toxic substance in products.

waste piles: Waste piled at industrial sites and then eventually disposed of in a landfill.

wastewater: Liquid wastestream primarily produced by five major sources: human and animal waste, household wastes, industrial wastes, stormwater runoff, and groundwater infiltration.

water content: In in situ volatilization, the influence water content has on rate of volatilization by affecting the rates at which chemicals can diffuse through the vadose zone. An increase in solid water content will decrease the rate at which volatile compounds are transported to the surface via vapor diffusion.

water pollutants: Unwanted contaminants that can pollute water.

water pollution: Any physical or chemical change in surface water or groundwater that can adversely affect living organisms.

water table: The upper surface of the saturation zone below which all void spaces are filled with water.

water vapor: The most visible constituent of the atmosphere (H_2O in vapor form).

waterborne pathogens: The transmission conduit for some pathogenic microorganisms.

watershed: The region draining into a river, river system, or body of water.

watershed divide: A ridge of high land dividing two areas drained by different river systems.

weather: The day-to-day pattern of precipitation, temperature, wind, barometric pressure, and humidity.

weathering: The chemical and mechanical breakdown of rocks and minerals under the action of atmospheric agencies.

weight: The force exerted on any object by gravity.

wet scrubber: A treatment device (a stacked tower, for example) in which the contaminant wastestream is passed through microorganism-laden media or through a chemical spray (such as caustic) to degrade and/or neutralize the harmful affects of the contaminant(s).

wetland: A lowland area, such as a marsh or swamp, saturated with moisture and usually thought of as natural wildlife habitat.

white goods: Large solid waste items such as household appliances (refrigerators, stoves, dish washers, washers and dryers, etc.)

wind: Horizontal air motion.

wind and breezes: Local conditions caused by the circulating movement of warm and cold air (convection) and differences in heating.

winter kill: A condition that can occur in a lake or pond when the entire water mass is frozen, thereby killing all inhabitants.

winter stratification: In a lake in winter, the condition that occurs when the epilimnion is ice-bound, is at the lowest temperature and thus lightest, the thermocline is at medium temperature and medium weight, and the hypolimnion is at about 4°C and heaviest.

worms: In stream ecology, certain species of worms present in bottom sediment indicates that the stream is polluted.

xenobiotics: Any chemical present in a natural environment, but which does not normally occur in nature, e.g., pesticides and/or industrial pollutants.

young pond: In the cycle of pond evolution, the initial or earliest phase.

zone of recent pollution: In streams or rivers, the point of pollution discharge.

Index

Government Institutes Mini-Catalog

PC #	ENVIRONMENTAL TITLES	Pub Date	Price
629	ABCs of Environmental Regulation: Understanding the Fed Regs	1998	$49
627	ABCs of Environmental Science	1998	$39
585	Book of Lists for Regulated Hazardous Substances, 8th Edition	1997	$79
579	Brownfields Redevelopment	1998	$79
4088	CFR Chemical Lists on CD ROM, 1997 Edition	1997	$125
4089	Chemical Data for Workplace Sampling & Analysis, Single User Disk	1997	$125
512	Clean Water Handbook, 2nd Edition	1996	$89
581	EH&S Auditing Made Easy	1997	$79
587	E H & S CFR Training Requirements, 3rd Edition	1997	$89
4082	EMMI-Envl Monitoring Methods Index for Windows-Network	1997	$537
4082	EMMI-Envl Monitoring Methods Index for Windows-Single User	1997	$179
525	Environmental Audits, 7th Edition	1996	$79
548	Environmental Engineering and Science: An Introduction	1997	$79
643	Environmental Guide to the Internet, 4rd Edition	1998	$59
560	Environmental Law Handbook, 14th Edition	1997	$79
353	Environmental Regulatory Glossary, 6th Edition	1993	$79
625	Environmental Statutes, 1998 Edition	1998	$69
4098	Environmental Statutes Book/CD-ROM, 1998 Edition	1997	$208
4994	Environmental Statutes on Disk for Windows-Network	1997	$405
4994	Environmental Statutes on Disk for Windows-Single User	1997	$139
570	Environmentalism at the Crossroads	1995	$39
536	ESAs Made Easy	1996	$59
515	Industrial Environmental Management: A Practical Approach	1996	$79
510	ISO 14000: Understanding Environmental Standards	1996	$69
551	ISO 14001: An Executive Repoert	1996	$55
588	International Environmental Auditing	1998	$149
518	Lead Regulation Handbook	1996	$79
478	Principles of EH&S Management	1995	$69
554	Property Rights: Understanding Government Takings	1997	$79
582	Recycling & Waste Mgmt Guide to the Internet	1997	$49
603	Superfund Manual, 6th Edition	1997	$115
566	TSCA Handbook, 3rd Edition	1997	$95
534	Wetland Mitigation: Mitigation Banking and Other Strategies	1997	$75

PC #	SAFETY and HEALTH TITLES	Pub Date	Price
547	Construction Safety Handbook	1996	$79
553	Cumulative Trauma Disorders	1997	$59
559	Forklift Safety	1997	$65
539	Fundamentals of Occupational Safety & Health	1996	$49
612	HAZWOPER Incident Command	1998	$59
535	Making Sense of OSHA Compliance	1997	$59
589	Managing Fatigue in Transportation, *ATA Conference*	1997	$75
558	PPE Made Easy	1998	$79
598	Project Mgmt for E H & S Professionals	1997	$59
552	Safety & Health in Agriculture, Forestry and Fisheries	1997	$125
613	Safety & Health on the Internet, 2nd Edition	1998	$49
597	Safety Is A People Business	1997	$49
463	Safety Made Easy	1995	$49
590	Your Company Safety and Health Manual	1997	$79

Government Institutes

4 Research Place, Suite 200 • Rockville, MD 20850-3226
Tel. (301) 921-2323 • FAX (301) 921-0264
Email: giinfo@govinst.com • Internet: http://www.govinst.com

Please call our customer service department at (301) 921-2323 for a free publications catalog.

CFRs now available online. Call (301) 921-2355 for info.

GOVERNMENT INSTITUTES ORDER FORM

4 Research Place, Suite 200 • Rockville, MD 20850-3226
Tel (301) 921-2323 • Fax (301) 921-0264
Internet: http://www.govinst.com • E-mail: giinfo@govinst.com

3 EASY WAYS TO ORDER

1. Phone: **(301) 921-2323**
Have your credit card ready when you call.

2. Fax: **(301) 921-0264**
Fax this completed order form with your company purchase order or credit card information.

3. Mail: **Government Institutes**
4 Research Place, Suite 200
Rockville, MD 20850-3226   USA
Mail this completed order form with a check, company purchase order, or credit card information.

PAYMENT OPTIONS

❏ **Check** (*payable to Government Institutes in US dollars*)

❏ **Purchase Order** (*This order form must be attached to your company P.O.* **Note:** *All International orders must be prepaid.*)

❏ **Credit Card**   ❏ *VISA*   ❏ *MasterCard*   ❏ *American Express*

Exp.____/____

Credit Card No. _____

Signature _____

(Government Institutes' Federal I.D.# is 52-0994196)

CUSTOMER INFORMATION

Ship To: (Please attach your purchase order)

Name: _____

GI Account # (*7 digits on mailing label*): _____

Company/Institution: _____

Address: _____
<small>(Please supply street address for UPS shipping)</small>
_____

City: _____ State/Province: _____

Zip/Postal Code: _____ Country: _____

Tel: (_____) _____

Fax: (_____) _____

Email Address: _____

Bill To: (if different from ship-to address)

Name: _____

Title/Position: _____

Company/Institution: _____

Address: _____
<small>(Please supply street address for UPS shipping)</small>
_____

City: _____ State/Province: _____

Zip/Postal Code: _____ Country: _____

Tel: (_____) _____

Fax: (_____) _____

Email Address: _____

Qty.	Product Code	Title	Price

Subtotal_____
MD Residents add 5% Sales Tax_____
Shipping and Handling (see box below)_____
Total Payment Enclosed_____

❏ **New Edition No Obligation Standing Order Program**
Please enroll me in this program for the products I have ordered. Government Institutes will notify me of new editions by sending me an invoice. I understand that there is no obligation to purchase the product. This invoice is simply my reminder that a new edition has been released.

15 DAY MONEY-BACK GUARANTEE
If you're not completely satisfied with any product, return it undamaged within 15 days for a full and immediate refund on the price of the product.

Within U.S:
1-4 products: $6/product
5 or more: $3/product

Outside U.S:
Add $15 for each item (Airmail)
Add $10 for each item (Surface)

SOURCE CODE: BP01